ESSAYS
ON
HISTORY OF GEOLOGY

See last pages of this volume
for a complete list of titles

ESSAYS ON HISTORY OF GEOLOGY

George W. White

With a Foreword
by Claude C. Albritton, Jr.

ARNO PRESS
A New York Times Company
New York / 1978

Editorial Supervision: ANDREA HICKS

Reprint Edition 1978 by Arno Press, Inc.

HISTORY OF GEOLOGY
ISBN for complete set: 0-405-10429-4
See last pages of this volume for titles.

Manufactured in the United States of America

Library of Congress Cataloging in Publication Data

White, George Willard, 1903-
Essays on history of geology.

(History of geology)
Reprint of essays originally published 1951-1976.
1. Geology--United States--History--Addresses,
essays, lectures. I. Title. II. Series.
QE13.U6W44 1977 557.3'09 77-6547
ISBN 0-405-10464-2

FOREWORD

No person has done more to excite interest in the history of American geology than George W. White. This he has accomplished not only through his own research and writing, but in a variety of ways less direct and less visible. First at Ohio State University and later at the University of Illinois he worked closely with the librarians in locating and acquiring treasured stores of primary materials vital to research in the history of geology. To graduate students and colleagues alike he has given over the years much assistance to their research as well as aid in finding outlets for their publications. He has been effective in persuading publishers to reissue classical geological writings long out of print.

Dr. White was the initial representative for North America on the International Comittee on the History of Geological Sciences, and the first Chairman of the U.S. National Committee on History of Geology. Largely through his efforts and influence the Geological Society of America established in 1976 a new Division for the History of Geology.

The wonder about all this is that Dr. White has been able to do so much else. Geohydrology and glacial geology have long been his scientific fields of special interest. At various times he has served as geologist to the Water Resources Branch of the U.S. Geological Survey, as State Geologist of Ohio, and as consultant to the State Geological Survey of Illinois. His professorial career includes appointments at the University of New Hampshire, Ohio State University, and the University of Illinois at Urbana. Presently he is Emeritus Research Professor of Geology at the University of Illinois, where for seven years he served as head of the Department of Geology.

Bringing together in one volume these historical writings was no easy task, even for the man who wrote them in the first place; but the result will be well worth the pains—to Dr. White's colleagues around the world, to his many former students, and to the future scholars and scientists concerned with the history of the geological sciences.

CLAUDE ALBRITTON
Southern Methodist University
February, 1977

PREFACE

The essays here reproduced have appeared over the past 25 years. They fall into several general groups, as shown in the Table of Contents.

My interest in history of geology began in graduate student days at Ohio State where Professor John Bownocker encouraged his students to read biographies of some of the men whose work they studied. From time to time I found copies of these biographies in second-hand book stores, sometimes in the bins in front of the stores where cheap books were displayed. Famous early works, as well as reference works and biographies, both American and European, were in the fine Edward Orton Memorial Geology Library at Ohio State. Professor Bownocker was an avid collector, not for himself, but for the Orton Library, which was generously supported by Edward Orton, Jr. I soon learned that there were such things as second-hand book catalogues, but I never put them to much use until some time later.

When I returned from 15 years in New Hampshire in 1941 to become a staff member of the Ohio State Geology Department, I found that my colleague, Dr. John W. Wells, was not only a most knowledgeable historian of geology, but was also an avid collector of early works. He generously shared with me as he worked on the Geology Library Committee his knowledge of importance, relative rarity, editions, states, and dealers.

Upon coming to Illinois in 1947 my interest in early works greatly increased. The University of Illinois Library and the Geology Library were even more complete than those at Ohio State and I was pleased that Dr. R.B. Downs, Director of Libraries, invited my assistance in building up the already excellent geology section of the history of science collection. A very generous special research fund was set up—"O happy day!" Within two years we were able to add the editions we needed of Hutton, Palissy, Steno, William Smith, and Werner, as well as many others. So the Illinois collection continued to increase and become ever more useful.

I had heard of the Lewis Evans 1755 map, but had never seen one—imagine my delight to find Illinois had a unique set of 10 copies of the various editions, with Evans and Pownall texts. I discovered that parts of Evans sounded strangely familiar—after 1755 writers had reprinted, usually without credit, Lewis Evans' descriptions! So some papers on Lewis Evans were written (two of these here reproduced) and his rightful place was asserted and is now recognized. Were there earlier geological observations? This led back to Thomas Hariot, 1588, then to Captain John Smith, and to other early references, so "Early American Geology" resulted.

My interests required my own reference books—the Illinois Library of course had these, but one needs tools at his own hand. Fortunately biographies and bibliographies and histories were available at low prices not too long ago. Bibliographic tools for early geology, especially American, seemed to need description and hence some of the bibliographic papers included here.

My own particular geological work and that of my graduate students has been in glacial geology. Some of the early observations on glacial geology we found seemed to deserve separate papers. Some of the early reports had interesting engravings; some annual "Brief Notes" were written to accompany the reproduction of the engravings for friends.

The geology in the early travel books is unexpectedly extensive. Its full exploration requires complete collections of these American travel narratives, fortunately available at Illinois. An extensive summary of geology in Midwestern travels is in *Annotated Bibliography of Observations on Illinois Water Resources 1673-1850,* by T.P. Schlunz, R.M. Sutton and G.W. White. It was printed in a form not suitable for good reproduction and is not included here, but the Water Resources Center, University of Illinois, Urbana, does have some copies still available.

It will be recognized that the list of references for the earlier papers are "dated." The bibliographies could now have extensive additions, but this is not the place or time for them. The recent great bibliography of early American geology by R.M. Hazen and M.H. Hazen(Geological Society of America, 1976) and Professor W.A.S. Sarjeant's forthcoming multi-volume bibliography of biographies and histories of geology will be the definitive works. The great history of petroleum geology by Edgar Owen (1976) has thousands of references and the history of the Geological Society of Canada by Morris Zaslow has exhaustive references.

History of geology, especially North American geology, now flourishes. The U.S. Committee for History of Geology, the Office of Historian of the Geological Survey, and a Geological Society of America Division of History of Geology have all been recently established. Biographies appear in increasing numbers and provide the basis for analytical history of geology.

Discoveries of little-known manuscript riches are now being investigated by Doskey on Maclure, Bray on Nicollet, Fleming on Cox, and Aldrich on Cook, as well as several other geologists by other historians. Their reports add wide new dimensions to our understanding of these men. It is an exciting time for history of geology, which is a field in which geologists and historians can assist each other.

GEORGE W. WHITE
University of Illinois
February 2, 1977

CONTENTS

REFERENCE BOOKS FOR HISTORY OF GEOLOGY (Reprinted from *Stechert-Hafner Book News,* Vol. XVIII, No. 3), November, 1963

ANNOTATED BIBLIOGRAPHY OF BOOKS FOR HISTORY OF GEOLOGY: Presented at the University of Nevada, Reno, Conference on History and Philosophy of Geology, August, 1964

EARLY STATE GEOLOGICAL SURVEY PUBLICATIONS (Reprinted from *Stechert-Hafner Book News,* Vol. XXIII, No. 2), October, 1968

THE HISTORY OF GEOLOGY AND MINERALOGY AS SEEN BY AMERICAN WRITERS, 1803-1835: A Bibliographic Essay (Reprinted from *Isis,* Vol. 64, No. 222), June, 1973

BIOGRAPHIES OF GEOLOGISTS. With John W. Wells (Reprinted from *The Ohio Journal of Science,* Vol. 58), September, 1958

HISTORY OF GLACIAL GEOLOGY

EARLY DISCOVERERS, XIX: Early Description and Explanation of Kettle Holes—Charles Whittlesey (1808-1886) (Reprinted from *Journal of Glaciology,* Vol. 5, No. 37), February, 1964

THE FIRST APPEARANCE IN OHIO OF THE THEORY OF CONTINENTAL GLACIATION (Reprinted from *The Ohio Journal of Science,* Vol. 67, No. 4), July, 1967

EARLY DISCOVERERS, XXVII: Announcement of Glaciation in Scotland—William Buckland (1784-1856) (Reprinted from *Journal of Glaciology,* Vol. 9, No. 55), 1970

HISTORY OF INVESTIGATION AND CLASSIFICATION OF WISCONSINAN DRIFT IN NORTH-CENTRAL UNITED STATES (Reprinted from *Geological Society of America, Memoir,* 136), 1973

GLACIAL GEOLOGY OF OHIO IN 1874—THE END OF AN ERA (Reprinted from *The Ohio Journal of Science,* Vol. 76), March, 1976

EARLY ENGRAVINGS OF GLACIERS AND FEATURES WITH BRIEF NOTES

THE GLACIAL THEORY COMES TO AMERICA (1841), 1966

THE FIRST ALPINE GEOLOGIST CLIMBS MONT BLANC (1787), 1968

THE RHÔNE GLACIER—Then and Now (1817), 1969

AGASSIZ'S MOBILE HOME ON THE AAR GLACIER (1840), 1970

THE PARALLEL ROADS OF GLEN ROY (1839), 1971

PETER MARTEL'S EARLY OBSERVATIONS ON ALPINE ANIMALS, ON HAY HARVESTING BY MARMOTS, AND ON GLACIERS (1742), 1972

PETER MARTEL'S OBSERVATIONS ON GLACIERS (1742), 1973

THE LAUTERBRUN VALLEY AND FALLS (1756), 1974

THE HUMBOLDT GLACIER (1854), 1975

REVIEWS

ASSAY OF THE WEST: Essay Review of *John Charles Frémont: The Expeditions of John Charles Frémont,* Vol. I, *Travels from 1838 to 1844,* Edited by Donald Jackson and Mary Lee Spence, 1970 (Reprinted from *Isis,* Vol. 62, Part 2, No. 212), 1970

REVIEW of *Philosophy of Geohistory: 1785-1970* (Benchmark Papers in Geology, Vol. 13), Edited by Claude C. Albritton, Jr., 1975 (Reprinted from *Quaternary Research,* Vol. 6), 1976

REVIEW of *Trek of the Oil Finders: A History of Exploration for Petroleum* by Edgar Wesley Owen (*American Association of Petroleum Geologists Memoir,* 6), 1975 (Reprinted from *Economic Geology,* Vol. 71), 1976

MEMORIALS

MEMORIAL TO JAMES WALTER GOLDTHWAIT (Reprinted from *Proceedings Volume of The Geological Society of America Annual Report For 1948*), April, 1949

MEMORIAL TO VINCENT CHESTER SHEPPS—1928-1967 Reprinted from *The Geological Society of America Proceedings for 1967*), 1967

MEMORIAL TO J. ERNEST CARMAN—1882-1966 (Reprinted from *The Geological Society of America Proceedings for 1966*), 1968

MEMORIAL TO HAROLD ROLLIN WANLESS—1898-1970 Reprinted from *The Geological Society of America Proceedings*), November, 1971

MEN
AND
OBSERVATIONS

EARLY AMERICAN GEOLOGY

Made in United States of America
Reprinted from THE SCIENTIFIC MONTHLY
Vol. LXXVI, No. 3, March, 1953

Early American Geology*

GEORGE W. WHITE

Until he went back to his native Ohio as professor of geology at Ohio State in 1941, George White had spent most of his academic career at the University of New Hampshire, which may account for his interest in the early history of geology in the United States. He left his alma mater (Ph.D., 1933) ***in 1947, to*** *become head of the staff of the Department of Geology, University of Illinois.*

THE early English voyagers sailed to America with certain information about the new continent, its aspect, and its resources. At first the information was vague and partly erroneous, and some of it was fantastic. They modified, amended, and added to the slim store of recorded facts on topography and mineral resources as Thomas Hariot[1] did so significantly in 1588.

After the first period of exploration, when actual settlement had begun, information which included geological data, both exact and inexact, was compiled for those in England contemplating settling in the new land—as, for example, by John Smith in his several works from 1608 to 1631, and by William Wood in *New Englands Prospect* in 1634.

After the settlements were well established, men of taste, thought, and observation commented on natural phenomena—including geological features —quite as penetratingly as their English cousins were doing in the late seventeeth and eighteenth centuries. A surprising number of the colonists were members of the Royal Society; one was a charter member.[2] Some of the American geological observations and theories, such as that on isostasy by Lewis Evans in 1743,[3] were fully as advanced as English ones of the same period.

This long period of geological observation, first by English visitors, but soon by colonial Americans, has been strangely neglected by historians of science and by historians of geology. Goode[4] discusses at length early biological observations; Osborn[5] and Simpson[6] have noted early observations on vertebrate fossils. The historians of geology from the time of the early outlines of the development of geology in America by Van Rensselaer[7] and Mather[8] to Merrill[9, 10] all agree that the beginning of geology in America was early in the nineteenth century and that William Maclure was its father. Samuel L. Mitchill is mentioned for his work of the same period, but the first book devoted entirely to American geology, that by Schoepf in 1787[11] is given only the barest mention. These early workers of the "Maclurean epoch," important as they certainly are, did not invent American geology out of thin air, or discover it in a vacuum. For more than two hundred years before 1800 facts had been accumulating upon which, in some cases, respectable theories had been constructed. It is my purpose to sketch the expanding knowledge of geologic fact recorded in English concerning the American colonies, and to cite examples from each of the colonial centuries.

The Earliest Voyagers and their Information

We actually know very little about the observations of John Cabot in his voyages to Newfoundland in 1497, but he is reported to have observed copper in possession of the natives.[12] Hakluyt in 1582[13] gave an English translation of the report of Giovanni de Verrazano of his sailing along the North Atlantic coast in 1524, in which copper possessed by the natives was noticed. As the reports of these voyages did not contain information on rich minerals, it is of interest to note just where the epic English voyagers of the late sixteenth and early seventeenth centuries—Frobisher, Gilbert, Lane, Grenville, Brereton, and Smith—did get their information about natural conditions, metals, and other products and their high hope of rich mineral resources.

Richard Eden[14] in 1555 brought together and translated a miscellany of travels by Peter Martyr, Oviedo, and others, and part of Biringuccio's book on metallurgy of precious metals, under the title of *The Decades of the New World*. Eden obviously is presenting these earlier reports on the richness of the New World in such a way as to encourage English expeditions and the support of such undertakings by those who could invest "venture capital."

Peter Martyr, as translated by Eden, refers to gold on almost every page by pounds, by tens of pounds, and by hundreds of pounds. Oviedo in his

* Address of the retiring vice president and chairman, Section E, AAAS, at St. Louis, December 29, 1952.

"Generall [natural] historie of the West Indies," which in Eden's compilation follows Martyr's work, begins with a very lucid description of placer gold mining, and makes it clear that the gold has come from veins ("mother lodes") in the mountains at the headwaters of the streams. Among other minerals described is bitumen used for caulking ships in Cuba and in other islands (Trinidad?), in what may be the earliest note on hydrocarbons in the New World and is almost certainly the first in English.

Oviedo tells of one "grayne [nugget] of golde of great weight" weighing 32 pounds, another of 7, and another of 5 pounds. He comments on a large amount of gold in other places; his detailed description of the golden armor of the Indians and of the "principall women" who wore brassieres of golden bars weighing more than 200 ducats must have fired the imagination of the English would-be voyagers. Eden's book of 1555 and the revision by Willis in 1577 were sources of information and of exciting stimulation to the English. When we realize the quarts of pearls, pounds of gold, and quantities of golden armor and lingerie the English expected to find, we can visualize their great disappointment when they found the Indians of Virginia and New England not only with no golden raiment, but without any armor or brassieres at all.

The Frenchman Jacques LeMoyne, reporting on the Laudonniere expedition to Florida in 1564–65 tells of the (placer) gold mining in the "Apalatcy Mountains" and painted a picture of this mining activity, an engraving of which was printed in 1591 by DeBry.[15] This appears to be the first picture of any part of the United States having to do with anything geological.

The "Minerall Men"

The personnel of the English expeditions of the latter part of the sixteenth and early part of the seventeenth centuries included economic geologists and mining and metallurgical experts. Some were English (Cornish?) and others were Germans; they are referred to as "minerall men" or "refiners." We even know the names of some of them: Captain Vaughan with Grenville, Lane, and Hariot; the Saxon Daniel with Sir Humphrey Gilbert; and William Callicut, William Dawson, and Abraham Ransacke, plus two goldsmiths with Captain John Smith. Frobisher, who explored in 1576–78 north of the present United States, appears to have had more than one such expert. They were of various abilities, even as present-day "minerall men."

Recent critical editions and translations of the sixteenth-century works of the mining geologist and engineer Agricola,[16] the metallurgist Biringuccio,[17] the assayer Ercker,[18] and the assayers of the *Probierbüchlein*[19] make it easier for us to examine the kind of knowledge that could have been available to the mineral men and to those others who had interest in metals and their ores in the New World. It is certain that not all the voyagers were equipped with even a small part of this knowledge, but the professional "minerall men" should have had at least a part of it. (Note, however, the repeated complaint of Captain John Smith about the ineptness of his mineral experts.)

Early Observations in Virginia

THOMAS HARIOT (1560–1621). Thomas Hariot's *A briefe and true report on the new found land of Virginia* of 1588[1] is the earliest publication on natural history of what is now the United States.[4, 20, 21] Biologists have heard of his accurate and lively observations on plants and animals, but geologists have only recently known of his considerable and respectable geologic observations.[22]

Hariot accompanied the Raleigh expedition of 1585 to Virginia (North Carolina) as historian, scientist, and adviser. Upon his return to England in 1586 he prepared the report which was published in 1588 and which is now one of the rarest and most desirable of Americana. In his list of "Marchantable commodities" Hariot mentions several minerals, including alum, and notes the occurrence of iron and copper. His observations are confirmed in the report of the governor, Ralph Lane,[23] who tells of the concurring opinion of Captain Vaughan, "minerall man" of the party.

In his "Conclusion" Hariot gives the first description of the changing character of the Coastal Plain toward the Piedmont. We are impressed by his geological knowledge when we summarize what Hariot knew from personal observation and actual experiment about the geology of Virginia:

1. There was a wide flat coastal region (Coastal Plain) without prominent stones.
2. Along a distinct line (Fall Line) hard (crystalline) rocks appeared and continued an unknown distance into the country (the Piedmont).
3. The character of the crystalline rock varied.
4. The rocky ground at places contained iron, which he had seen and knew on professional advice to be ore.
5. Copper occurred someplace beyond the limit of exploration but supposedly not far away. The material was positively copper on the basis of its melting character.
6. The copper was silver-bearing, as determined by assay by Hariot and the "minerall man."
7. Excellent and plentiful clay for brick making existed in the Coastal Plain.

8. The fossil shells in the Coastal Plain sediments may have been recognized as remains of actual organisms. We know, from other sources, of Hariot's interest in fossils.[24]

JOHN BRERETON (fl. 1602). After the voyages supported by Sir Walter Raleigh in 1584–91 there was little exploration by the English along the Atlantic Coast until 1607, except for that of John Brereton, who passed along the New England Coast in 1602, and the same year published a little book on his "discoverie."[25] Brereton's observations on natural history consist mainly of those on minerals and topography of southern New England, Martha's Vineyard, and Nantucket. He was much impressed by what we today call industrial minerals —especially by building stone and clay.

CAPTAIN JOHN SMITH (1580–1631). Captain John Smith's claim to notice in any history of early American natural history[4] and in any history of early American geology rests upon the scientific and geographic observations recorded in his several books on Virginia and on New England which appeared between 1608 and 1631.

It is hard to realize that the Captain John Smith who was the actual and later the titular leader of the Jamestown colonists was in 1607 a man not yet twenty-seven years old.[26] All his incredible adventures as seaman, soldier, shipwrecked mariner, ambassador, Turkish slave, escapee, and traveler in Europe, Turkey, North Africa, and Ireland took place between his thirteenth and his twenty-fourth birthdays. When he sailed to Virginia with the settlers it was as one of those who had invested in the project and not as a hired member of the party.

From Smith's voluminous writings and his compilations of the writings of others, it appears that Smith himself was from the outset skeptical about finding precious metals or a route to the Indies. One gets the impression that he felt from the first that the future of Virginia lay in colonization and the development of trade. His attention to soil types confirms his interest in agricultural resources. Indeed, he is honest enough on one occasion to write that the hope of gold was used only to secure backing for his voyage to New England.

The principal source of information about the Virginia seen by Smith and his associates is the book[27] first published in 1612, *A Map of Virginia, with a description of the Countrey. . . .* The map, whether made by Smith or by members of his party under his direction[18, 28] is a remarkable production considering the conditions under which it was made. It is tolerably accurate geographically, shows the Indian territories and towns, and—a point I have not seen emphasized—clearly indicates which parts are based on actual exploration of the ground and which based on reports. The flat Coastal Plain is differentiated from the more hilly Piedmont. Falls are indicated where the various rivers cross the boundary from the harder rocks of the Piedmont to the softer ones of the Coastal Plain (the Fall Line).

Smith describes crystalline rock at "the head of the Bay" near Baltimore. His description of soil—probably the first one in English for America—clearly recognizes the variation of vegetation with soil types. After describing the rivers, their valleys, and their inhabitants, Smith passes on to a lengthy description of trees, fruits, plants, animals, fishes, and birds. He concludes this summary of the natural history of Virginia by a paragraph on "The Rocks":

> Concerning the entrails of the earth little can be saide for certainty. There wanted good Refiners: for these that tooke upon them to have skill this way, tooke up the washings from the mounetaines and some moskered [disintegrated] shining stones and spangles which the waters brought down; . . . The crust also of these rocks would easily perswade a man to beleeve there are other mines than yron and steele, if there were but meanes and men of experience that knew the mine from spar [i.e., knew ore from worthless gangue].

Smith's lack of certainty "Concerning the entrails of the earth" is refreshingly modern; we are still uncertain about them. Smith is again contemptuously sure that the "moskered shining stones" are not gold and is disdainful of the "refiners," of whom he had three.

Captain John Smith left Virginia for England in 1609, never to return to the colony that could not have survived the rigors of the winters of 1607 and 1608 without his guidance. After several years in England, during which he wrote his famous book on Virginia and defended himself from the charges of his enemies, he returned to British America, this time to a more northeasterly part of it. Smith briefly explored the New England shore in 1614, but his return in 1615 was prevented by storms and his capture by the French. Quite characteristically, during his captivity and upon his return he wrote a book: *A Description of New England . . .* ,[29] in the first page of which he names New England for the first time. He contrasts the rocky coast of Maine with the sandy shores of Massachusetts, and generalizes on mineral resources.

The description of the geography and natural history of New England is not quite so smoothly organized as the earlier one of Virginia, but nevertheless is full of information. He apparently did not have an assayer ("minerall man" or "refiner") along on this voyage to New England, but acted as his own assayer. Again the common attitude "there must be gold in them thar hills" is expressed by Smith, although we get the impression he himself did not really believe it. Always one to emphasize the more immediate and known resources, such as fish and furs, he knew at firsthand about iron ore and how important it could be to a new colony.

John Smith—soldier, sailor, explorer, administrator, colonizer, writer—was less credulous than most; he realized that in America the English had a world for settlement as well as for exploitation of any precious metals. In contrast to fertile soil and fundamental resources of iron and fuel to smelt it, he felt that precious metals were unimportant. He had the keen eye of the successful soldier for topography, but not the scholar's contemplative consideration of its continuity and origin, as had Hariot. He had the scientific spirit of true report of data, but left to others analysis and theory.

Observations in New England

WILLIAM WOOD (fl. 1629–39). William Wood, in his famous 1634 description,[30] *New Englands Prospect,* devotes Chapter 4 to "the nature of the Soyle," which he says "is for the generall a warme Kinde of earth," and after describing meadows and hay crops, tells of soil derived from clay, sand, and gravel.

Wood did not take "great notice" of "commodities as lie underground," and gives scant space to meager secondary material on minerals. The attitude toward precious metal had changed markedly in the twenty-five years since the settlement of Jamestown, when the settlers, fired by the tales of Martyr and Oviedo as translated by Eden, expected to find gold and silver at every bend in the rivers. Now, although he asserts that precious metals may yet be discovered in the "barren mountaines," Wood says "nobody dare confidently conclude" that they are present. The New Englanders had turned to more obvious resources—slate, building stone, and clay for bricks—which were to be of real use for their comfort.

Wood had considerable firsthand information of water supply and discusses in a fascinating manner the quality of water. His remarks on water supply are the earliest in America so far known to me. He comments in detail on the ease of finding wells and asserts that the quality of the water is so superlative that it is almost as good as "good Beere" and that "any man will choose it before bad Beere, wheay, or Buttermilk."

THOMAS MORTON (d. 1646). Thomas Morton was a man who flaunted his impertinent actions before his Puritan neighbors and satirized them in his book *New English Canaan.*[31] His book, published in 1637, makes clear his high approval of the country and the Indians, but also—and not guardedly—his disapproval of the Puritan settlers.

Morton's contribution is more to our amusement than to our geological knowledge. Slight though his geological observations were, his book does show that some minerals were known, and that industrial minerals—lime, whetstones, and building stones—were coming into use.

He notes the excellent supply of ground water and tells of the ease of digging satisfactory wells. He overestimates the curative value of certain medicinal waters, but in so broad a way that there is little doubt he does so facetiously.

JOHN JOSSELYN (fl. 1630–75). Over the next forty years about a dozen publications contain material on American natural history and include geological and topographic observations. Among these is *New Englands rarities discovered . . . ,* by John Josselyn,[32] in 1672. He describes the White Mountains and the topography of Mount Washington in considerable detail, especially the valleys cut into that mountain. He ascribes at least part of the cutting of these "Gullies" to running water, the first record I have found of such assumption in the New World.

Josselyn's chapter on "Stones, Minerals, Metals, and Earths" is disappointing in its brevity, merely listing several kinds of clay, metal ores, and gems (two of them mistakenly identified). The concluding paragraph of the mineral chapter on a strange cure for cancer is as out of context as it is startling.

JOHN CLAYTON (fl. 1688–93). About twenty years later John Clayton in 1693 wrote in the *Philosophical Transactions* on several American natural history subjects.[33] He also was impressed with the excellent ground-water resources of Virginia. He says that the waters in the springs are "somewhat more eager" than those in England and require "some quantity more of malt to make strong beer."

Clayton's observations on both invertebrate and vertebrate fossils in Virginia and Maryland are the equal of those across the Atlantic and were quoted approvingly in England.[34] He describes the concentration of fossil oysters in certain strata, and their association with huge teeth (shark's teeth). His description of a fossil whale is noteworthy. Clayton says the shells may be remains of animals or may be figured stones, exactly the attitude of Ray, Lister, and Lhwyd, who at this time described English fossils at length, but could not bring themselves to decide their origin.[35, 36]

First Half of the Eighteenth Century

ROBERT BEVERLEY (1673?–1722). Robert Beverley, the Virginia planter, was the first native American to write on Virginia.[37] In 1705 he published *History and present State of Virginia* in which he tells of the expedition of Captain Henry Batt to the Appalachian Mountains. There appears in this record a suggestion of the recognition of the parallel ridges of the folded Appalachians, the wide valleys between the ridges, and of the Appalachian Plateau, where a "Rivulet descended backwards."

In the chapter "Of the Earths, and Soil," Beverley shows a keen discrimination of soil varieties based on parent material—clay, sand, gravel, stones, or marl; and on topographic situation—lowland or upland. He is particularly specific in his references to the various shrubs and trees that grow on the different soils.

He is thorough in his listing of several varieties of clay and pigments, and mentions coal, slate, and building stone. He differentiates between Coastal Plain and Piedmont, and contrasts the region of "lower parts," flat, and free from stones (Coastal Plain) with the higher country where, especially near the "Falls of the Rivers," are found "vast quantities of Stone." He recognizes the continuity of the regular boundary between the Coastal Plain and Piedmont, and indicates that the Piedmont gives way farther west to mountains.

Beverley's reference to Colonel Byrd's exploration for the extension and reopening of the iron mine at Falling Creek by "boring, and searching after the richest Veins, near the Place of the former Work" is almost certainly the first record of exploratory drilling in America.

Beverley's sense of appreciation of amount and quality of water supply is repeatedly evident. His consciousness of rivers for navigation and springs for water supply and mills reflects his needs as a planter with large interests and operations. He also knows of "several Mineral Springs" but only briefly describes them, for he says he is "not Naturalist skilful enough, to describe them with the Exactness they deserve."

His account is a valuable one because it tells us what an educated man of 1705 knew of the topography and geology of Virginia. There is no speculation about causes or origins, but we are glad to have his literate and pleasant observations.

JOHN BARTRAM (1699–1777) and PETER KALM (1716–1779). At least 40 more books and papers on natural history were written in the next fifty years. The quality and amount of geologic observation vary, but the bulk is considerable. These must be passed over, for reasons of time and space, to discuss the observations and publications of three men who show great advance, not only in observation, but also in synthesizing data and proposing explanations. It is difficult to understand how their work has been neglected, for it provided a foundation—both acknowledged and unacknowledged—for those better-known men who followed.

These three are John Bartram, Philadelphia botanist and natural historian; Peter Kalm, Swedish traveler and scientist; and Lewis Evans, surveyor, cartographer, and geologic observer. Bartram's book on natural history of 1751[38] describes topography, rocks, fossils, and raised beaches near Lake Ontario. Of Kalm's several publications, the chief is the account of his travels first published in Swedish in 1753, in English in 1770, and available in a recent translation.[39] His more than 150 excellent geological observations and interesting speculations aggregate tens of pages and cannot be discussed adequately in a short space. It should be noted, however, that he speculates that Ireland and North America were once either united or else an island chain existed between them. He repeatedly mentions erratics and at one place till, but does not theorize on their origin.

LEWIS EVANS (1700–56). Lewis Evans' work[40] will serve as an example of the very respectable American geologic knowledge of the middle of the eighteenth century. Evans was born in Wales in 1700, but his entire professional career as surveyor, cartographer, and pamphleteer was spent in America.[41]

Evans traveled widely and observed acutely. In 1743 he made a trip from Philadelphia to Onondaga and Lake Ontario in the company of John Bartram and Conrad Weiser, interpreter and ambassador to the Indians. Evans carefully recorded his observations in a journal,[42] from which we note his certainty that fossil shells are remains of marine organisms, his recognition of the exhumation of the present Appalachian ridges from a former plain (peneplain!), the erosion of valleys to leave mountains, the linear character of the Appalachian ridges, the Coastal Plain, the former greater extent of the Great Lakes, and the isostatic uplift consequent on unloading the earth's crust by partial draining of the lakes.

Evans' maps[43] of Pennsylvania and adjacent states are very rare, but facsimiles are now available.[41] In these maps Evans filled in blank spaces with notes on weather, roads, streams, and geology. His "Remarks on the Endless [Appalachian] Mountains, etc." are excellent geologic descriptions.

The 1755 Evans' *Map of the Middle British Colonies*[44] is so full of geographic data that, except for spot indication of "Coals" and "Freestone," all geologic notes are presented in a separate booklet of 36 pages: *An analysis of a general map of the Middle British Colonies.*[45] In the preface Evans regrets not being able to include on the map profiles and geologic sections, stating:

> But Want of Room in the Plate, has obliged me to leave out what would have very much assisted my Explanation of the Face of the Country, I mean a Section of it in several Directions; such would have exhibited the Rising and Falling of the Ground, and how elevated above the Surface of the Sea; what Parts are level, what rugged; where the Mountains rise and how far they spread. Nor is this all that a perpendicular Section might be made to represent; for, as on the upper Side, the Elevations, Depressions, outer Appearances and Names of Places may be laid down; on the lower, the Nature of the Soil, Substrata and particular Fossils may be exprest. It was with Regret I was obliged to omit it. But in some future Maps of Separate Colonies, I hope to be furnished with more Room.

We share Evans' regret for the omission, for it was more than fifty years before Maclure[46] presented geologic sections, and these did not show all the features Evans proposed to include had there not been "Want of Room in the Plate." Neither these sections, nor "Maps of the Separate Colonies," were published, for Evans died under tragic circumstances the following year.

In the *Analysis* is a clear statement of the physiographic and geologic provinces of the eastern United States.[47] It is the first attempt at delineating physiographic provinces of the whole country and is valid today. Although his description of New England is brief, for the country west of the Hudson he describes in very considerable detail what we now call the Coastal Plain, the Fall Line, the Piedmont Plateau, the Blue Ridge, the Folded Appalachians, the Allegheny Front, and the Allegheny Plateau.

Evans deserves an important place in the history of American geology because of his maps with their geological notes; his *Analysis* with geological descriptions; and his journal, with theories on Great Lakes drainage, isostasy, and stream origin. He is perhaps equally important because he furnished additional geological information to his contemporaries John Bartram, Kalm, and others.

We can trace Evans' geologic contributions for sixty years after his death. Kalm widely disseminated them. Books of travel display knowledge of them as, for example, that of J. F. D. Smyth.[48] The idea of deposition of sedimentary rock and fossils in inland bodies of water, often attributed to Volney[49] and to Mitchill,[50] who developed the concept in great detail, traces back to Evans.

It is unfortunate that Evans did not live to publish the planned maps of separate colonies with geologic cross sections. If he had done so his just place in the history of American geology would have been earlier recognized.

Last Half of the Eighteenth Century

It is necessary to pass over many other men of the eighteenth century who contributed geological observations, descriptions, and even hypotheses, for their consideration would require a lengthy paper in itself. We can only note that more or less extensive geologic references are to be found in the writings of John Winthrop (Hollis professor at Harvard), Benjamin Franklin and several of his circle, William Bartram, Peter Collinson, Jonathan Carver, Thomas Hutchins, Thomas Jefferson, Jeremy Belknap, Thomas Pownall, and many others. I pass over the first geological map of North America by Jean Ettienne Guettard in 1752[51] because it was made by a man who was never on this continent, but who compiled his map from information from French correspondents.

The extensive observations in 1784 of the very little-known J. F. D. Smyth[48] occur in a two-volume book of travels. In addition to describing the Appalachian and Cumberland mountains he pays much attention to waters, streams, and valleys, noting in detail natural levees in some of the valleys.

The two-volume travel book[52] of 1787 by the philosopher-soldier François Jean, Marquis de Chastellux (1734–88), second in command of the French forces in the Revolution, provides unexpectedly rich geologic references. His approach is genetic, and he seeks and proposes many clever explanations to account for geologic phenomena. He is not the first to notice erratics, but he is the first I have found who speculates on their origin; although he could produce no hypothesis that satisfied him, he was thoroughly aware of the problem.

The first book entirely devoted to the geology of the United States is by the Hessian physician Johann David Schoepf (1752–1800) in 1787.[11] Much of the material in it is included in a more general travel book published the next year.[53]

With the mention of Schoepf we make contact with already known history of American geology. In the first decade of the nineteenth century appeared the works of Maclure,[46] Mitchill, Cleaveland,[54] and others with whose names we are at last familiar,[9] and my project of pointing out the hitherto unsuspected colonial interest in, and knowledge of, geologic phenomena comes to an end. Before 1800 the scores of publications by more than 40 different men contain important geological descriptions and observations, and include respectable hypotheses. A list of the first observations of various phenomena or of first proposal of concepts may provide a useful summary (Table 1).

TABLE 1

1524	First reference to copper—Verrazano
1588	First book on natural history and geology—Hariot
1591	First picture of mining (Appalachian region)—DeBry from LeMoyne
1612	First maps differentiating Piedmont from Coastal Plain—John Smith
1612	First description of soil types—John Smith
1634	First ground-water report—Wood
1672	First recognition of valley cutting by water—Josselyn
1705	First exploratory drilling—Byrd
1705	Physiographic description of Virginia and ground-water report—Beverley
1743	Great Lakes more extensive (shown by beach ridges) and isostatic rebound on their draining—Evans
1752	First geologic map of any part of the United States—Guettard
1753	Land bridge from North America to Europe—Kalm
1753	First mention of erratics—Kalm
1755	First division of America into physiographic provinces—Evans
1784	Description of natural levees—J. F. D. Smyth
1787	Speculation on origin of erratics—Chastellux
1787	First book on geology of United States—Schoepf

While recognizing the importance of the early nineteenth-century work of Maclure, Cleaveland, Mitchill, and their contemporaries, we must realize that these men really had a very considerable foundation of earlier observation and hypothesis upon which to build.

References

1. Hariot, T. *A briefe and true report on the new found land of Virginia. . . .* London. (1588); (facsimile ed., Ann Arbor: [1931]).
2. Stearns, R. P. *Notes and Records of the Royal Society of London,* **8,** 178 (1951).
3. White, G. W. *Science,* **114,** 302 (1951).
4. Goode, G. B. *Rept. U. S. Natl. Museum for 1897,* Pt. 2, 357 (1901).
5. Osborn, H. F. *Cope: Master Naturalist.* Princeton, N. J.: Princeton Univ. Press (1931).
6. Simpson, G. G. *Proc. Am. Phil. Soc.,* **86,** 130 (1942).
7. Van Rensselaer, J. *Lectures on geology. . . .* New York: Bliss and White (1825).
8. Mather, W. W. *Elements of geology. . . .* Norwich: Wm. Lester (1833).
9. Merrill, G. P. *Contributions to the History of American Geology.* Washington, D. C.: GPO (1906).
10. ———. *The First One Hundred Years of American Geology.* New Haven, Conn.: Yale Univ. Press (1924).
11. Schoepf, J. D. *Beiträge zur mineralogischen Kenntniss des östlichen Theils von Nord Amerika und seiner Gebürge.* Erlangen: J. J. Palm (1787).
12. Rickard, T. A. *A History of American Mining.* New York: McGraw-Hill (1932).
13. Fite, E. D., and Freeman, A. *A Book of Old Maps Delineating American History. . . .* Cambridge, Mass.: Harvard Univ. Press (1926).
14. Eden, R. *The Decades of the Newe Worlde . . . with particular description of the Moste ryche and large landes . . . by Peter Martyr of Angleria. . . .* London: Guilhelmi Powell (1555); (another ed., 1577; also Arber, E., Ed. Birmingham: [1885]).
15. Lorant, S. *The New World.* New York: Duell, Sloan & Pearce (1946).
16. Hoover, H. C., and Hoover, L. H. *Georgius Agricola, De re metallica, translated from the first Latin edition of 1556. . . .* London: *Mining Magazine* (1912); New York: Dover (1950).
17. Smith, C. S., and Gnudi, M. T. *The Pirotechnia of Vannoccio Biringuccio. . . .* New York: Am. Inst. Mining Met. Engrs. (1942).
18. Sisco, A. G., and Smith, C. S. *Lazarus Ercker's Treatise on Ores and Assaying, translated from the German edition of 1580.* Chicago: Univ. Chicago Press (1951).
19. ———. *Bergwerk- und Probierbüchlein.* New York: Am. Inst. Mining Met. Engrs. (1949).
20. Meisel, M. *A Bibliography of American Natural History,* 3 vols. Brooklyn: Premier (1924–29).
21. Jaffe, B. *Men of Science in America.* New York: Simon & Schuster (1946).
22. White, G. W. *Trans. Illinois State Acad. Sci.,* **45,** 116 (1953).
23. Hakluyt, R. *The principal navigations. . . .* London: (1598–1600); Vol. 8, Glasgow: (1903).
24. Parks, G. B. *Richard Hakluyt and the English Voyages.* New York: Am. Geogr. Soc. (1930).
25. Brereton, J. *A briefe and true relation of the discoverie of the north part of Virginia [New England] . . . made this present yeere 1602. . . .* London: Geor. Bishop (1602).
26. Arber, E. *Travels and Works of Captain John Smith,* 2 vols. Edinburgh: John Grant (1910).
27. Smith, Captain J. *A Map of Virginia. With a description of the Countrey. . . .* Oxford: Joseph Barnes (1612).
28. Ford, W. C. *Geogr. Rev.,* **14,** 433 (1924).
29. Smith, Captain J. *A Description of New England. . . .* London: Humfrey Lownes (1616).
30. Wood, W. *New Englands Prospect. A true, lively, and experimental description. . . .* London: Iohn Bellamie (1634); (also Boston: [1865]).
31. Morton, T. *New English Canaan. . . .* Amsterdam: J. F. Stam (1637).
32. Josselyn, J. *New-Englands rarities discovered. . . .* London: G. Widdowes (1672).
33. Clayton, J. *Philos. Trans.,* **17,** 941 (1693); **18,** 121 (1694).
34. Whitehurst, J. *An Inquiry into the original state and formation of the earth. . . .* London: J. Cooper (1778).
35. Gunther, R. T. *Life and Letters of Edward Lhwyd.* Oxford: Univ. Press (1945).
36. Zittel, K. A. v. *Geschichte der Geologie und Paläontologie.* München: R. Oldenbourg (1899).
37. Beverley, R. In L. B. Wright (Ed.), *The History and Present State of Virginia.* Chapel Hill: Univ. North Carolina Press (1947); (1st ed. [1705]).
38. Bartram, J. *Observations on the inhabitants, climate, soil, rivers. . . .* London: J. Whiston and B. White (1751).
39. Kalm, P. In A. B. Benson (Ed.), *Peter Kalm's Travels in North America. . . .* New York: Wilson-Erickson (1937).

40. White, G. W. *Trans. Illinois State Acad. Sci.* **44**, 152 (1951).
41. Gipson, L. H. *Lewis Evans.* Philadelphia: Historical Society of Penna. (1939).
42. Pownall, T. *Topographical description. . . .* London: J. Almon (1776).
43. Evans, L. *A map of Pensilvania, New-Jersey, New-York. . . .* Philadelphia: (1749, 1752).
44. ———. *A general map of the middle British colonies. . . .* Philadelphia: (1755).
45.———. *An analysis of a general map. . . .* Philadelphia: B. Franklin and D. Hall (1755).
46. Maclure, W. *Observations on the geology of the United States. . . .* Philadelphia: A. Small (1817).
47. Mather, K. F., and Mason, S. L. *A Source Book in Geology.* New York: McGraw-Hill (1939).
48. Smyth, J. F. D. *A tour in the United States of America,* 2 vols. London (1784).
49. Volney, C. F. *View of the climate and soil of the United States. . . .* London: J. Johnson (1804).
50. Mitchill, S. L. *Observations on the Geology of North America . . . ,* pp. 319–431 of Jameson's translation Cuvier, *Essay on the theory of earth.* New York: Kirk and Mercein (1818).
51. Ireland, H. A. *Bull. Geol. Soc. Am.,* **54,** 1227 (1943).
52. Chastellux, F. J. de. *Travels in North America . . . ,* 2 vols. London: Robinson (1787); (French ed., 1786).
53. Schoepf, J. D. *Reise durch einige der mittlern und sudlichen Vereinigten nordamerikanischen Staaten . . . ,* 2 vols. Erlangen: J. J. Palm (1788); (English trans., Philadelphia: [1911]).
54. Cleaveland, P. *An elementary treatise on mineralogy and geology.* Boston: Cummings and Hilliard (1816).

LEWIS EVANS' CONTRIBUTIONS TO EARLY AMERICAN GEOLOGY 1743-1755

LEWIS EVANS' CONTRIBUTIONS TO EARLY AMERICAN GEOLOGY 1743-1755

GEORGE W. WHITE
University of Illinois, Urbana

LEWIS EVANS' CONTRIBUTIONS TO EARLY AMERICAN GEOLOGY—1743-1755

GEORGE W. WHITE
University of Illinois, Urbana

Lewis Evans' important work in cartography of the American colonies is now recognized, but his very considerable contributions to early American geology are almost unknown. His observations and theories formed the basis for much better known work of later writers, such as Volney (1804), Mitchill (1818), and others. The subject of this paper is Evans' geological contributions, which have not been noticed by historians of geology, except for the quotation of his physiographic descriptions of the eastern colonies by Mather and Mason (1939, pp. 55-59).

Stevens' (1905, 1920, 1924) monographic studies of Evans' maps have made known his importance to colonial cartography, and recently the details of his life have been given by Wroth (1934, pp. 48-56; 148-66) and by Gipson (1939). The biographical data in this paper are summarized from Gipson.

Lewis Evans was born in 1700 at Llangwnadl, near Pwllheli, Carnarvonshire, Wales. He is known to have been in Pennsylvania as early as 1736. He was a surveyor, draftsman, and map maker. His earliest known map, a manuscript map of a small part of Pennsylvania (The Walking Purchase), is signed and dated 1738. He was well known to Benjamin Franklin, often purchased books from him, and may have on occasion worked for him as draftsman.

Evans travelled widely and observed acutely. In 1743 he made a trip from Philadelphia to Onandoga and Lake Ontario in the company of John Bartram, botanist and natural historian, and Conrad Weiser, interpreter and ambassador to the Indians (Bartram, 1751). He carefully recorded his observations in a journal which was not published until 1776—and then only in part—by Thomas Pownall.

From the journal of 1743 we note Evans' certainty that fossil shells are remains of marine organisms,[1] his recognition of the exhumation of the present Appalachian ridges from a former plain (peneplain!), the erosion of valleys to leave mountains, the linear character of the Appalachian ridges, the Coastal Plain, the former greater extent of the Great Lakes, and isostatic uplift consequent on unloading the earth's crust by partial draining of the lakes (White, 1951). The salient observations are (Pownall, 1776, pp. 29, 30; unchanged in Mulkearn, 1949, pp. 112-116):

> The Stones in all Parts of these Mountains are full of Sea Shells: It is not in the loose Stones scattered through the Vales that these Shells abound only, but

[1] This is all the more noteworthy when we remember that less than 40 years earlier Edward Lhwyd of Oxford, author of the first book on fossils published in England, after 20 years studying and collecting, could not bring himself to believe them to be organic remains (Gunther, 1945).

they are found at the Tops of the Mountains also . . . there could be no Doubt but that the Shells were genuine, and where the Shells were actually of the same Sort as those which I had observed in the Mountains, convinced me that those Shells of the Mountains were real, and had been mixed with and finally incrusted in the Stones where they were found . . .

Various Systems and Theories of the present Earth have been devised in order to account for this Phaenominon. One System supposes that the Whole of this Continent, the highest Mountains themselves, as they now appear, were formerly but one large Plain, inclining with a considerable Slant towards the Sea; that this has been worn into its present Appearance of Ridges, with Vales between them, by the Rains of the Heavens and Waters of the Earth washing away the Soil from the upper Parts, and carrying it down to Seawards. That the Soil thus carried down and lodged in various Places hath in a Series of Ages formed the lower Plains of the Jerseys, Pennsylvania, Maryland, Virginia, and the Carolinas . . . The Downfall of Waters from the melting of the Snow, the Rains, and the swollen Springs is such amongst the Mountains, and the Discharge from thence so great, that the Freshes on the Susquehanna River, where it is a Mile broad, rise 20 Feet, though they are discharged with a violent and precipitate Current. These Freshes carry down with them immense Quantities of Soil which they begin to drop as the Velocity of their Course slackens in gliding over the lower Plains, and which they finally lodge in Bars and Islands at the Mouths of the Rivers where they meet the Sea. Thus have been many very extensive Countries formed at the Mouths of all the great Rivers in the World, and thus at the several Mouths of the many great Rivers ranging so near one another along this Coast may that long continued Range of flat country, which is herein before called the *Lower Plains*, be formed . . .

But we must have recourse to some other Explanation in order to account for the Situation of the Shells on the Tops of the Mountains.

It is easy to shew the Earth and Sea may assume one another's Places, but positively to assert how that hath actually happened in Times past, is hazardous; we know what an immense Body of Water is contained in the great Lakes at the Top of the Country, and that this is damm'd and held up by Ridges of Rocks: Let us suppose these Ridges broken down by any natural Accident, or that in a long Course of Ages a Passage may be worn through them, the Space occupied by the Water would be drained: This Part of America, disburthened of such a Load of Waters, would of course rise, as the immediate Effect of the shifting of the Center of Gravity in the Globe at once or by Degrees, much or little, accordingly as the Operation of such Event had Effect on that Center. . . .

Some such Changes may have come gradually and advanced by such slow Degrees, as that in a Period of a few Ages would not be perceptible; History therefore could take no Notice of them.

Evans' maps (1749, 1752) of Pennsylvania and adjacent states are now vary rare, but facsimiles are included in Gipson (1939). In these maps Evans filled in blank spaces with notes on weather, roads, streams, and geology. His "Remarks on the Endless [Appalachian] Mountains, etc."[2] follow:

These Mountains are about 900 Miles in Length and back of Pensilvania from 70 to 100 Miles right Across; scarce an Acre of 10 of which is capable of Culture. They are not confusedly scatter'd, and here and there in lofty Peaks over topping one another but stretch in long uniform Ridges, scarce half a Mile Perpendicular in any Place I saw them. They furnish endless Funds for Systems and Theories of the World; but the most obvious to me was, That this Earth was made of the Ruins of another; at the Creation Bones and Shells which escaped the Fate of softer animal Substances are find mixt with the old Materials and elegantly preserved in the loose Stones and rocky Bases of the highest of these Hills. These Mountains existed in their present elevated Heights before the Deluge, but not so bare of Soil as now. The further ridges which are much the largest and highest, proceeding from the Inclination of the whole towards the Sea, are of very rich Land, even on the Tops; while the very Vallies, on the hither Side, seem swept of all the Soil.

[2] His statement that "this Earth was made of the Ruins of another" contains the germ of Huttonianism—"For having, in the natural history of this earth, seen a succession of worlds . . . ." (Hutton, 1788, p. 304). It would be interesting to know if Hutton ever saw Evans' maps or "Analysis."

Their Height rendered them no doubt less exposed to that general Devastation, and preserv'd them unhurt, while the Soil and the loose Parts of the lower Hills and Vallies, agitated by a greater Weight of Water, were borne away, suspended in the dashing Waves, and thrown downwards in Stratas of different Kinds, as the Billows roll'd from different Parts; still obvious in our lower Lands Northward and Westward of the Raritan and Delaware. But in N. Jersey on the other Side of these Rivers, the Land is made by an Accumulation of Sand from the Ocean: Digging there ab. 18 foot thro white worn Sand, you come to a Stratum, of Sea Mud mixt with Shells and other Drift Trash; and in some Places vast Beds of Shells of all Sorts, in Pairs, entire 30 Miles from the Sea. Dr. Woodward from infinite examples discover'd, that the World had been in a State of Dissolution. But the Power he ascribes to the Water of the Deluge is too much of a Miracle to obtain Belief. We have glaring Marks of a Deluge of far more recent Date, in which the Compass of Britain might not perhaps have furnished the Dr. with.

In 1755 Evans published his "Map of the Middle British Colonies." It was reprinted with and without acknowledgment for the succeeding 50 years (Stevens, 1904, 1920, 1924). It was used by both sides in the Revolutionary War. The map was so full of geographic data that, except for spot indication of "Coals" and "Freestone," all geologic notes were presented in a separate booklet of 36 pages, "An Analysis of a General Map of the Middle British Colonies . . . ," which could be purchased with the map. The Map and "Analysis" are bibliographic treasures. Fortunately, facsimiles of the map and the "Analysis" are now available (Gipson, 1939). In the preface Evans regrets not being able to include on the map profiles and geologic sections, stating:

But Want of Room in the Plate, has obliged me to leave out, what would have very much assisted my Explanation of the Face of the Country, I mean a Section of it in several Directions; such would have exhibited the Rising and Falling of the Ground, and how elevated above the Surface of the Sea; what Parts are level, what rugged; where the Mountains rise and how far they spread. Nor is this all that a perpendicular Section might be made to represent; for, as on the upper Side, the Elevations, Depressions, outer Appearances and Names of Places may be laid down; on the lower, the Nature of the Soil, Substrata and particular Fossils may be exprest. It was with Regret I was obliged to omit it. But in some future Maps of Separate Colonies, I hope to be furnished with more Room.

We share Evans' regret for the omission. It was not until more than 50 years later that Maclure (1817) presented geologic sections, which do not show all the features Evans proposed to include had there not been "Want of Room in the Plate." Neither these sections, nor "Maps of the Separate Colonies," were published, for Evans died under tragic circumstances the following year. It is likely that at least working drawings of sections were made by Evans. The best chance for their preservation would have been by Sir Thomas Pownall, who had the Evans' diary (or a copy?) and took it to England. The reader can carry as far as he likes any further conjecture on the possible existence of the sections.

In the "Analysis" is a clear statement of the physiographic and geologic provinces of the eastern United States. It is the first attempt at delineating physiographic provinces of the country and is valid today. The description of New England (p. 6) is brief, but west of the Hudson Evans was on surer ground. He describes what we now call the Coastal Plain, the Fall Line, the Piedmont Plateau, the Blue Ridge, the Folded

Appalachians, the Allegheny Front, and the Allegheny Plateau (pp. 6-9) (the modern terms are inserted in brackets):

The Land, South Westward of Hudson's River, is more regularly divided, and into a greater Number of Stages than the other. The first Object worthy Regard, in this Part, is a Rief or Vein of Rocks, of the Talky or Isinglassy Kind, some two or three, or Half a Dozen Miles broad; rising generally some small Matter higher than the adjoining Land; and extending from New-York City South Westerly by the Lower Falls of Delaware, Schuylkill, Susquehanna, Gun-Powder, Patapsco, Potomack, Rapahannock, James River and Ronoak. This was the antient maritime Boundary of America, and forms a very regular Curve. The Land between this Rief and the Sea, and from the Navesink Hills South Westward as far as this Map extends, and probably to the Extremity of Georgia, may be denominated the *Lower Plains* [Coastal Plain], and consists of Soil washt down from above, and Sand accumulated from the Ocean. . . .

From this Rief of Rocks, over which all the Rivers fall [Fall Line], to that Chain of broken Hills, called the South Mountain, there is the Distance of 50, 60 or 70 Miles of very uneven Ground, rising sensibly as you advance further inland; and may be denominated the *Upland* [Piedmont]. This consists of Veins of different Kinds of Soil and Substrata, some Scores of Miles in Length; and in some Places overlaid with little Ridges and Chains of Hills. The Declivity of the whole gives great Rapidity to the Streams; and our violent Gusts of Rain have washt it all into Gullies, and carried down the Soil to enrich the Borders of the Rivers in the *Lower Plains*. These Inequalities render half the Country not easily capable of Culture; and impoverishes it, where torne up with the Plough, by daily washing away the richer Mould that covers the Surface.

The *South* Mountain is not in Ridges like the *Endless* Mountains, but in small, broken, steep, stony Hills; nor does it run with so much Regularity. In some Places it gradually degenerates to Nothing, not to appear again for some Miles, and in others spreads several Miles in Breadth. Between the South Mountain and the hither Chain of the Endless Mountains, (often for Distinction called the North Mountain, and in some Places the Kittatinni, and Pequilin,) there is a Valley of pretty even, good Land, some 8, 10 or 20 Miles wide, and is the most considerable Quantity of valuable Land that the English are possest of; and runs through New-Jersey, Pensilvania, Mariland and Virginia. It has yet obtained no general Name, but may properly enough be called Piemont, from it Situation. Besides Conveniencies always attending good Land, this Valley is every where enriched with Limestone.

The *Endless Mountains* [Folded Appalachians], so called from a Translation of the Indian Name, bearing the Signification, come next in Order. They are not confusedly scattered, and in lofty Peaks overtopping one another, but stretch in long uniform Ridges, scarce Half a Mile perpendicular in any Place above the intermediate Vallies. Their Name is expressive of their Extent, though no Doubt, not in a literal Sense. . . .

There are several Chains of the Endless Mountains which have not come to my Knowledge, and had they been so, might have filled several Places which lie vacant in the Map. But so far as we are acquainted with them, we observe that each Chain consists of a particular Kind of Stone, and each different from the rest; and these Differences continue for their whole Extent, as far as I can learn. . . .

To the North Westward of the Endless Mountains is a Country of vast Extent, and in a Manner as high as the Mountains themselves. To look at the abrupt Termination of it, near the Sea Level, as is Case on the West Side of Hudson's River, below Albany, it looks as a vast high Mountain; for the Kaats Kills, though of more lofty Stature than any other Mountains in these Parts of America, are but the Continuation of the Plains on the Top; and the Cliffs of them, in the Front they present towards Kinderhook. These UPPER PLAINS [Appalachian Plateau] are of extraordinary rich level Land, and extend from the Mohocks River, through the Country of the Confederates. Their Termination Northward is at a little Distance from Lake Ontario; but what it is Westward is not known, for those most extensive Plains of Ohio are Part of them; which continue to widen as they extend further Westward, even far beyond the Mississippi; and its Boundary Southward is a little Chain of broken Hills, about 10 or 15 Miles South of the Ohio River. 'Tis an odd Phaenomenon to observe how near the Tide comes up Hudson's River to the Heads of Delaware and Susquehanna; when these two Rivers are obliged to go so far to meet it in their own Channels. The Reason is Delaware and Susquehanna have their Heads in the *Plains*, and Hudson's River the Tide at the Foot of them.

Evans' description of the rivers, lakes, and portages is in sufficient detail to be useful as a guide to navigation (pp. 16-31). He described at length the Indian tribes and the territory they occupied (pp. 14-16). He was alarmed over the encroachment of the French in the Ohio Valley and insisted on the necessity of English colonization of that region and of proper treatment of Indian allies (p. 16).

In some conservative quarters his views were regarded as "expansionist" and "war-mongering," which might lead to war with the French. Arguments with the opposition, whom he regarded not only as misguided but even as traitorous to the Colonial interest (pp. 31-32), led to his disfavor in certain quarters. Efforts were made to silence him (Gipson, pp. 75-78) and for a time he was imprisoned. After his release from jail he died on June 11, 1756, "after a lingering Indisposition" (Gipson, p. 77). The "Analysis" closes with a statement of the resources of the Ohio country (p. 32), which he believes through lack of knowledge and governmental inertia (he suggests—and not too delicately—even stupidity) may be allowed to fall to the French:

Ohio is naturally furnished with Salt, Coal, Limestone, Grindstone, Millstone, Clay for Glass-houses and Pottery, which are of vast Advantage to an Inland Country, and well deserving the Notice I take of them in the Map.

Evans deserves an important place in the history of American geology because of his maps with their geological notes; his "Analysis" with geological descriptions; and his journal, with theories on Great Lakes drainage, isostasy, and stream origin. He is perhaps equally important because he furnished geological information in other ways to his contemporaries. He and John Bartram had travelled together to Lake Ontario in 1743 and must have then and often later discussed geology. A comparison of the extracts from Evans' journal with Bartram's published journal (1751) shows that geologically Evans was the more penetrating and that Bartram was the more naive. It seems to me, on the basis of comparison of the journals, that Bartram gained more from Evans than he contributed to him. (Study of Bartram's work in geology is now in progress and my opinion may be changed by fuller investigation.)

An example of the information Evans had, but did not formally publish, is contained in an extensive manuscript account of Pennsylvania drawn in answer to some "Queries proposed by a Gentleman in Europe to his friend in Philadelphia" (Gipson, 1939, pp. 87-137). He described natural resources, the animals and plants, and tells of iron, copper, lead, mineral water, marble, millstones, amethyst, clays, and other mineral resources (Gipson, pp. 104-108).

Evans was certainly familiar with at least some foreign geologic literature for he refers to Woodward (1695) and discussed Burnet (1684) with Bartram (1751). A further interesting study will be to determine what other scientific literature of the day was known to him.

Evans supplied topographical and geological information to Peter Kalm, the Swedish scientific visitor (Kalm, 1937, pp. 61, 106, 351), and much of the geological information given Kalm by John Bartram (Kalm, 1937, pp. 36, 64, 65, 71-73, 173) may have come originally from Evans, or at any rate certainly must have often been discussed by Evans and Bartram. Kalm's "Travels" was first published in Swedish in 1753 and went through many editions in several European languages. His work had great influence for several decades and is often quoted by travellers such as Schoepf (1787; 1911, vol. 1, p. 300).

We can trace Evans' geologic contributions for ~60 years after his death. Kalm widely disseminated them. Books of travels display knowledge of them, as for example that of Smyth (1784). The idea of deposition of sedimentary rock and fossils in inland bodies of water, often attributed to Volney (1804) and to Mitchill (1818), who developed the concept in great detail, appears to trace back to Evans.

It is unfortunate that Evans did not live to publish the planned maps of separate colonies with geologic cross sections. If he had done so his just place in the history of American geology would have been earlier recognized. His belief in a deluge as an explanation of certain geologic features is not to his discredit—this explanation was accepted almost into the middle of the 19th century for more phenomena than even Evans explained by it. His observations were acute and his explanations were, in part, in advance of their day. His theories were presented as such and were not insisted upon, except as observation supported them. His ideas were based on field observations. These are astonishingly correct and his theories no worse, and often better, than those of many who came long after him.

REFERENCES

BARTRAM, JOHN, 1751, Observations on the inhabitants, climate, soil, rivers . . .: London, J. Whiston and B. White.

BURNET, THOMAS, 1684, The Theory of the Earth . . .: London, R. Norton for W. Kettilby (several later editions).

EVANS, LEWIS, 1749, 1752, A map of Pensilvania, New-Jersey, New-York . . .: Philadelphia.

———, 1755, A general map of the middle British colonies . . .: Philadelphia.

———, 1755, An analysis of a general map of the middle British colonies . . . : Philadelphia, B. Franklin and D. Hall.

GIPSON, L. H., 1939, Lewis Evans: Philadelphia, Historical Society of Pennsylvania.

GUNTHER, R. T., 1945, Early Science in Oxford, vol. 14; Life and letters of Edward Lhwyd: Oxford, University Press.

HUTTON, JAMES, 1788, Theory of the Earth . . .: Royal Soc. Edinburgh Trans., vol. 1, pt. 2, pp. 209-304.

KALM, PEHR, 1937, Peter Kalm's travels in North America . . .: edited by A. B. Benson, New York, Wilson-Erickson.

MACLURE, WILLIAM, 1817, Observations on the geology of the United States of America . . .: Philadelphia, Abraham Small; also in Am. Philos. Soc. Trans., 1818, new series., vol. 1, pp. 1-91.

MITCHILL, L. L., 1818, Observations on the geology of North America . . .: pp. 319-431 of Jameson's translation of Cuvier, Essay on the theory of the earth: New York, Kirk and Mercein.

MATHER, K. F., AND MASON, S. L., 1939, A source book in geology: New York, McGraw-Hill.

MULKEARN, LOIS, ED., 1949, POWNALL, T., A topographical description of the United States of America (being a revised and enlarged edition): Pittsburgh, University of Pittsburgh Press.

POWNALL, THOMAS, 1776, A topographical description of such parts of North America as are contained in the (annexed) map of the middle British colonies in North America: London, J. Almon.

SCHOEPF, J. D., 1788, Reise durch einige der mittlern und sudlichen Vereinigten nordamerikanischen Staaten . . .: 2 vols., Erlangen, J. J. Palm. 1788; English translation by A. J. Morrison, 1911, Travels in the Confederation: 2 vols., Philadelphia, W. J. Campbell.

SMYTH, J. F. D., 1784, A tour in the United States of America: 2 vols., London.

STEVENS, H. N., 1905, 1920, 1924, Lewis Evans, his map of the middle British colonies . . .: London, H. Stevens, Son, and Stiles; three editions.

VOLNEY, C. F., 1804, View of the climate and soil of the United States of America . . .: London, J. Johnson.

WHITE, G. W., 1951, Lewis Evans' early American notice of isostasy: Science, vol. 114, pp. 302-303.

WOODWARD, JOHN, 1695, An essay toward a natural history of the earth . . .: London, R. Wilkin.

WROTH, L. C., 1934, An American bookshelf 1755: Philadelphia, University of Pennsylvania Press.

Lewis Evans' Early American Notice of Isostasy

George W. White
Department of Geology, University of Illinois, Urbana

Reprinted from SCIENCE, September 21, 1951, Vol. 114, No. 2960, pages 302–303.

Lewis Evans' Early American Notice of Isostasy

George W. White

Department of Geology, University of Illinois, Urbana

The term "isostasy" was first proposed by Dutton in 1889 (*1, 2*). As early as about 1500 Leonardo da Vinci (*3, 4*) recognized that change of load causes movement of the earth's crust. The earliest recognition in America of what we now call isostatic adjustment appears to have been in 1743 by Lewis Evans, colonial surveyor, cartographer, and geological observer.

Evans was born in Wales about 1700 and died in New York in 1756 (*5*). He is best known for his *Map of the Middle British Colonies* (*6*). He was a keen observer of geologic phenomena and topography and inserted geologic notes on his earlier maps (*7*) and included much more extended geologic notes in *An Analysis of a General Map . . .*, a booklet of 36 pages which accompanied his 1755 map (*8*). Evans' surprisingly modern geological observations, almost completely unnoticed by historians of geology, are now the subject of study, which will be reported more extensively elsewhere.

Evans' remarks on isostasy were recorded in a journal he kept while on an expedition in 1743 from Philadelphia to Onondago and Lake Ontario with John Bartram (*9*), the pioneer botanist, and Conrad Weiser, interpreter and "ambassador" to the Indian tribes. The journal was not published by Evans and is now lost, but a copy was in possession of Thomas Pownall, colonial administrator and patron of Evans, who published parts of it. In 1776 Pownall brought out a revised edition of Evans' 1755 map and *Analysis* (*10*), which he supplemented with his own observations, especially of New England, a territory not well known to Evans. Pownall is careful to indicate exactly the material copied from Evans' *Analysis* and from Evans' 1743 journal. Pownall's original publication is now rare, but a new edition based on Pownall's notes for a revised edition was published in 1949 (*11*).

Pownall, after a discussion of the Appalachian ridges and valleys, in which he mainly quotes Evans' *Analysis,* introduces Evans' unpublished journal observation of 1743 on fossils, the drainage of once much more extensive Great Lakes, and the consequent effect on crustal elevation and concomitant eustatic movement of sea level. The material on isostasy from Evans' journal is on pages 29 and 30 of the 1776 edition (*10*) and on pages 112–116 of that of 1949 (*11*):

> Amidst the Detail of these dry Descriptions, it may perhaps relieve and amuse the Reader to insert some Observations and Opinions which I found in and extracted from Mr. Evans' Journal.

"The Stones in all Parts of these Mountains are full of Sea Shells: . . .

"Various Systems and Theories of the present Earth have been devised in order to account for this Phaenominon. One System supposes that the Whole of this Continent, the highest Mountains themselves, as they now appear, were formerly but one large Plain, inclining with a considerable Slant towards the Sea; that this has been worn into its present Appearance of Ridges, with Vales between them, by the Rains of the Heavens and Waters of the Earth washing away the Soil from the upper Parts, and carrying it down to Seawards. That the Soil thus carried down and lodged in various Places hath in a Series of Ages formed the lower Plains of the Jerseys, Pennsylvania, Maryland, Virginia, and the Carolinas. . . .

"But we must have recourse to some other Explanation in order to account for the Situation of the Shells on the Tops of the Mountains.

"It is easy to shew the Earth and Sea may assume one another's Places, but positively to assert how that hath actually happened in Times past, is hazardous; we know what an immense Body of Water is contained in the great Lakes at the Top of the Country, and this is a damm'd and held up by Ridges of Rocks: Let us suppose these Ridges broken down by any natural Accident, or that in a long Course of Ages a Passage may be worn through them, the Space occupied by the Water would be drained: This part of America, disburthened of such a Load of Waters, would of course rise, as the immediate Effect of the shifting of the Center of Gravity in the Globe at once or by Degrees, much or little, accordingly as the Operation of such Event had Effect on that Center. . . .

". . . Some such Changes may have come gradually and advanced by such slow Degrees, as that in a Period of a few Ages would not be perceptible; History therefore could take no Notice of them."

Evans recognized the late Pleistocene beach ridges near Lake Ontario as former shore lines of a once more extensive lake, but he assumed also that the fossils in the Paleozoic rocks of the northern Appalachian mountains and Allegheny plateau were remains of forms which once lived in the very much larger body of water which spread to the eastern margin of the mountains. There is some evidence that Evans realized the consolidated Paleozoic rocks were much older that the lacustrine Pleistocene deposits near Lake Ontario. The belief in a great water body extending to the eastern margin of the mountains was held at least until 1818, when Mitchill set forth the idea in great detail (*12*). A few years earlier Volney (*13*) had proposed the same theory. Their source was Evans, but neither followed him in ascribing land uplift to reduction in crustal load.

References

1. DUTTON, C. E. *Phil. Soc. Washington,* **11**, 51 (1889).
2. BOWIE, W. *Isostacy.* New York : Dutton (1927).
3. DELANEY, J. P. *Science,* **91**, 546 (1940).
4. MACCURDY, E. *The Notebooks of Leonardo da Vinci,* 2 vol., New York : Reynal & Hitchcock (1938).
5. GIPSON, L. H. *Lewis Evans.* Philadelphia : Historical Society of Pennsylvania (1939).
6. EVANS, L. *A General Map of the Middle British Colonies . . .* Philadelphia : (1755).

7. ———. *A Map of Pennsylvania, New Jersey, New York* . . . Philadelphia: (1749, 1752).

8. ———. *An Analysis of a General Map of the Middle British Colonies* . . . Philadelphia: B. Franklin and D. Hall (1755).

9. Bartram, J. *Observations on the Inhabitants, Climate, Soil Rivers* . . . London: J. Whiston and B. White (1751).

10. Pownall, T. *A Topographical Description of such Parts of North America as are Contained in the (annexed) Map of the Middle British Colonies in North America.* London: 1776).

11. ———. In Mulkearn, L. (Ed.), *A Topographical Description of the United States of America (being a revised and enlarged edition).* Pittsburgh: University of Pittsburgh Press (1949).

12. Mitchill, S. L. *Observation on the Geology of North America* . . . pp. 319–431 of Jameson's translation of Cuvier, *Essay on the Theory of the Earth.* New York: Kirk & Mercein (1818).

13. Volney, C. F. *View of the Climate and Soil of the United States of America* . . . London: (1804).

THOMAS HARIOT'S OBSERVATIONS ON AMERICAN GEOLOGY IN 1588

GEORGE W. WHITE
University of Illinois, Urbana

Illinois Academy of Science Transactions, Vol. 45, 1952

THOMAS HARIOT'S OBSERVATIONS ON AMERICAN GEOLOGY IN 1588

GEORGE W. WHITE
University of Illinois, Urbana

Thomas Hariot's "A Briefe and True Report on the New Found Land of Virginia" of 1588 is sometimes regarded as the earliest publication on the natural history of what is now the United States (Meisel, 1929, vol. 3, p. 329; Goode, 1901, p. 363). Biologists have heard of his accurate and lively observations on plants and animals, but historians of American geology do not record his considerable and respectable geologic observations.

Observations of topography and mineral resources were made in the first half of the sixteenth century by the French in Canada and by the Spanish in Central and South America (Hornberger, 1937). English observations on minerals, especially of Newfoundland (Rickard, 1932, p. 3), were made before Hariot and were known to him; but to Thomas Hariot must go the credit for first writing in a widely available book form about mineral resources and about geological observations in what is now the United States.

Thomas Hariot (1560-1621) was born in Oxford. He graduated B.A. in 1580 and later became a fellow of Trinity College. Soon after attaining his degree he became mathematical tutor and adviser to Sir Walter Raleigh. Raleigh was long his patron, as was also Henry Percy, Earl of Northumberland, who provided him with a very adequate pension and a house suitable for a gentleman and scholar. Hariot spent most of his life in mathematical and astronomical studies. He is credited with independently inventing the telescope (he appears to have had some such device in Virginia in 1585), with modernizing algebra, and with inventing the signs for inequality. Details on his life are given by Goode (1901), Stevens (1900), Lorant (1946) and Dictionary of National Biography.

Hariot accompanied the Raleigh expedition of 1585 to Virginia[1] as historian, scientist, and adviser. Upon the return of the colonists to England in 1586 he prepared a report which was published in 1588. "A Briefe and True Report of the New Found Land of Virginia" is one of the rarest and most desirable items of Americana, there being but 5 known copies in the world, of which two are in the United States (Adams 1931, p. viii). De Bry republished it in Latin, French, German, and English in 1590, with engravings of water colors made by John White (Sabin, 1870, v. 3, p. 20). Facsimile reprints have been published, and recently it has appeared in an edition with modernized spelling and grammar and with the De Bry illustrations (Lorant, 1946).

[1] The settlement was in the present state of North Carolina on Roanoke Island. The inland explorations were mainly in Albemarle Sound and the Roanoke River. In Hariot's time "Virginia" was the indefinite region north of Florida.

Hariot's well-organized, straightforward, and compact report presents a favorable summary of the new country, its people, its natural history and its resources. He is contemptuous of those who issued "malicious, and slanderous reports" and dwelt on the hardships and discomforts they had suffered, dealing with them as follows (Hariot, 1588, p. A4 verso):

The cause of their ignorance was, in that they were of that many that were neuer out of the Iland where wee were seated, or not farre, or at the leastwise in few places els, during the time of our aboade in the countrey; or of that many that after golde and siluer was not so soone found, as it was by them looked for, had little or no care of any other thing but to pamper their bellies; or of that many which had little vnderstanding, lesse discretion, and more tongue then was needfull or requisite. Some also were of a nice bringing vp, only in cities or townes, or such as neuer (as I may say) had seene the world before. Because there were not to bee found any English cities, nor such faire houses, nor at their owne wish any of their olde accustomed daintie food, nor any soft beds of downe or fethers; the countrey was to them miserable, & their reports thereof according.

Because my purpose was but in briefe to open the cause of the varietie of such speeches; the particularities of them, and of many enuious, malicious, and slaunderous reports and deuises els, by our owne countrey men besides; as trifles that are not worthy of wise men to bee thought vpon, I meane not to trouble you withall: but will passe to the commodities, the substance of that which I haue to make relation of vnto you.

In his list of "Marchantable commodities" Hariot mentions several minerals, including alum, of which he says (p. B2):

Allum: There is a veine of earth along the sea coast for the space of fourtie or fiftie miles, whereof by the iudgement of some that haue made triall heere in England, is made good *Allum*, of that kinde which is called *Roche Allum*. The richnesse of such a commoditie is so well knowne that I neede not to saye any thing thereof. The same earth doth also yeelde *White Copresse*, *Nitrum*, and *Alumen plumeum*, but nothing so plentifully as the common *Allum;* which be also of price and profitable.

Hariot noted the occurrence of iron and copper. His observations are confirmed in the report of the governor, Ralph Lane (Hakluyt, v. 8, p. 332), which will be quoted later. This passage is of great importance because of the mention of a "minerall man," upon whom we shall remark again. Hariot wrote (p. B3) as follows:

Iron: In two places of the countrey specially, one about fourescore and the other sixe score miles from the Fort or place where wee dwelt: wee founde neere the water side the ground to be rockie, which by the triall of a minerall man, was founde to holde yron richly. It is founde in manie places of the countrey else. I knowe nothing to the contrarie, but that it maie bee allowed for a good marchantable commoditie, considering there the small charge for the labour and feeding of men: the infinite store of wood: the want of wood and deerenesse thereof in England: & the necessity of ballasting of shippes.

Copper: A hundred and fiftie miles into the maine in two townes wee founde with the inhabitaunts diuerse small plates of copper, that had beene made as wee vnderstood, by the inhabitantes that dwell farther into the countrey: where as they say are mountaines and Riuers that yeelde also whyte graynes of Mettall, which is to bee deemed *Siluer*. For confirmation whereof at the time of our first arriuall in the Countrey, I sawe with some others with mee, two small peeces of siluer grosly beaten about the weight of a Testrone [testone, shilling], hangyng in the eares of a *Wiroans* or *chiefe Lorde* that dwelt about fourescore myles from vs; of whom thorowe enquiry, by the number of dayes and the way, I learned that it had come to his handes from the same place or neere, where I after vnderstood the copper was made and the white graynes of mettall founde. The aforesaide copper wee also founde by triall to holde siluer.

In addition to the first section on merchantable commodities Hariot included a second section on "commodities for victuall and sustenance," which includes no minerals, and a third section on "things as is behoofull for those which shall plant settle and inhabit to know of . . . for building and other necessary uses." After discussing timber for houses and ships he says (p. D4 verso, El):

Now for Stone, Bricke and Lime, thus it is. Neere vnto the Sea coast where wee dwelt, there are no kinde of stones to bee found (except a fewe small pebbles about foure miles off) but such as haue bene brought from farther out of the maine. In some of our voiages wee haue seene diuers hard raggie stones, great pebbles, and a kinde of grey stone like vnto marble, of which the inhabitants make their hatchets to cleeue wood. Vpon inquirie wee heard that a little further vp into the Countrey were of all sortes verie many, although of Quarries they are ignorant, neither haue they vse of any store whereupon they should haue occasion to seeke any. For if euerie housholde baue [have] one or two to cracke Nuttes, grinde shelles, whet copper, and sometimes other stones for hatchets, they haue enough: neither vse they any digging, but onely for graues about three foote deepe: and therefore no maruaile that they know neither Quarries, nor lime stones, which both may bee in places neerer than they wot of.

In the meane time vntill there bee discouerie of sufficient store [stone?] in some place or other conuenient, the want of you which are and shalbe the planters therein may be as well supplied by Bricke: for the making whereof in diuers places of the countrey there is clay both excellent good, and plentie; and also by lime made of Oister shels, and of others burnt, after the maner as they vse in the Iles of Tenet and Shepy, and also in diuers other places of England: Which kinde of lime is well knowne to bee as good as any other. And of Oister shels there is plentie enough: for besides diuers other particular places where are abundance, there is one shallowe sounde along the coast, where for the space of many miles together in length, and two or three miles in breadth, the grounde is nothing els beeing but halfe a foote or a foote vnder water for the most part.

This much I can say further more of stones, that about 120. miles from our fort neere the water in the side of a hill was founde by a Gentleman of our company, a great veine of hard ragge stones, which I thought good to remember vnto you.

The description of Hariot and the independent narrative of Lane show that part of the colonists crossed the Coastal Plain to the Fall Line and entered the Piedmont. In his "Conclusion" Hariot gives the first description of the changing character of the Coastal Plain toward the Piedmont. Americans would now say that an affirmative answer has been given to his optimistic final question (p. F2 verso, F3):

Now I haue as I hope made relation not of so fewe and smal things but that the countrey of men that are indifferent & wel disposed maie be sufficiently liked: If there were no more knowen then I haue mentioned, which doubtlesse and in great reason is nothing to that which remaineth to bee discouered, neither the soile, nor commodities. As we haue reason so to gather by the difference we found in our trauails; for although all which I haue before spoken of, haue bin discouered & experimented not far from the sea coast where was our abode & most of our trauailing: yet somtimes as we made our iourneies farther into the maine and-countrey; we found the soyle to bee fatter; the trees greater and to growe thinner; the grounde more firme and deeper mould; more and larger champions [fields]; finer grasse and as good as euer we saw any in England; in some places rockie and farre more high and hillie ground; more plentie of their fruites; more abundance of beastes; the more inhabited with people, and of greater pollicie & larger dominions, with greater townes and houses.

Why may wee not then looke for in good hope from the inner parts of more and greater plentie, as well of other things, as of those which wee haue alreadie discouered?

Ralph Lane, governor of the colony left in Virginia over the winter of 1585-86, gives more details about the journey on which the iron and copper were found. The journey appears to have been made primarily to search for a reported copper mine, which was presumably the source of the copper plates possessed by the local Indians. He did not find the mine. In making his excuses to Raleigh it is evident that importance of the discovery of metals was well recognized. It is in this passage that we find a clue to the identity of Hariot's "minerall man" (Hakluyt, v. 8, p. 331-332; modernized in Lorant, p. 741, 742):

> I have set downe this Voyage somewhat particularly, to the ende it may appeare unto you (as true it is) that there wanted no great good will from the most to the least amongst us, to have perfited this discoverie of the Mine: for that the discovery of a good Mine, by the goodnesse of God, or a passage to the Southsea, or some way to it, and nothing els can bring this Countrey in request to be inhabited by our nation. And with the discovery of either of the two above shewed, it will bee the most sweete and healthfullest climate, and therewithall the most fertile soyle (being manured) in the world: . . .
>
> For this river of Moratico promiseth great things, and by the opinion of M. Hariots the head of it by the description of the Countrey, either riseth from the bay of Mexico, or els from very neere unto the same, that openeth out into the South sea.
>
> And touching the Minerall, thus doeth M. Youghan affirme, that though it be but copper, seeing the Savages are able to melt it, it is one of the richest Minerals in the world.

This Master "Youghan" is undoubtedly the "minerall man" of Hariot (1588, p. B3). But no Youghan appears in the list "of those as well Gentlemen as others, that remained one whole yeere in Virginia, under the Government of Master Ralph Lane" as given in Hakluyt (v. 8, p. 317-318). "Captaine Vaughan" is listed eighth in the list of 107 men. It is logical to assume that "Youghan" is a printer's error for "Vaughan." If so, who was "Captaine Vaughan"? He is the only man listed as "Captaine." Fourteen are listed as "Master," including Hariot, but in the text several are referred to as "Captaine" (p. 316).

The late sixteenth century "mineral man" appears to have been a combination geologist, mining engineer, assayer, and refiner. The profession was recognized and the title not uncommon (Hoover and Hoover, 1912, footnote, p. 283). Master Vaughan appears to have been the earliest English one in what is now the United States, but he was not the earliest attached to an English expedition to North America. That honor goes to "the minerall man and refiner . . . Saxon borne, honest and religious, named Daniel," who is reported by Edward Hayes to have been attached to the expedition of 1583 of Sir Humphrey Gilbert to Newfoundland and neighboring regions (Hakluyt, v. 8, p. 60). Along with Sir Humphrey, both Daniel and his ore specimens were lost on the return to England (*ibid.*, p. 67, 71).

It is possible that Hariot knew more than one mineral man. His patron, Raleigh, was a Devonshire man, who is reputed to have been a favorite of the miners of Devon and Cornwall. Raleigh became Warden of the Stannaries in 1585 and was active in using his new powers over the mining districts he already knew so well (Encycl. Brit., article Raleigh).

It is easy to speculate that after Hariot's return to England in 1586 he came into contact with the mine captains, mineral men, and refiners of Cornwall. Perhaps Captain Vaughan himself was a Cornish mining man. In the two and one-half years between his return and the publication of "Briefe and True Report," Hariot may have had many opportunities to increase his knowledge of mineral resources. But at any rate we can be impressed by his geological knowledge when we summarize what Hariot knew, from personal observation and actual experiment, about the geology of Virginia:

1. There was a wide flat coastal region (Coastal Plain) without prominent stones.

2. Along a distinct line (Fall Line) hard (crystalline) rocks appeared and continued an unknown distance into the country (the Piedmont).

3. The character of the crystalline rock varied and included a "hard ragge stone" (see "rag," Fay, p. 554; Rice, p. 336).

4. The rocky ground at places contained iron, which he had seen and knew on professional advice to be ore.

5. Copper occurred beyond the limit of exploration, but supposedly not far away. The material was positively copper on the basis of its melting character.

6. The copper was silver-bearing, as determined by trial by Hariot and the "minerall man."

7. Excellent and plentiful clay for brick existed in the Coastal Plain.

8. Probably the fossil shells in the Coastal Plain sediments were recognized as remains of actual organisms. Hariot knew enough about fossils[2] and was interested enough to discuss them in 1589 with Jacob Cole, nephew of the geographer Ortelius (Parks, p. 168, 251). Almost certainly he had made note of them in Virginia only three years before. He had noted the abundance of oyster shells there and it would have been strange if he had not noted similar, but fossil, shells in the Coastal Plain sediments. Certainly other observers who came after him did so and made the proper interpretation (Evans, 1749; also quoted in White, 1951, p. 154).

[2] Until the eighteenth century shells, bones, teeth, crystals, stalactites, and other objects were all called fossils. Much of the early insistence on the inorganic origin of fossils is thus more easily understood.

REFERENCES

ADAMS, R. A., 1931, Introductory Essay; to A brief and true report of the new found land of Virginia, by Thomas Hariot, A facsimile reproduction of the 1588 quarto: Ann Arbor, Edwards Brothers.

DICTIONARY OF NATIONAL BIOGRAPHY, article on Hariot.

ENCYCLOPAEDIA BRITANNICA, 1947, Article on Raleigh: Chicago.

EVANS, LEWIS, 1749, A map of Pensilvania, New-Jersey, New-York . . . ; Philadelphia.

FAY, A. H., 1920; 1947, A Glossary of the Mining and Mineral Industry: U. S. Bur. Mines, Bull. 95.

GOODE, G. B., 1901, The beginnings of natural history in America: Rept. U. S. Nat. Mus., 1897, pt. 2, 1901, pp. 357-406.

HAKLUYT, RICHARD, 1598-1600, The Principal Navigations Voyages, Traffiques & Discoveries of the English Nation . . . : References are to the readily available Glasgow edition of 1903-1905.

Hariot, Thomas, 1588 [1589 new style], A briefe and true report of the new found land of Virginia . . . : London; the facsimile reproduction, Ann Arbor, 1931, was used for this study.

Hoover, H. C. and Hoover. L. H., 1912, Georgius Agricola, De Re Metallica, translated from the first Latin edition of 1556 . . . : London, Mining Magazine; reprinted 1950, New York, Dover.

Hornberger, Theodore, 1937, Science and the New World (1526-1800): Huntington Library, San Marino, California.

Lorant, Stefan, 1946, The New World: New York, Duell, Sloan and Pearce.

Meisel, Max, 1929, A bibliography of American natural history: vol. 3, Brooklyn, Premier.

Parks, G. B., 1930, Richard Hakluyt and the English voyages: Am. Geogr. Soc. Spec. Pub. No. 10, New York.

Rice, C. M., 1948, Dictionary of geologic terms: Ann Arbor, Edwards Brothers.

Rickard, T. A., 1932, A history of American mining: A.I.M.E. series, New York, McGraw-Hill.

Sabin, Joseph, 1870, A dictionary of books relating to America: vol. 3, New York, J. Sabin and Sons.

Stevens, Henry, 1900, Thomas Hariot the mathematician the philosopher and the scholar . . . : London, privately printed.

White, G. W., 1951, Lewis Evans' contributions to early American Geology —1743-1755: Ill. Acad. Sci. Trans., vol. 44, pp. 152-158.

GEOLOGICAL OBSERVATIONS OF CAPTAIN JOHN SMITH IN 1607-1614

GEORGE W. WHITE
University of Illinois, Urbana

Illinois Academy of Science Transactions, Vol. 46, 1953

GEOLOGICAL OBSERVATIONS OF CAPTAIN JOHN SMITH IN 1607-1614

GEORGE W. WHITE
University of Illinois, Urbana

The fabulous and famous John Smith (1580-1631), always referred to as "Captain John Smith," has been more highly regarded for veracity in this century than he was in the preceding one (McIlwaine, 1947). There is now a tendency, on the basis of confirmation from various sources, to take Smith's autobiography (1630), including the story of Pocahontas, quite seriously. His claim to notice in any history of early American natural history (Goode, 1901, p. 371) or of early American geology rests upon his observations on natural history, mineral resources, and physical features recorded in his several books on Virginia and New England which appeared between 1608 and 1631.[1]

It is hard to realize that the Captain John Smith who was the actual and later the titular leader of the Jamestown colonists was in 1607 a man not yet 27. All his incredible adventures as seaman, soldier, shipwrecked mariner, ambassador, Turkish slave, escapee and traveler in Europe, Turkey, North Africa, and Ireland took place between his thirteenth and twenty-fourth birthdays! When he sailed to Virginia with the settlers it was as one of those who had invested in the project and not as a hired member of the party.

From Smith's voluminous writings and his compilations of the writings of others, especially of his old companions in arms, Phettiplace and Todkill, it appears that Smith himself early became skeptical about finding precious metals or a route to the Indies. One gets the impression that he felt from the first that the real future of Virginia lay in colonization and the development of trade. His attention to soil types confirms his interest in agricultural resources. One almost feels that he realized that the type of settlement of the Spaniards, based on extraction of precious metals and its necessary exploitation of the natives, was not to be desired for the English. Indeed, he is honest enough on one occasion to write that the hope of finding gold was used only to secure backing for his voyage to New England. He was obviously thoroughly familiar with Thomas Hariot's (1588; White, 1952) sane and sensible report of Virginia (actually North Carolina) resources and appears to have relied more on it than on the rosy compilations of Eden (1555) on Spanish gold.

SMITH'S MAP AND DESCRIPTION

Smith's first book, *A True Relation . . .*, published in 1608 (the first printing does not have Smith's name as author) contains little on obser-

[1] I am much more sympathetic to the viewpoint of Arber (1910) and McIlwaine (1947) about the veracity of Smith than to the carpingly critical view of Brown (1891). I have come to this attitude on topographic and geologic grounds—his geologic and physiographic observations "stand up." To professional historians must be left the assessment of the Smith-Newport administrative antagonism, but there can be no doubt that Smith was right and Newport wrong on the precious metal resources.

vations of natural history and nothing that is not in later, more full accounts. The principal source of information on the Virginia seen by Smith and his associates is the book first published in 1612, *A Map of Virginia, with a description of the Countrey. . . .* The map, whether made by Smith or by Nathaniel Powell or by other members of his party under Smith's direction (Fite and Freeman, 1926, pp. 116-119; Ford, 1924), is a remarkable production considering the conditions under which it was made.[2] It is tolerably accurate geographically and shows the Indian territories and towns. It clearly indicates which parts are based on actual exploration of the ground and which based on reports. I have not seen this interesting point taken into account elsewhere—it is, however, of importance when discussing Smith's observation of topography and geology.[3] The Coastal Plain is shown with almost no relief, but landward from the Fall Line, little hills are shown to mark the more hilly Piedmont. This distinction between Coastal Plain and Piedmont, recognized by Hariot (1588, p. F2; White, 1952, p. 118), is well shown although not described in the text.[4]

The *Description* starts by giving the location of Virginia, landfalls by which it may be reached, the climate, and then a description of the topography, soil, and the various rivers. In describing the kinds of rocks Smith makes very clear that glistening particles are not gold but may easily be so mistaken (1612, p. 3; 1910, p. 49):[5]

> The mountains are of diverse natures, for at the head of the Bay the rocks are of a composition like milnstones. Some of marble, &c. And many peeces of christall we found as throwne downe by water from the mountaines. For in winter these mountaines are covered with much snow, and when it dissolveth the waters fall with such violence, that it causeth great inundations in the narrow valleyes which yet is scarce perceived being once in the rivers. These waters wash from the rocks such glistening tinctures that the ground in some places seemeth as guilded, where both the rocks and the earth are so splendent to behold, *that better judgments then ours might have been perswaded, they contained more then probabilities.*
>
> The vesture of the earth in most places doeth manifestly prove the nature of the soile to be lusty and very rich. The colour of the earth we found in diverse places, resembleth *bole Armoniac, terra sigillata ad lemnia*, Fullers earth, marle, and divers other such appearances. But generally for the most part the earth is a black sandy mould, in some places a fat slimy clay, in other places a very barren gravell. But the best ground is knowne by the vesture it beareth, as by the greatnesse of trees or abundance of weedes, &c.
>
> The country is not mountanous nor yet low but such pleasant plaine hils and fertle valleyes, one prettily crossing an other, and watered so conveniently with their sweete brookes and christall springs, as if art it selfe had devised them.

[2] Brown's (1891) denial of Smith's authorship of the map carries little conviction. The reproduction of the manuscript map he exhibits to show how crude Smith's work probably was is not at all convincing. Many an excellent map is produced by a competent cartographic draftsman from just such rough notes. This is especially true if the author of the map is at hand to confer with cartographer and engraver as the map develops.

[3] Reproductions of the map are given by Arber (1910), Ford (1924) and Fite and Freeman (1926).

[4] Smith's map is not the first to show Coastal Plain and Piedmont. Hariot's map of "Virginia" of 1588 (Lorant, 1946, pp. 274-275) shows a hilly region inland from the coastal region. Le Moyne's map of "Florida" of 1565 (?) (Lorant, pp. 34-35) shows hills beyond a coastal region, placed on the map probably from hearsay.

[5] Each reference to Smith's writings will be given both to the first edition and to the much more accessible Arber edition of 1910.

The spelling and punctuation of the original are retained, except for the long "s" which not even Arber reproduced; and, when required, the substitution of initial "u" for "v", of medial "v" for "u" and of "j" for "i." I feel justified in making these changes because these quotations are for geologists rather than for bibliographers who will consult the original anyway. (The generally rigid, rather than capricious, usage of these letters by early printers is described by McKerrow, 1928, p. 310). Those few comments I have thought necessary to make within the quotations are shown by square brackets [].

The crystalline rocks at "the head of the Bay" are those near Baltimore. The description of soil is probably the first one in English for America. In it Smith clearly recognizes the variation of vegetation with soil types. The *bole Armoniac* and *terra sigillata* are varieties of clay used for pigment and medicine; they are often referred to in writings of the time (Hoover and Hoover, 1912, p. 31, note 10).

After describing the rivers, their valleys and their inhabitants, Smith passes on to a lengthy description of trees, fruits, plants, animals, fishes, and birds. He concludes this summary of the natural history of Virginia by a paragraph on "The Rocks" (1612, p. 15; 1910, p. 61):

> Concerning the entrailes of the earth little can be saide for certainty. There wanted good Refiners: for these that tooke upon them to have skill this way, tooke up the washings from the mounetaines and some moskered shining stones and spangles which the waters brought down; flattering themselves in their own vaine conceits to have bin supposed that they were not, by the meanes of that ore, if it proved as their arts and judgments expected. Only this is certaine, that many regions lying in the same latitude, afford mines very rich of diverse natures. The crust also of these rocks would easily perswade a man to beleeve there are other mines then yron and steele, if there were but meanes and men of experience that knew the mine from *spare* [spar].[6]

We are still uncertain "concerning the entrailes of the earth," although we do know more about rocks at the earth's surface than was known in 1612. Smith is again contemptuously sure that the "shining stones" are not gold. He is disdainful of the "refiners," of whom he had three besides Captain Newport who also had such pretensions.

It is interesting to note that Smith concludes his own part of the book (1612, pp. 37-38; 1910, pp. 82-83) by castigating those who have spread slanderous stories of hardships in Virginia in terms so similar to those used by Hariot in 1588 (p. A4; White, 1952, p. 117) that it is certain he had Hariot's book (or the Debry or Hakluyt reprints) before him when composing this section.

The "Refiners"

It is necessary to distinguish Part 1 of the 1612 *Description* by Smith himself, which is well written and well organized, from Part 2, which is a collection of partly repetitive writings of seven of the Virginia colonists, of varying length, expression, detail, and literacy. They give first-hand information of events and locations at which Smith was not present. They are in a sense the raw data which Smith in part used for his summary in Part 1.

In Chapter III of Part 2, Anas Todkill and Thomas Studly tell of the arrival (Jan. 8, 1608) of a ship commanded by Captain Newport. Todkill and Studly write disapprovingly of Captain Newport and his sailors. Instead of delivering the supplies and returning, the sailors stayed on for 14 weeks, eating up the supplies they had brought for the settlers, interfering with the indian trade, but to quote Todkill (1612, p. 21, 22; 1910, p. 104; the poetry, not found in the 1612 *Description* but added in the *Generall Historie* [1624, p. 53; 1910, p. 408], shows Smith's continuing bitterness toward Captain Newport):

[6] Smith's last sentence is more comprehensible if we remember "mine" meant "deposit or ore mineral [of]," whence "tell mine from spar" means "tell ore mineral from worthless rock."

But the worst mischiefe was our gilded refiners, with their golden promises, made all men their slaves in hope of recompence. There was no talke, no hope, nor worke, but dig gold, wash gold, refine gold, load gold. Such a brute of gold, as one made fellow [a wag] desired to bee buried in the sandes, least they should by their art make gold of his bones. Little need there was and lesse reason, the ship should stay, their wages run on, our victuall consume 14 weekes, that the Marriners might say, they built such a golden Church, that we can say, the raine washed neare to nothing in 14 daies.

Were it that Captaine *Smith* would not applaud all those golden inventions, because they admitted him not to the sight of their trials, nor golden consultations I knowe not: but I [? Anas Todkill] heard him question with Captaine *Martin* and tell him, except he would shew him a more substantiall triall, hee was not inamored with their durtie skill. Breathing out these and many other passions, never any thing did more torment him, than to see all necessarie businesse neglected, to fraught such a drunken ship with so much gilded durt.

Till then wee never accounted Captaine *Newport* a refiner. Who being fit to set saile for England, and wee not having any use of Parliaments, plaies, petitions, admirals, recorders, interpreters, chronologers, courts of plea, nor Iustices of peace, sent Maister *Wingfield*, and Captaine *Archer* with him, for England, to seeke some place of better imploiment.

Oh cursed gold, those hunger-starved movers,
To what misfortunes lead'st thou all those lovers!
For all the China *wealth, nor Indies can*
Suffice the minde of an av'ritious man.

At the end of the Todkill-Studly section (1612, p. 23-29; 1910, p. 106-108) is a list of those who came on April 20, 1608, in the *Phoenix*. This vessel, which Captain Martin wished to load with "phantasticall gold," finally sailed for England with a cargo of cedar which was what Smith intended from the outset she carry home.[7] There were landed from the *Phoenix* 33 "Gent.," 21 "Labourers," 6 "Tailers," 2 "Apothecaries," a "Jeweller," a "Gunsmith," a "Perfumer," 2 "Goldsmiths," 2 "Refiners," 4 other artisans, and "divers others to the number of 120."

The refiners were William Dawson and Abraham Ransacke. Already William Callicut (1612, p. 47; 1910, p. 125) was present and listed as a "refiner." Captain Newport also had pretentions as such and Captain Martin (Brown, 1898, p. 46) was the son of Sir Richard Martin, master of the mint. Yet Newport and Martin, the four refiners, and the two goldsmiths were deluded by the "phantasticall gold" which appears to have been only sand with glittering mica flakes. Throughout his books Smith is always contemptuous of his "refiners" and surely with good reason. They seem to have been far less expert, cautious, conservative, and respected than Captain Vaughan, who was refiner ("minerall man") with Lane and Hariot 20 years earlier, or than the Saxon Daniel, who perished with Sir Humphrey Gilbert a few years before that (White, 1952, p. 119). These refiners with whom Smith was plagued could not have been serious students of Ercker (Sisco and Smith, 1951), Biringuccio (Smith and Gnudi, 1942), or Agricola (Hoover and Hoover, 1912), and knew nothing of the *Probierbuchlein* (Sisco and Smith, 1949).

[7] But see Beverley ([1705], 1947, p. 31), who reports that this ship did carry micaceous dirt—"a yellow sort of Dust-Isinglass"—with a part load of cedar, and that soon after a second ship sailed with this "supposed Gold-Dust, . . . filling her up with Cedar and Clap-board." The report by Todkill may be interpreted either way. Brown (1898, p. 59) asserts that only "samples of minerals" were taken back in these voyages.

Even a touchstone, used alike by assayers, goldsmiths, and jewellers of that day, unreliable as this instrument is for careful assaying, would have shown them that mica is not gold. Had they been resourceful they could even have used a piece of gold coin for a touch needle (Sisco and Smith, 1951, p. 122). An interesting early confirmation of Newport's "gold," followed by more thorough assays in which the worthless samples "turned to vapore," is quoted from a contemporary letter by Brown (1898, p. 45, 46).

The report of Dr. Walter Russell and Anas Todkill on the "Accidents that hapned in the Discoverie of the bay Chesapeake" makes up chapter 5 of Part 2 of *A Map of Virginia.* They relate some details of a voyage up the Potomac River (1612, p. 33; 1910, p. 113):

> The cause of this discovery was to search a glistering mettal, the Salvages told vs they had from *Patawomeck* (the which *Newport* assured that he had tryed to hold halfe silver), also to search what furres, metals, rivers, Rockes, nations, woods, fishings, fruits, victuals, and other commodities the land afforded; and whether the bay were endlesse, or how farre it extended.
>
> The mine we found 9 or 10 myles up in the country from the river; but it proved of no value.

Much greater detail of the same report is given in the 1624 *Generall Historie,* which includes a bit more geological observation on the travel to the falls of the Potomac and then later up the Quiyough (Occoquan) River to the Fall Line. They found no precious metals, but the unusual method they used to try to catch fish is amusing (1624, p. 58; 1910, p. 417-418):

> Hauing gone so high as we could with the bote, we met divers Salvages in Canowes, well loaden with the flesh of Beares, Deere and other beasts; whereof we had part. Here we found mighty Rocks, growing in some places above the grownd as high as the shrubby trees, and divers other solid quarries of divers tinctures: and divers places where the waters had falne from the high mountaines they had left a tinctured spangled skurfe, that made many bare places seeme as guilded. Digging the growne above in the highest clifts of rocks, we saw it was a claie sand so mingled with yeallow spangles as if it had beene halfe pindust.
>
> In our returne inquiring still for this *Matchqueon,* the king of *Patawomeke* gave us guides to conduct us up a little river called Quiyough, up which we rowed so high as we could. Leaving the bote; with six shot and divers Salvages, he marched seven or eight myle before they came to the mine: leading his hostages in a small chaine they were to have for their paines, being proud so richly to be adorned.
>
> The mine is a great Rocky mountaine like Antimony; wherein they digged a great hole with shells and hatchets: and hard by it, runneth a fayre brooke of Christal-like water, where they wash away the drosse and keepe the remainder, which they put in little baggs and sell it all over the country to paint there bodyes, faces, or Idols; which makes them looke like Blackmores dusted over with silver. With so much as we could carry we returned to our bote, kindly requiting this kinde king and all his kinde people.
>
> The cause of this [voyage of] discovery was to search [for] this mine, of which *Newport* did assure us that those small baggs (we had given him), in England he had tryed [assayed] to hold halfe silver; but all we got proved of no value: also to search what furrs, . . . and what other mineralls, rivers, rocks, nations, woods, fishings, fruites, victuall, and what other commodities the land afforded: and whether the bay were endlesse or how farre it extended.
>
> Of mines we were all ignorant, but a few Bevers, Otters, Beares, Martins and minkes [skins] we found, and in divers places that aboundance of fish, lying so thicke with their heads above the water, as for want of nets (our barge driving amongst them) we attempted to catch them with a frying pan: but we found it a bad instrument to catch fish with: neither better fish, more plenty, nor more variety for smal fish, had any of us ever seene in any place so swimming in the water, but they are not to be caught with frying pans.

Some of Captain Newport's actions after January, 1608, are understandable when we realize that he had secret orders "not to returne without a lumpe of gold, a certainty of the south sea, or one of the lost company of Sir Walter Rawley" (1612, pt. 2, p. 42; 1910, p. 121). He also brought a "5 pieced barge" which was to bear him to the south sea after "we had borne her over the mountaines (which how farre they extend is yet unknown)" (*ibid*). The troublesome Captain Newport was unable to carry out any of his secret orders, but he did disrupt the organization and development of the infant settlement.

The narrative by Richard Wiffen, William Phettiplace, and Anas Todkill also recounts the journey of Captain Newport (Smith stayed behind) to Monacan 40 miles beyond the falls of the Powhatan (James) River at the present site of Richmond and the search on the return for minerals. They also recount how William Callicut, "a refiner, fitted for that purpose . . . perswaded us to beleeve he extracted some smal quantitie of silver (and not unlikely better stuffe might bee had for the digging)" (1612, p. 47; 1910, p. 125, 126). That these men had read of the great riches won by the Spaniards in Richard Eden's *Decades of the New World* in the 1555 or 1577 edition or in one of the editions of Hakluyt is indicated by the conclusion of their report (1612, pt. 2, p. 77, 78; 1910, p. 148):

> But to conclude, against all rumor of opinion I only say this for those that the three first yeares began this plantation: notwithstanding al their factions, mutenies, and miseries, so gently corrected and well prevented, peruse the Spanish *Decades*, the relations of *Master Hacklut;* and tell mee how many ever, with such smal meanes as a barge of 2 Tunnes, sometimes with 7, 8, 9, or but at most 15 men, did ever discover so many faire and navigable rivers, subject so many severall kings people and nations to obedience and contribution, with so little bloud shed.
>
> And if in the search of those Countries, wee had hapned where wealth had beene, we had as surely had it, as obedience and contribution; but if wee have overskipped it, we will not envy them that shall chance to finde it. Yet can wee not but lament it was our ill fortunes to end, when wee had but only learned how to begin, and found the right course how to proceed.

Eight Dutchmen and Poles came with Newport in 1608 (1612, p. 53; 1910, p. 129) to make pitch, tar, glass, and "sope ashes." On his return Newport took "tryals" of these (1612, p. 51; 1910, p. 128). Some of the Dutchmen sold arms to the Indians and joined them for a time (1612, p. 65; 1910, p. 139). "The randavus for all their unsuspected villany" was "the glasse-house, a place in the woods neere a myle from *James* Towne . . ." (1612, p. 80; 1910, p. 150). Thus early in American history came the use of industrial minerals in manufactures.

New England

Captain John Smith left Virginia for England in 1609, never to return to the colony which would not have survived the rigors of the winters of 1607 and 1608 without his guidance. After several years in England, during which he wrote his famous book on Virginia (1612) and defended himself from stories of his enemies, he returned to British America, but to a more northeasterly part of it. Smith explored briefly the New England coast in 1614, but his return in 1615 was prevented by storms and his capture by the

French. Quite characteristically, during his captivity and upon his return he wrote a book: *A Description of New England* . . . (1616). In the first page of his book, in which he names New England for the first time, Smith gives the reasons for this voyage to America (1616, p. 1; 1910, p. 187):

> In the moneth of Aprill, 1614. with two Ships from *London*, of a few Marchants, I chanced to arrive in *New-England*, a parte of *Ameryca;* at the Ile of *Monahigan*, in 43½ of Northerly latitude; our plot was there to take Whales and make tryalls of a Myne of Gold and Copper. If those failed, Fish and Furres was then our refuge . . .
>
> For our Golde, it was rather the Masters device to get a voyage that projected it, then any knowledge hee had at all of any such matter.

His succinct summary of the explorations, and his estimation of the extent of knowledge of America from Florida northward, is very useful (1616, pp. 3, 4; 1910, p. 189). He mentions the character of the coast south of Pennobscot and contrasts the rocky coast of Maine with the sandy shores of Massachusetts. He generalizes on mineral resources from quite insufficient direct evidence, but his statement shows that he and his party did pay some attention to these matters (1616, p. 9; 1910, p. 193):

> From *Pennobscot* to *Sagadahock* [cf. Sagadahoc County at mouth of Kennebec River] this Coast is all Mountainous and Iles of huge Rocks, but overgrowen with all sorts of excellent good woodes . . .
>
> Betwixt *Sagadahock* and *Sawocatuck* [? Cape Porpoise] there is but two or three sandy Bayes, but betwixt that and *Cape Cod* very many: especialy the Coast of the *Massachusets* is so indifferently mixed with high clayie or sandy cliffes in one place, and then tracts of large long ledges of divers sorts, and quarries of stones in other places so strangely divided with tinctured veines of divers colours: as, Free stone for building, Slate for tiling, smooth stone to make Fornaces and Forges for glasse or iron, and iron ore sufficient, conveniently to melt in them. But the most part so resembleth the Coast of *Devonshire*, I think most of the cliffes would make such lime stone. If they be not of these qualities, they are so like, they may deceive a better judgment then mine. All which are so neere adjoyning to those other advantages I observed in these parts, that if the Ore prove as good iron and steele in those parts, as I know it is within the bounds of the Countrey, I dare engage my head (having but men skilfull to worke the simples there growing) to have all things belonging to the building [and] the rigging of shippes of any proportion, and good marchandize for the fraight, within a square of 10 or 14 leagues. . . .

The description of the geography and natural history of New England is not so smoothly organized as the earlier one of Virginia, but nevertheless is full of information. He apparently did not have an assayer ("mineral man" or "refiner") along on this voyage to New England, but acted as his own assayer. He apparently had had instruction in assaying and probably had instructions along with him. However, we get the impression that Smith really did not expect to find precious metals. Always one to emphasize the more immediate and known resources, such as fish and furs, he knew at first hand about iron ore and knew it could be important to a new colony. At any rate he was sure of iron if not of gold (1616, pp. 20, 21; 1910, p. 201):

> Of Mynes of Golde and Silver, Copper, and probabilities of Lead, Christall and Allum, I could say much if relations were good assurances. It is true indeed, I made many trials, according to those instructions I had, which doe perswade mee I need not despaire, but there are metalls in the Countrey: but I am no Alchymist, nor will promise more then I know: which is, Who will undertake the rectifying [? erecting] of an Iron forge, if those that buy meate, drinke,

coals, ore, and all necessaries at a dear rate gaine; where all these things are to be had for the taking up, in my opinion cannot lose.

In his *Description of New England* Smith included a map which is as remarkable as his earlier map of Virginia. It is the first real map of this part of North America. It does not show the interior as well as the map of Virginia, but it must be remembered that he spent far less time in New England than in Virginia. No special geological implication can be read from the map of New England as can be from that of Virginia.

Conclusion

Smith wrote still other books, which, however, are mainly restatements of his earlier works on Virginia and New England. The most important is *The Generall Historie of Virginia, New England and the Summer Isles* of 1624. In a complete study, they must be examined because there is an occasional amplification, amendment, or addition.

In his maps and books about the new land, Smith supplied clear and factual information for English readers. Intelligent Englishmen could no longer hold the extravagant and fantastic ideas of the New World portrayed in the well-known play of 1605, *Eastward Hoe.*

Smith's report and maps indicate he was aware of the existence of a low region of sand and clay (Coastal Plain) beyond which lay a more hilly belt (Piedmont) with rivers passing from hilly belt to the lower part by falls. Crystalline rock outcropped at falls and upstream from the falls. Mineral resources included clay, building stone, iron ore, limestone, and materials for glass making. Smith's careful and repeated denial of precious metals was a factor in bringing the English to their senses about the prospects of wealth in the new land; already by 1634 Wood, and by 1637 Morton, accepted the very probable nonexistence of precious metals in New England.

John Smith was a keen observer of natural history but, unlike Hariot, his scientific observations were only incidental to his major activities as soldier, sailor, explorer, administrator, colonizer, and writer. Less credulous than most, he realized that in America the English had a world primarily for settlement rather than for exploitation of any precious metal. He shrugged off precious metals as unimportant, as compared to fertile soil and fundamental resources of iron and fuel to smelt it. He had the keen eye of the successful soldier for topography, but not the scholar's contemplative consideration of its continuity and origin, as did Hariot. He had the scientific spirit of true report of data. He left to others analysis and theory.

REFERENCES

Arber, Edward, Travels and works of Captain John Smith: Edinburgh, John Grant, 2 vols., 1910.

Beverley, Robert, The history and present state of Virginia: (1947, L. B. Wright, Ed., Chapel Hill, Univ. North Carolina Press), 1705.

Brown, Alexander, The genesis of the United States: Boston, Houghton, Mifflin and Co., 1891.

———, The first republic in America: Boston, Houghton, Mifflin and Co., 1898.

CHAPMAN, GEO.; IONSON, BEN; MARSTON, IOH., Eastward Hoe: London, William Apley, 1605.

EDEN, RICHARD, The decades of the New World . . . with the particular description of the moste ryche and large landes . . . by Peter Martyr of Angleria . . .: London, Guilhelmi Powell, 1555.

FITE, E. D., and FREEMAN, ARCHIBALD, A book of old maps delineating American History . . .: Cambridge, Harvard Univ. Press, 1926.

FORD, W. C., Captain John Smith's map of Virginia, 1612: Geogr. Review, vol. 14, pp. 433-443, 1924.

GOODE, G. B., The beginnings of natural history in America: Rept. U. S. Nat. Mus. for 1897, pt. 2, pp. 357-406, 1901.

HARIOT, THOMAS, A briefe and true report of the new found land of Virginia . . .: London, 1588.

HOOVER, H. C., and HOOVER, L. H., Georgius Agricola, De Re Metallica, translated from the first Latin edition of 1556 . . .: London, Mining Magazine, 1912; reprinted 1950, New York, Dover.

LORANT, STEFAN, The New World: New York, Duell, Sloan and Pearce, 1946.

MCILWAINE, H. R., John Smith: Encyclopedia Brittanica, vol. 20, pp. 831-833, 1947.

MCKERROW, R. B., An introduction to bibliography for literary students: Oxford, Clarendon Press, 1928.

MORTON, THOMAS, New English Canaan . . .: Amsterdam, J. F. Stam, 1637.

SISCO, A. G., and SMITH, C. S., Bergwerk- und Probierbüchlein: New York, Am. Inst. Mining Engrs., 1949.

————, Lazarus Ercker's Treatise on Ores and assaying, translated from the German edition of 1580: Chicago, Univ. of Chicago Press, 1951.

SMITH, CAPTAIN JOHN, A true relation of such occurrences and accidents of noate as hapned in Virginia . . .: London, John Tappe, 1608.

————, A map of Virginia, with a description of the countrey, the commodities, people, government and religion: Oxford, Joseph Barnes, 1612.

————, A description of New England . . .: London, Humprey Lownes for Robert Clerke, 1616.

————, The general historie of Virginia, New-England, and the Summer Isles . . .: London, Michael Sparkes, 1624.

————, The true travels, adventures and observations of Captain J. Smith: London, T. Slater, 1630.

SMITH, C. S., and GNUDI, M. T., The Pirotechnia of Vannoccio Biringuccio . . .: New York, Am. Inst. Mining Engrs., 1942.

WOOD, WILLIAM, New Englands prospect, a true, lively, and experimental description . . .: London, John Bellamie, 1634.

WHITE, G. W., Thomas Hariot's observations on American geology in 1588: Trans. Ill. Acad. Sci., vol. 45, pp. 116-121, 1952.

————, Early American geology: Scientific Monthly, vol. 76, pp. 134-141, 1953.

JOHN JOSSELYN'S GEOLOGICAL OBSERVATIONS IN 1672-1674

Illinois Academy of Science Transactions, Vol. 48, 1956

JOHN JOSSELYN'S GEOLOGICAL OBSERVATIONS IN 1672-1674

GEORGE W. WHITE
University of Illinois, Urbana

In seventeenth-century England the stirrings of interest in science culminated in the founding of the Royal Society. The new Society eagerly accepted for publication observations of facts and of phenomena in the New World and reviewed approvingly publications issued independently. From a slightly lower level came the support of travel books which contained more or less of natural history and on a still lower level even the common people eagerly paid to see exhibitions of curiosities from across the sea (Goode, 1901: 68, 378).

John Josselyn was one of the men in Restoration England who was influenced by those who were encouraging each other in observation and discussion of "rarities" and natural phenomena. That Josselyn, an educated man and possibly a physician, was stirred by this influence to write on American natural history is indicated by his statement that he made it his "business to discover all along the Natural, Physical, and Chyrurgical Rarities of this New-found World" (Mood, 1935: 26).

John Josselyn (*ca.* 1608-1675)[1] was the son of Sir Thomas Josselyn of Willingdale-Doe, Essex. He spent 15 months in 1638 and 1639 partly in Massachusetts, but mostly at Black Point (now Scarborough), Maine, where his older brother Henry was an important official. Many years later he returned to New England where he lived from 1663 to 1671. The intervening years in England correspond closely to the period of the Protectorate and must have been difficult ones for Josselyn, who was a staunch Royalist. Josselyn, a bachelor with no family ties, could travel extensively and appears to have done so in New England.

Upon his return to England in 1671 Josselyn wrote a general description of New England, which was published in 1672 as *New-England Rarities*. So well was this received and so favorably reviewed by the Royal Society that he put together his journals and added more material of local color and in 1674 dedicated to the Royal Society *An Account of Two Voyages to New-England. . . .* This was so successful that a second edition was necessary the following year.

Commentators on Josselyn have usually written approvingly of him. *New-England Rarities* was immediately and favorably reviewed (by Henry Oldenburg ?) in the *Philosophical Transactions of the Royal Society* (1672:5021-5023). Edward Tuckerman in 1865 reprinted Josselyn's books and wrote approvingly of him and of the importance of his botanical writings. Tyler (1885: 180-185), the colonial literary his-

[1] [Biographical information on Josselyn is not extensive; my summary is derived from Felter (1927), Mood (1933, 1935), and Goodwin (no date).]

torian, wrote of Josselyn in a delightfully sympathetic way. Tyler's little essay is full of quotable remarks, perhaps the best of which is the comment about "this author, whose scientific methods had in them so little severity. . . ." The historian of early American biology (Goode, 1901: 376) spoke approvingly of Josselyn's moderate tone and "picturesque style", but had only a moderately favorable opinion of his biological observations. The historian of American materia medica, Dr. Felter (1927: 8), referred to "the cheerful author, Josselyn" and quoted extensively and approvingly his lengthy sections on plants and medical observations. Dr. Felter reproduced Josselyn's illustrations of plants, supplied the botanical names for them as well as for most of the plants listed, and gave the present day names of the diseases so quaintly but recognizably described. Mood (1935) in a critical and perceptive essay analyzed Josselyn's writings against the background of the time in which he worked, speaking of his "timeless" and "freshness of appeal".

These writers recognized Josselyn's claim to important and very early contributions to American botany, zoology, materia medica, anthropology, politics, and medical observations, but no one of them, or any one else writing on history of science, as far as I can determine, has noted that the two books contain geological observations—of uneven quality to be sure—but of considerable importance.

Josselyn's writing is delightful and should be made available to present day readers. Josselyn was a man of "polite reading" and good family, was acquainted with physic, if not actually a physician, and displayed a continuous interest in matters of the kitchen and cellar.

Josselyn's *Rarities,* after an introduction, is divided into six sections descriptive of the "Natural, Physical and Chyrurgical Rarities": (1) Birds, (2) Beasts, (3) Fishes, (4) Serpents and Insects, (5) Plants, (6) Stones, Minerals, Metals and Earths. The section on plants and their culinary and individual uses is much the most extensive one. The first geological observation is in the introduction where Josselyn describes the aspect of the country and especially of the White Mountains and Mt. Washington (p. 3-4):[2]

The Country generally is Rocky and Mountanous, and extremely overgrown with wood, yet here and there beautified with large rich Valleys, wherein are Lakes ten, twenty, yea sixty miles in compass, out of which our great Rivers have their Beginnings.

Fourscore miles (upon a direct line) to the Northwest of *Scarborow* [Maine] a Ridge of Mountains run Northwest and Northeast an hundred Leagues[3], known by the name of the *White Mountains,* upon which lieth, Snow all the year, and is a Land-mark twenty miles off at Sea. It is rising ground from the Sea shore to these Hills, and they are inaccessible but by the Gullies which the dissolved Snow hath made; in these Gullies grow *Saven* Bushes, [dwarf conifers] which being taken hold of are a good help to the climbing Discoverer;

---

[2] [References are to the 1672 edition of the *Rarities* and to the 1674 edition of the *Account of Two Voyages.* I was able to consult these editions in the Grenville Kane Collection of the Princeton University Library. It is a pleasure to record the helpfulness of Miss Julie Hudson, Curator of Rare Books, and of Dr. Howard Rice, Jr., Chief of the Division of Rare Books and Manuscripts, in my studies in the Princeton Library in 1954.

Quotations from Josselyn are printed with the spelling and punctuation of the original, except for the medial long "s". An occasional comment or explanation within a quotation is indicated by square brackets [ ].]

[3] [The Grenville Kane copy, now in the Princeton Library, has "Leagues" struck out and "miles" substituted in what appears to be a seventeenth or eighteenth century hand.]

upon the top of the highest of these Mountains is a large Level or Plain of a day's journey over, whereon nothing grows but Moss; at the farther end of this Plain is another Hill called the *Sugar-Loaf*, to outward appearance a rude heap of massie stones piled one upon another, and you may as you ascend step from one stone to another, as if you were going up a pair of stairs, but winding still about the Hill till you come to the top, which will require half a day's time, and yet it is not above a Mile, where there is also a Level of about an Acre of ground, with a pond of clear water in the midst of it; which you may hear run down, but how it ascends is a mystery. From this rocky Hill you may see the whole Country round about; it is far above the lower Clouds, and from hence we beheld a Vapour (like a great Pillar) drawn up by the Sun Beams out of a great Lake or Pond into the Air, where it was formed into a Cloud. The Country beyound these Hills Northward is daunting terrible, being full of rocky Hills, as thick as Mole-hills in a Meadow, and cloathed with infinite thick Woods.

This description, the earliest printed of the White Mountains reads like one written by a man who had climbed Mt. Washington; the description of Bigelows Lawn, Lake of the Clouds, and the "ravines" seem to be derived from first-hand knowledge.[4] The country northward from Mt. Washington is still "full of rocky Hills, . . . cloathed with infinite thick wood." Josselyn was not the first to climb Mt. Washington, that was Darby Field in 1642 (Tuckerman, 1860: 35).

The most important physiographic observation of the introduction is that the "gullies" on Mt. Washington, now known as gulfs or ravines, were eroded by "dissolved [melted] snow". This appears to be the first recognition in America of the cutting of valleys by running water. It was to be almost 150 years until this view was generally accepted in England. Anyone who has made his way up Mt. Washington by way of Tuckerman Ravine on a warm late spring day when the water from the rapidly melting snow is running down the valley sides will recognize why Josselyn implies that the water in the brooks comes from melting snow and does not mention rainfall. It is interesting to note that the major valleys on Mt. Washington are in part due to glacial modification of earlier stream valleys (Goldwait, *et al.* 1951: 12).

The reference to the "pond of clear water" near the top of Mt. Washington is to the Lake of the Clouds. At the same time that Josselyn was a century and a half in advance of his time in recognizing (at least in part) the cutting of the valleys by running water, he was much less advanced in his thinking about the origin of the water in the lake. He could see the water run down the mountain from the lake but "how it ascends is a mystery". This is all the more remarkable when in the very next passage he describes cloud formation visible from this spot. His reference to water ascending is to the ancient and medieval belief in origin of springs and "sources" of streams in the hills and mountains from upward subterranean movement of water from the sea into the mountains where the water, desalted in some unspecified way, broke forth to the surface.[5]

[4] [See Tuckerman's note (1865: 36) for an analysis of routes and possibility of Josselyn's actual or hearsay information.]

[5] [For a discussion of these beliefs see Adams (1938: 426-460) who reproduced several diagrams illustrating the principle. Several of these diagrams are from the 1678 edition of Kircher's *Mundus Subterraneus;* the edition of 1665 seems not to have been available to Adams. Some of the diagrams of Kircher (1665) showing subterranean passages for upward movement and storage of underground water are on pp. 71, 148, 233, and 234. Kircher's books were widely distributed over Europe, and perhaps Josselyn may have seen a

Josselyn's sixth section, on "Stones . . ." is less than three pages long and half of that is irrelevant. The whole section is reproduced (1672: 92-94):

Of Stones, Minerals, Metals and Earths.
As first, the *Emrald* which grows in flat Rocks, and is very good.

Rubies, which here are very watry. I have heard a story of an *Indian,* that found a stone, up in the Country, by a great Pond as big as an Egg, that in a dark Night would give a light to read by; but I take it to be but a story.

Diamond, which are very brittle, and therefore of little worth.

Crystal, called by our West Country Men the *Kenning Stone;* by *Sebegug* Pond is found in considerable quantity, not far from thence is a Rock of Crystal called the *Moose* Rock, because in shape *Moose,* and

Muscovy Glass, both white and purple of reasonable content.
Black Lead.
Bole Armoniack.
Red and Yellow *Oker.*
Terra Sigilla.
Vitriol.
Antimony.
Arsnick, too much.
Lead.
Tin.
Tin Glass.
Silver.

Iron, in abundance, and as good bog Iron as any in the World.

Copper. It is reported that the French have a *Copper* Mine at *Port Royal,* that yieldeth them twelve Ounces of pure *Copper* out of a Pound of *Oar.*

I shall conclude this Section with a strange Cure effected upon a Drummers Wife, much afflicted with a Wolf in her Breast; the poor Woman lived with her Husband at a Town called by the *Indians, Casco,* but by the *English, Famouth;* where for some time she swaged the Pain of her Sore, by bathing it with strong Malt Beer, ... When she could come by no more Beer, (for it was brought from *Boston,* along the Coasts by Merchants,) she made use of Rhum, a strong Water drawn from Sugar Canes, with which it was lull'd a sleep; at last, (to be rid of it altogether) she put a quantity of *Arsnick* to the *Rhum,* and bathing of it as formerly, she utterly destroyed it, and Cured her self; . . .

His "Emrald" is probably beryl, sometimes found in schist; his "Rubies" garnets; "Diamond" probably quartz; "Muscovy glass" muscovite; the purple probably biotite; and "Bole armoniack" and "Terra sigilla" varieties of clays. The "Arsnick" was probably arsenopyrite, which is not uncommon in New England. To this day arsenopyrite in northern New England is sometimes believed by the uninformed to carry important precious metal values; perhaps that is what Josselyn meant by "too much". A small amount of tin has been found in New England and Josselyn's informant may have seen a bit of the ore from Jackson, New Hampshire.

It is difficult to explain the shortness of this section; Josselyn had travelled widely about New England and diligently, if indiscriminately, collected notes on plants and animals. It is noteworthy that all through the *Rarities* the author exhibits little sense of location; indeed, after the introduction, in which location is passably dealt with, he rarely mentions any locality. Josselyn's almost disdain of the value of the minerals is in direct contrast to the reports of earlier travellers who still had hope of finding mineral treasure and who describe, often overly optimistically, every observation or rumor about minerals. By Josselyn's time the settlers were convinced that, aside from iron ore, there were few minerals of value—the rubies were watery and the diamonds brittle and "of little worth". We leave to medical historians the

copy. At any rate, the concept was so "classical" and so much a part of "general knowledge" that Josselyn would have included it in his mental equipment.]

evaluation of Josselyn's report on the cure for "wolf" or cancer.

Josselyn exhibited a high regard for the Indians; his description of their physical features, while having nothing to do with geological observation, may interest the reader (1672: 99-100):

> The Men are somewhat Horse Fac'd, and generally Faucious, *i.e.* without Beards; but the Women many of them have very good Features; seldome without a *Come to me*, or *Cos Amoris*, in their Countenance; all of them black Eyed, having even short Teeth, and very white; their Hair black, thick and long, broad Breasted; handsome streight Bodies, and slender, considering their constant loose habit: Their limbs cleanly, straight, and of a convenient stature, generally, as plump as Partridges, and saving here and there one, of a modest deportment.

In his second book, *An account of two voyages to New-England* . . . published in 1674, Josselyn reproduced a large part of his diary of the voyages, in some parts without much editing. He recorded that on his first voyage, somewhere on the banks, on the 14th of June, 1638 (1674: 9):

> . . . we sailed by an Island of Ice . . . three leagues in length mountain high, in form of land, with Bayes and Capes like high clift land and a River pouring off it into the Sea. We saw likewise two or three Foxes or Devils skipping upon it. These Islands of Ice are congealed in the North, and brought down in the spring-time with the Current to the banks on this side New-found-land, and there stopt, where they dissolve at last to water.

Josselyn knew the fate of icebergs even if he was ignorant of their origin. Even men who had been in higher latitudes at first failed to recognize the origin of icebergs from glaciers, as may be seen in the early reports of the Frobisher voyages.

A day or two before entering Boston Harbor, on June 29th, Josselyn's ship had some difficulty, possibly at the Isles of Shoals, but the delay enabled them to hear news, partly geological, from Boston (1674: 11):

> The Nine and twentieth day, sounded at night, and found 120 fathome water, the head of the Ship struck against a rock; At 4 of the clock we descryed two sail bound for *New-found-land*, and so for the *Streights*, they told us of a general Earth-quake in *New-England*, of the Birth of a Monster at *Boston*, in the *Massachusets-Bay* a mortality, and now we are two leagues off *Cap Aun*.

The report of the earthquake is of that of June 1, 1638, the earliest recorded in the list of earthquakes in the United States (Heck, 1938: 11). Brigham (1871: 2) quoted several more extensive contemporary reports of this apparently widespread and at least moderately violent shock. New Englanders were always impressed by the score or more of earthquakes in the colonial period; they provided the occasion for many stirring sermons (Leet, 1948: 21-23).

Josselyn went on to describe the location of New England and its boundaries. Here he discussed the sea which forms part of the boundary and passed on to a description of the tides. His statements about depths of the sea are very early ones, perhaps the first which referred to the region off the American coast (1674: 42-43):

> . . . the Sea lyes East and South from the land, and is very deep, some say that the depth of the Sea being measured with line and plummet, seldom exceeds two or three miles, except in some places near the *Swevian*-shores, and about *Pontus*, observed by *Pliny*. Sir *Francis Drake* threw out 7 Hogsheads of line near *Porto-bello* and found no bottom, but whether this be true or no, or that

they were deceived by the Currants carrying away their lead and line, this is certainly true, that there is more Sea in the Western than the Eastern *Hemisphere*, on the shore in more places than one at spring-tides, that is at the full or new of the moon, the Sea riseth 18 foot perpendicular, the reason of this great flow of waters I refer to the learned, onely by the way I shall acquaint you with two reasons for the ebbing and flowing of the Sea; the one delivered in Common conference, the other in a Sermon at *Boston* in the *Massachusets-Bay* by an eminent man; The first was, *that God and his spirit moving upon the waters caused the motion;* the other, *that the spirit of the waters gathered the waters together; as the spirit of Christ gathered Souls.*

The spirit of the time is well illustrated in this passage in which in one sentence the author refers to Pliny, the classical authority, and in the next to Drake, the man who had actually attempted to measure depth by a sounding line. The inductive attitude, very often lost sight of by Josselyn, is exhibited here in the shrewdly expressed possibility of erroneous record of sounding because of currents acting on the lead and line.

Josselyn would have seen tides at least 18 feet high in Maine and heard of those much higher in the Bay of Fundy. The explanation of tides in a sermon in Boston "by an eminent man" would have been taken at face value by most of his readers, but it seems to me that Josselyn actually wrote this with tongue in cheek.

Josselyn described the New England shore and the territory back from it; the prominent mountain mentioned in the following passage is of course Mt. Washington (1674: 43-44):

The shore is Rockie, with high cliffs, having a multitude of considerable Harbours; many of which are capacious enough for a Navy of 500 sail, one of a thousand, the Countrie within Rockie and mountanious, full of tall wood, one stately mountain there is surmounting the rest, about four score mile from the Sea: The description of it you have in my rarities of *New-England*, between the mountains are many ample rich and pregnant valleys as ever eye beheld, beset on each side with variety of goodly Trees, the grass man-high, unmowed uneaten and uselesly withering; within these valleys are spacious lakes or ponds well stored with Fish and Beavers; the original of all the great Rivers in the Countrie, of which there are many with lesser streams (wherein are an infinite of fish) manifesting the goodness of the soil which is black, red-clay, gravel, sand, loom, and very deep in some places, as in the valleys and swamps, which are low grounds and bottoms infinitely thick set with Trees and Bushes of all sorts for the most part, others having no other shrub or Tree growing, but spruce, under the shades whereof you may freely walk two or three mile together; being goodly large Trees, and convenient for masts and sail-yards.

Josselyn's remarks on the intervales of northern New England and on the soils in them are early comments on pedology, but not the earliest in America; John Smith in 1612 mentioned the soils in Virginia (Smith, 1612: 3; White, 1953a: 136; 1953b: 125) and William Wood had a short chapter in *New Englands Prospect* of 1634 on "Soyle" (White, 1953a: 137). Like Smith, Josselyn realized the relation of plants to soil types and thus early were ecological observations made in America. Only a few years later in 1705 Robert Beverley in *History and Present State of Virginia* was to write in great detail and extent about Virginia soils and the relation of soils to vegetation.

In a time when the settlers were at first dependent on springs and streams for water supply and later on shallow wells, it is natural that considerable attention would be paid to these springs and to the quality of their water. Josselyn left no doubt about abundance and quality of water from springs, calling on the aid of poetry to reinforce his enthusiasm (1674: 44):

> The whole Countrie produceth springs in abundance replenished with excellent waters, having all the properties ascribed to the best in the world.
>
> *Swift is't in pace, light poiz'd, to look in clear,*
> *And quick in boiling (which esteemed were)*
> *Such qualities, as rightly understood*
> *Withouten these no water could be good.*
>
> *One Spring there is, at* Black-point *in the Province of* Main, *coming out of muddy-clay that will colour a spade, as if hatcht with silver, it is purgative and cures scabs and Itch. &c.*

The spring at Black-point was probably of sulfur water, which would have the effects that Josselyn ascribed to it. The curative properties of various mineral waters were usually described in writings of the time on natural resources in America as well as in England. An English example is the long chapter on the water of Oxfordshire by Robert Plot (1677: 18-50).

Josselyn is at his poorest in the remarks on metal deposits and on minerals and rocks in general. His credulity shows up in his reports of precious metal resources. In the *Rarities* he had said little about these, but allowed himself to speculate at greater length two years later in the *Two Voyages* (1674: 44-47):

> The Mountains and Rocky Hills are richly furnished with mines of Lead Silver, Copper, Tin, and divers sorts of minerals, branching out even to their summits, where in small Crannies you may meet with shreds of perfect silver; yet have the *English* no maw to open any of them, whether out of ignorance or fear of bringing a forraign Enemy upon them, or (like the dog in the manger) to keep their Soveraign from partaking of the benefits, who certainly may claim an interest in them as his due, . . .
>
> The stones in the Countrey are for the most *mettle*-stone, freestone, pebble, slate, none that will run to lime, of which they have great want, of the slate you may make Tables easie to be split to the thickness of an inch, or thicker if you please, and long enough for a dozen men to sit at. Pretious stones there are too, but if you desire to know further of them, see the Rarities of *New-England;* onely let me add this observation by the way, that Crystal set in the Sun taketh fire, and setteth dry Tow or brown Paper on fire held to it. There is likewise a sort of glittering sand, which is altogether as good as the glassie powder brought from the *Indies* to dry up ink on paper newly written.

The reference to mineral deposits branching out even to the summits of the mountains is a survival of the medieval idea of the tree of ores growing in the earth (Adams, 1938: 286). Perhaps it is a conscious or unconscious remembrance of the extensively developed exposition of Peter Martyr of Angleria translated into English by Eden in 1555, and reappearing in later editions.

The continued mixture of fanciful foolishness and accurate observation is illustrated by the jumble of lore about crystal and the accurate report on cleavable slate and on the lack of stone "that will run to lime", still a serious lack in New England's natural resources.

Josselyn discussed weather and climate at some length and concluded with a page on earthquakes in which his reference to the raising of the White Mountains by that agency is of interest. His proof for the hollowness of these mountains is startling, but of course they had to be hollow so the subterranean water could ascend and so the "tree of metals" could grow (1674: 58):[6]

In *Anno Dom.* 1667. March, appeared a sign in the Heavens in the form of a Sphear, pointing directly to the West: and in the year following on the third day of *April* being *Friday*, there was a terrible Earthquake, before that a very great one in 1638, and another in 58 and in 1662/3. January 26, 27, & 28. (which was the year before I came thither) there were Earthquakes 6 or 7 times in the space of three dayes. Earthquakes are frequent in the Countrie; some suppose that the white mountains were first raised by Earthquakes, they are hollow as may be guessed by the resounding of the rain upon the level on the top. The *Indians* told us of a River whose course was not only stopt by an Earthquake in 1668. (as near as I can remember) but the whole River swallowed up. And I have heard it reported from credible persons, that (whilst I was there in the Countrie) there happened a terrible Earthquake amongst the *French*, rending a huge Rock asunder even to the center, wherein was a vast hollow of an immeasurable depth, out of which came many infernal Spirits.

Josselyn next went on to extensive lists and description of trees and other plants (1674: 59-122; the famous panegyric on tobacco is on p. 76); to mammals, including the egg-laying porcupine, birds, fish, mollusks, snakes, and insects (including the amusing tale of the farmer and the bumble bee). Much of the description is from rumor and hearsay and some of it worth just about as much as the reward Josselyn believed would be his in the ribald quotation from Ben Jonson (1674: 123).

After a long description of towns and rivers Josselyn presented the final observation of geological interest on the "clay bullets cast up by a mineral vapor" near the "Kenibunck River" which " is by the Town of Wells" (1674: 204). This is no doubt in reference to the concretions in the late Pleistocene marine clay.

Josselyn was like many other writers of the time who touched on scientific subjects and even like some who devoted most of their time to scientific activities—he could write accurate description and could also report outrageous rumor as probable fact. He had some classical knowledge and liked to exhibit it. John Lederer (1672), the German physician, who at almost the same time reported on a part of the colonies much farther south, was far less chatty and perhaps less credulous than Josselyn, but Lederer's report tells us much less of what was being seen, conjectured, and written about at the time.

The writer on geological subjects in England at about the same time who invites comparison with Josselyn is Robert Plot, whose *Natural History of Oxfordshire* appeared in 1677. Dr. Plot was Professor of Chemistry, first Keeper of the Ashmolean Museum at Oxford, and secretary of the Royal Society. His book on Oxfordshire was much longer than both little books of Josselyn's which deal only incidentally with natural history. However, it should be noted that the Oxford professor was not sure that streams were supported by rainfall, gives no indication of any thought of valley cutting by streams, and, although he publishes many excellent plates of fossils, cannot bring himself to be

[6] [The listing of earthquakes corresponds fairly well with the listing of Brigham (1871: 2-4). If we read "1663" for "1668" for the Indians' story, it then also refers to the great 1663 St. Lawrence valley earthquake "amongst the *French*", which is known from several sources to have been violent even in New England (Brigham, 1871: 3, 4) and was the most famous earthquake of the colonial period (Heck, 1936: 135).]

sure they are actual remains of real organisms. Plot wrote more voluminously, observed more acutely, repeated wondrous tales, and quoted more classical references, but considering his position and advantages he was not nearly as much better than Josselyn as he should have been in development of explanations and exhibition of the inductive method.

Josselyn should not be criticized for credulity and inaccuracy — he was quite abreast of his time. Although his books were general reports on New England and not primarily on natural history, they stand up surprisingly well when compared with the productions of the professionals in England at about the same time—those of Plot, Ray, or Lhuyd, for example. Furthermore, it is probable that some of Josselyn's information is from his 1638 notes of a third of a century earlier.

In summary, Josselyn's major contributions were on:

1. The sea off New England, and its depth, determined by actual measurements.
2. Tides and their (probably incorrect) causes.
3. Icebergs, accurate as to their fate, but inaccurate as to their origin.
4. Mt. Washington, its topography and its valleys, believed to have been cut, at least in part, by running water; and on the lake near the summit with unknown origin of its "ascending water".
5. Mineral resources, but in a disappointingly brief and partly inaccurate way, except for iron ore, building stone, and clay.
6. Earthquakes and their possible cause of mountain building, as well as in producing "hollowness" of the mountains.
7. The quality and character of groundwater.

Because of his worthwhile observations, especially on topography, John Josselyn deserves a place in the history of geology in America.

Literature Cited

Adams, F. D. 1938. Birth and development of the geological sciences. Baltimore, Williams & Wilkins. 506 pp.

Brigham, W. T. 1871. Volcanic manifestations in New England: being an enumeration of the principal earthquakes from 1638 to 1869. Mem. Boston Soc. Nat. Hist., 2: 1-28.

Eden, Richard. 1555. The decades of the New World. . . . London, Guilhelmi Powell. 362 pp.

Felter, H. W. 1927. The genesis of the American materia medica including a biographical sketch of "John Josselyn, Gent". . . . Cincinnati, Bull. Lloyd Library Reprod. Ser. No. 83: 64 pp.

Goldthwait, J. W., L. Goldthwait, and R. P. Goldthwait. 1951. The geology of New Hampshire, Part I—Surficial Geology. N. H. State Plan. and Devel. Comm., 83 pp.

Goode, G. B. 1901. The beginnings of natural history in America. Rept. U. S. Nat. Mus., 1897, pt. 2, 515 pp.

Goodwin, Gordon. No date. John Josselyn. Dict. Nat. Biog., 10: 1096.

Heck, N. H. 1936. Earthquake. Princeton, Princeton Univ. Press. 222 pp.

.... 1938. Earthquake history of the United States. U. S. Coast and Geodetic Surv. Ser. No. 609: 85 pp.

Josselyn, John. 1672. New-England rarities discovered. . . . London, G. Widdowes. 114 pp.

.... 1674. An account of two voyages. . . . London, G. Widdowes. 279 pp.

Kircher, Athanasius. 1665. Mundus Subterraneus. Amstelodami, J. Jansonium & E. Weyerstraten. 346 pp.

Lederer, John. 1672. The discoveries of John Lederer . . . begun in March

1669, and ended in September 1670 . . . translated out of the Latin . . . by Sir William Talbot. London, J. C. for Samuel Heyrick. 27 pp.

LEET, L. D. 1948. Causes of catastrophe. New York, McGraw-Hill Book Co. 232 pp.

MOOD, FULMER. 1933. John Josselyn. Dict. Amer. Biog., 10: 219-220.

.... 1935. Notes on John Josselyn, Gent. Colon. Soc. Mass. Public., 28: 24-36.

[?] OLDENBERG, HENRY. 1672. [Review of New-England rarities]. Phil. Trans. Roy. Soc., No. 85: 5021-5023.

PLOT, ROBERT. 1677. Natural history of Oxford-shire. . . . Oxford, Oxford Univ. 358 pp.

SMITH, CAPTAIN JOHN. 1612. A map of Virginia, with a description of the countrey. . . . Oxford, Joseph Barnes. 153 pp.

TUCKERMAN, EDWARD. 1860. History of exploration of the White Hills. In T. S. King, *The White Hills*, Boston, Crosby, Nichols and Co. pp. 34-36.

.... 1865. Introduction and notes [to] New-England's rarities. . . . by John Josselyn. Boston, William Veazie. 169 pp.

TYLER, MOSES C. 1885. History of American literature. New York, G. P. Putnam's Sons, 1: 180-185.

WHITE, G. W. 1953a. Early American geology. Sci. Monthly, 76: 134-141.

.... 1953b. Geological observations of Captain John Smith in 1607-1614. Trans. Ill. Acad. Sci., 46: 124-132.

WOOD, WILLIAM. 1634. New Englands prospect. London, Tho. Cotes for Iohn Bellamie. 83 pp.

ENGINEERING GEOLOGY IN CONSTRUCTION OF THE NATIONAL ROAD 1825

From THE ENGINEERING GEOLOGIST (Engineering Geology Division of the Geological Society of America), Volume 1, No. 3, October 1966, pp. 11-14.

- -

ENGINEERING GEOLOGY IN CONSTRUCTION OF THE NATIONAL ROAD - 1825

George W. White

Engineering geology is a relatively new term, but the connection of engineers and geology is very ancient, as shown by the great canal engineers Vitruvius, Leonardo da Vinci, and William Smith, "the father of stratigraphic geology". The way in which geology was taken into account in canal building would be a monographic study for a historian of geology and of technology. It is also known that geology was taken into account in the building of turnpikes and other great new roads of the early United States. The kinds of geologic information used in planning and construction of a part of the famous National Road, which eventually ran from Cumberland, Maryland to St. Louis, are illustrated in a 32 page report, House of Representatives, War Dept. Document No. 51, 19th Congress, First Session; Road from Wheeling to Missouri..... report of the chief engineer in relation to the survey, location, and progress of the road from the right bank of the Ohio, opposite to Wheeling, to the state of Missouri, Washington: 1826. This report actually deals only with that part in Ohio from Bridgeport, opposite Wheeling, to Columbus, of the road always called the "National Road" or "National Pike", and much later to become U. S. Highway 40.

Major General Alex. Macomb, Chief Engineer, U. S. Army, transmitted to Secretary of War James Barbour copies of the correspondence with Jonathan Knight, Commissioner for the Continuation of the Cumberland Road, and with Caspar W. Wever, Superintendent of the Road, together with the reports of the commissioner and superintendent. In this present day of elaborate instructions couched in legal phraseology it is refreshing to read the simple instructions to Knight and Wever to build a road from the Ohio River to Columbus.

Jonathan Knight (1787-1858) was already known in 1825 as a competent civil engineer and was later to be known as "one of the notable engineers of his time" (Dict. Am. Biog.). In letters of March 29 and March 30, 1825, covering less than three printed pages, General Macomb instructed Knight to review the preliminary location from Wheeling to Zanesville, change the route if he thought advisable, locate a route from Zanesville to Columbus and thence to Indianapolis, Vandalia (Illinois) and St. Louis. The Commissioner was to receive $6.00, each surveyor $3.00, and each other necessary assistant $1.00 per day. To defray "the expense of surveying and marking the road" $10,000 was appropriated and Knight was instructed to draw on the department for required sums using "due economy".

Knight was also instructed, on the plat of the road he was to submit, to "designate the hills and mountains, towns, villages, and settlements; also, the rivers, runs, marshes, woods and rocks, and any other remarkable objects......The quality of the stones or rocks, as well as of the soil, will demand your particular notice and examination. You will also state whether there is.... the proper materials for making permanent roads, to be found on the route, or near enough to be applied for that purpose......It is unnecessary to be more particular, having entire confidence in your abilities to execute the same." In other words -- "build the road".

The route was to be planned for a width of 80 feet. The road bed was to be 30 feet with "hills to be cut down, the earth, rocks and stumps removed". The cuts and fills were to be such that no grade was to be greater than four degrees. On the fills the sides were to slope not exceeding 30 degrees and proper allowance was to be made for settling in the fills and "no stumps, logs or wood of any kind be permitted in the filling". In necessary places "side walls must be built for sustaining the earth" where the "nature of the ground" would require. Special precautions were to be taken to prevent "the earth from falling or slipping" in cuts and fills. The roadway was to be carefully drained by sufficient ditches. The surface was to be "covered 20 feet in width with stone broken into pieces weighing not more than 4 ounces". The first layer of stone was to be 3 inches, and "no stones to be used but such as hard or granite, flint or limestone". An additional cover of 3 inches was to be put on in the following year, but it appears the engineers were allowed to use their own judgment about the time of the second layer. The road was to be made according to the principles of McAdam "that you are particularly referred to".

The instructions further stated that "all the masonry must be executed in the best manner, that is, durable in character, not extravagant in embellishment, but plain and strong and of the best materials". Particular attention was to be paid to the mortar. It appears that General Macomb had been informed that careless mixing of mortar and even use of clay was characteristic of the frontier. It appears that Mr. Wever examined the ground in the next few weeks and wished to modify the instructions for one three-inch layer of stone now and another after a year to two three-inch layers immediately. General Macomb informed Mr. Wever that he would like to have a fair trial of the original instructions, but "after the first year three inches more will be put on and if that quantity does not answer three inches more the next year, which will give nine inches and so on and so on, until the road shall have attained a complete and impervious cover". No doubt Mr. Wever was correct, because over half of the route from the Ohio River to Zanesville lies upon Pennsylvanian Conemaugh rocks, which include much fine-grained shale and the resulting clay loam soils; three inches of crushed stone is not sufficient for a road bed on these soils.

Within the following few weeks Knight and Wever had decided between two alternate routes, choosing the more direct one through Fairview rather than a more southerly one through Barnesville. Although they had been instructed to let the contracts in two mile or larger units they actually let the contracts for units as little as one-half mile in length. One reason for doing so was that if larger units were let, only a few people could bid and these would then proceed to let smaller units to sub-contractors, so it was decided to let the routes directly to those who would otherwise be sub-contractors. Another reason was that "each section would include throughout ground as nearly alike as possible", so that estimation and bidding could be more precise, thus saving money.

The final seven pages of the little document is a "Memoir" which is the interim report of Commissioner Knight to December 22, 1825. It is in this report that we discover much of the geological reasoning used by Knight and Wever in the location and construction of the road. The route from the Ohio River to Zanesville was over the unglaciated Allegheny Plateau with a relief of several hundred feet. The route, consistent with directness, generally followed the ridges. Where it followed favorably situated valleys it was cleverly located far enough above the streams to be generally above flood level. It was recognized that the route westward from Zanesville to Columbus presented more variety: "As we proceed from Zanesville, the country, for the first seven miles, is literally cut up with drains and tributaries .... West of that for about thirteen miles, the country is uneven...... The hills thus formed, are not so high as those east of Zanesville, but they are, if possible, more irregular." The hills terminated 20 miles west of Zanesville and "west of the termination of the hills....... the ground to be surveyed" was on the "Great Plain" which continued far beyond Columbus. Knight took into account these changes in topography, relief and soils, but of course did not realize their cause. The boundary of the Illinoian glacial drift runs approximately north-south about 7 miles west of Zanesville and the 13 mile wide area of Illinoian drift consists of more rounded bedrock hills than those in the unglaciated area to the east. Also other "irregular hills" (kames) in the Illinoian tract required unusually detailed exploration of alternate routes in order to find the most suitable one.

Twenty miles west of Zanesville is the very sharp and noticeable Wisconsin glacial boundary which almost coincides with one of the escarpments of the plateau, to the west of which the rocks are Mississippian shales and siltstones, with very little real sandstone. From this point on it was noted that the main problem in location would be to avoid the swamps, particularly the large one which was later flooded to become Buckeye Lake, and to find suitable places to cross several of the major north-south streams tributary to the Scioto River, whose floodplains were incised below the level of the plains. Particular attention was paid to the materials for constructing the road: "Throughout the hilly country, good materials are convenient and abundant for the mason work. For the metal or paving of the road, not so much so. There is plenty of sandstone;but this description should only be used where limestone, or freestone of a hard quality, cannot be procured under reasonable expense." There was indeed sufficient sandstone for dimension stone located close to the road from Wheeling to Zanesville. Some of the beautiful sandstone bridges still remain in roadside parks or near the now relocated U. S. Highway 40, attesting to the skill in design and construction from local stone. Limestone and "freestone" were reasonably available for road metal.

West of Zanesville it was stated that "stone can be had in sufficient abundance....... (for)....... a distance of 20 1/4 miles, after which, stone becomes very scarce and the mason work is rated accordingly". This was the area of Mississippian shale and thick glacial drift. Knight noted that outcrops of sandstone were many miles away from the route and in a restricted quantity, but that in the Scioto valley at Columbus, limestone (Devonian) would be available. It was therefore recommended "that the remainder of the road to Columbus (that is beginning 20 miles west of Zanesville) be not stoned, but gravelled", because gravel is available throughout this area. Indeed, it was discovered that "about three miles of the route, between the system of swamps in the south fork of the licking", the road would pass "on a sub-stratum of clay about one foot deep and underneath it gravel. Here the grading of the road would also gravel it". In other places it was thought that gravel could be found within a distance of not more than two miles. They clearly recognized the vast quantity of gravel although they had no idea that its origin was glacial.

The material for bridge construction was taken into account; the bridges in the hilly part were "near to good rock. They are estimated to be constructed altogether of stone." On the route from 20 miles west of Zanesville to Columbus, "where rock is scarce", it was suggested that the bridges "be constructed of stone underneath and a superstructure of wood to be weather-boarded and roofed for preservation from decay". Such construction was successful -- one of the largest of these covered bridges, at Hebron, in use for more than one hundred years, testified to the satisfactory structure of the abutments on rather difficult foundations.

It is apparent from reading the documents that Knight, the Commissioner, and Wever, the Superintendent, worked harmoniously together and that each had the complete confidence and respect of the other. They used the latitude and discretion allotted them and they unquestionably selected the best route and made the best use of local materials in construction. They had examined the character of the soil and the possible construction materials far enough in advance so that the road building could proceed expeditiously. Thus, the greatest road of the young Republic was brought to quick completion and became an important factor in the development of the western country.

- -

ACCOUNTS OF ILLINOIS GEOLOGY
BY JOHN BRADBURY
IN 1817

Reprinted from the Illinois State Academy of Science
Volume 60, Number 4—1967

ACCOUNTS OF ILLINOIS GEOLOGY BY JOHN BRADBURY IN 1817

GEORGE W. WHITE
Department of Geology, University of Illinois, Urbana

ABSTRACT. — John Bradbury (1765-1823), an English botanist well skilled in mineralogy and geology, traveled in Illinois in 1809, 1810, and 1811. The important geological observations made in Illinois and included in his report of 1817 have never been recorded and deserve attention. He recognized the flat-lying strata of limestone, shale, sandstone and coal and asserted they had been deposited after a long period of time and had not been affected by later "convulsions." His description of Illinois fluorspar is probably the earliest known.

One of the early accounts of geological features in Illinois is contained in *Travels in the Interior of America,* 1817, by John Bradbury (1765-1823), the English botanist. He came to America primarily to collect plants, but he was well informed in chemistry, was a competent mineralogist and had geological interests. He let little of scientific interest escape his keen observation (True, 1929). He traveled in the United States from 1809 to 1811 and was in Illinois in 1809 and 1810. In 1811 he traveled up the Missouri River almost to the Montana state line before returning to St. Louis and thence to New Orleans. His book contains an exciting account of this part of his travels. His travels in Illinois were along the Ohio and Mississippi rivers from which he made some excursions inland, as he traveled leisurely and was not in haste to reach a destination. From his base at St. Louis in 1810 he made many trips of 80 or 100 miles, and it is apparent that on some of these he crossed the Mississippi and examined the Illinois Territory for many miles east of the river.

Bradbury's book was widely read, and a second edition appeared in 1819. His literate and perceptive descriptions of the "western country" became part of the basis of accumulating general knowledge of its physical geography and geology. He had keen appreciation of the high value of Illinois soils and of the mineral resources of coal, limestone, salt and lead. But Bradbury was more than an accurate describer of natural features—he must be credited with some important analyses as well. He not only recognized that the strata westward from the Appalachian Mountains, including those in Illinois, were essentially horizontal, but also realized that this indicated quiet conditions of marine deposition and that later "convulsions" had not affected them.

At a time when it was not generally agreed that valleys were cut by the streams now flowing in them,

Bradbury was sure that the valleys he saw were stream eroded and the strata that matched on both sides of the valleys had originally been continuous. He regarded some of the level surfaces as having been produced by erosion from once higher levels. He realized that a long period of time was necessary for the deposition and later erosion of the rocks he saw. Although he did not mention Hutton or Playfair, his opinions on length of geologic time and slow action of geological forces are in harmony with their views.

One of the important geological events of pioneer Illinois history was the series of great earthquake shocks which began on December 16, 1811, and continued for many months. The whole state was affected, but most strongly in the southern part, which was very close to the center of the disturbance across the Mississippi River at New Madrid, Missouri. Bradbury had just passed the southern tip of Illinois and had reached New Madrid on December 16 when at 2:00 a.m. the first of the great earthquake shocks took place. Bradbury was the closest to the earthquake of any man of scientific experience and his factual report (p. 199-207) is a most valuable eye-witness account (Fuller, 1912, p. 9, 18).

Bradbury's book is divided into three main parts: the first deals with the Missouri Valley journey in 1811 and is not germane to the present discussion. The second part, "Description of the Missouri Territory" (p. 234-278) and the third, "Remarks on the States of Ohio, Kentucky, and Indiana, with the Illinois and Western Territory. . ." (p. 279-364), contain the Illinois references. His recognition of the difference between the surface and strata in Illinois and those of the eastern part of the country is illustrated by the statement that:

> In . . . geological formation, this country . . . is much more level, and the strata more regular and undisturbed. In general the order of the strata is sand lying on sand-stone, afterwards lime-stone, beneath which is argillaceous schist lying on coal (p. 308).

His recognition of the origin of the strata, of the surface and of the valleys, together with the long time required is well summarized in this passage:

> Several geological facts tend to prove that this portion of the globe has been peculiarly exempted from the operation of local and disorganizing convulsions, and that it has remained for a vast length of time in its present state. The most prominent of these facts is the undisturbed uniformity of the strata, and their general parallelism to the surface of the earth . . . and in the limestone rocks, wherever they occur. The depth and extent of the valleys of the river, together with the peculiar formation of the hills, tend to confirm the opinion, that whatever changes have taken place on the surface have been effected by the operations of a slow, but continually acting cause (p. 241-242).

Bradbury's reference to Illinois minerals includes one of the earliest on record of oolitic limestone, galèna and fluorite. His reference to Illinois fluorite is earlier than that of Jessup (Cleaveland, 1822, p. 202-203) or of Joseph Baldwin of Shawnee Town (1818, p. 52):

> About fourteen miles west of the Ohio Saline, in the Illinois Territory, there is a lead mine, which was discovered by a gentleman from Tennessee, of the name of Guest. It is not yet worked, but seems to promise well. Some small excavations have been made, and a quantity of galena found. It appears to have no connection or affinity with the

mines of St. Genevieve, not only on account of the distance being about 150 miles, but from the marked difference in the rock which is the matrix of the ore, and in the substances which are concomitant with it. The rock in this mine is of that species of limestone called kettonstone, or compact limestone of Kirwan, and consists of very small accreted round granulations. The ore is mixed with very beautiful fluor spar, of several colours, as blue, brown, yellow, and pellucid (p. 287).

Bradbury's recognition of why little of the geology can be known in the early stages of settlement of a "fertile country" with low relief, such as Illinois, is included in the following passage describing the surface and some of the mineral resources:

From the Alleghanies to the Lakes there are no mountains, or scarcely an elevation deserving the name of a hill, the bluffs which border the rivers excepted. It is nevertheless relieved from the dull monotony of a level plain by numberless valleys, through which the streams flow, and by small elevations, termed ridges. The soil is much superior to that of the countries east of the Alleghanies. . . . In the early stage of the transition of a fertile country from a state of nature to that of improvement and the arts, but little can be known as respects its geological formation, or, in other words, what it may contain as regards minerals. In general the first settlers are only cultivators of the soil, and never examine to any considerable depth below the surface, except by digging wells. All that is known of this country on those subjects has been discovered in that way, or has manifested itself on the surface, and is confined, as regards useful articles, to coal, salt, iron, lead, and nitre. . . . Coal . . . may be presumed to be very abundant. In examining the beds of most of the rivers, rounded nodules of coal may be found mixed with the stones and gravel, and beds of argillaceous schist, containing vegetable impressions, are frequent: in some instances these beds contain masses of pyrites. Coal is actually found at Pittsburgh, at Zanesville, on Green River, in the Illinois, and in the western territories. It is uniformly bituminous, and highly charged with that substance. In all these instances it has manifested itself on the surface of the earth, and indicates almost inexhaustible beds.

Salt, the most useful article at present, is found in various places, but as yet only in a state of solution, and has mostly been indicated by the excavations made by wild animals before the country was discovered by the whites (p. 283-284).

We wish that John Bradbury, who saw so much of the geology of southern and of west-central Illinois, had written a narrative account of his travels in the state. However, we are grateful for his record of observations of Illinois geology, which led to such respectable general conclusions about continuity of strata and the great length of geologic time during which slow and long-continued forces operated to produce the surface forms and rock structures of Illinois.

Literature Cited

Baldwin, Joseph. 1818. New locality of fluor spar, or fluate of lime, and of galena, or sulfuret of lead. Am. Jour. Science. 1:52-53.

Bradbury, John. 1817. Travels in the interior of America, in the years 1809, 1810, and 1811; including a description of Upper Louisiana, together with the states of Ohio, Kentucky, Indiana, and Tennessee, with the Illinois and western territories. . . . Printed for the author, Liverpool. 364 pp. (Second edition, with map, 1819.)

Cleaveland, Parker. 1822. An elementary treatise on mineralogy and geology. . . . Second edition. Cummings and Hilliard, Boston. 819 pp.

Fuller, M. L. 1912. The New Madrid earthquake. U. S. Geol. Survey Bull. 494. 119 pp.

True, R. H. 1929. A sketch of the life of John Bradbury including his unpublished correspondence with Thomas Jefferson. Proc. Am. Phil. Soc. 68:133-150.

Manuscript received February 3, 1967.

EARLY GEOLOGIC OBSERVATIONS IN THE CENTRAL INTERIOR OF THE UNITED STATES

EARLY GEOLOGIC OBSERVATIONS

IN THE

CENTRAL INTERIOR OF THE UNITED STATES

George W. White
University of Illinois

Presented at the New Hampshire Conference on History of Geology, Rye Beach, September 7-12, 1967. Cecil J. Schneer, Convenor and Editor.

The first half of this paper was published in a shortened form as "Early Geological Observations in the American Midwest:, in Toward a History of Geology, C. J. Schneer, Editor, Cambridge, M.I.T. Press, 1969, pp. 415-425.

INTRODUCTION

Geologic observations in America were made along the Atlantic seaboard by some of the first explorers, Verrazano in 1524, Thomas Hariot in 1588, and Captain John Smith in 1612 (White, 1953). There was almost no travel by the English across the Appalachian Mountains until about 1700, although one Englishman may have reached the Ohio River as early as 1669 (Banta, 1949, p. 49). In the first half of the 18th century more and more travelers and traders moved beyond the folded rocks of the mountains into western New York and western Pennsylvania and on into the eastern margin of the "Ohio Country."

Until about 1763 the "Western Country" meant western Pennsylvania and New York, but after 1763, and more especially after 1783, it would mean the whole central interior west to the Mississippi River and northwest to its headwaters and the whole Great Lakes basin as well--the region which was to become the Northwest Territory soon after the establishment of the new United States. In this study I shall use the term "Central Interior" to mean this region west of the folded Appalachian Mountains, essentially the Allegheny Plateau, the Lake Plains, and the Interior Plains. This is almost coincident with upper Mississippi basin, plus the southern part of the Great Lakes basin. The Missouri basin does not enter the discussion until after 1800 because until 1803 the region west of the Mississippi was Spanish and later French territory and not part of the United States. I will use practically synonymously "Mississippi Basin," "trans-Allegheny Region," and "Central Interior." This is the region of almost flat-lying Paleozoic sedimentary rocks in the eastern two-thirds of the region, but consisting of Cretaceous rocks in the upper Missouri Basin. Most of this region (except for most of Kentucky and parts of Missouri) is glaciated, but this received only incidental recognition until after the period of the present discussion.

The present essay attempts to summarize the geological observations in English to 1804. A further study will be concerned with extending the record to 1820 and another until 1840, when the period of "modern geology" begins. It is hoped that after the observations are recorded, further study can be made of the development of concepts and this influence of the geological features of the Central Interior in development of geological thinking in the United States, and indeed, in other parts of the world. Here was the first area of really extensive horizontal Paleozoic sedimentary rocks that was studied. It was distinctly different from the folded Appalachians with the parallel valleys (some of them notably prosperous) and barren mountain ridges, the hills of the metamorphic and igneous Piedmont and New England, and even from the gently dipping rocks of the Coastal Plain, which were unconsolidated and more like "alluvium" than "real rocks." The only comparable area of "platform" (cratonic) rocks is in eastern Europe and perhaps some day someone interested in development of history of geology in the North American craton and someone similarly interested in developments of history of geology of the eastern European craton can compare the two areas.

This paper does not deal with the many observations by the French when they controlled the northern and central parts of North America (e.g., coal near Joliet, Illinois, was described as early as 1673). Nevertheless, one publication should be specifically mentioned, that of Jean Etienne Guettard (1715-1786) who presented to the Royal Academy in Paris in 1752 a memoir on the minerals and rocks of North America which was published in 1756. Guettard, who was never in North America, compiled a map from French publications and through personal correspondence. He had access to many French maps, one as early as 1703 (Tucker, 1942, pl. 13, 15, 17), which showed mineral locations in the upper Mississippi Valley. The Guettard map is not only the earliest geological map of this continent but one of the oldest geological maps of any continent (Wells, 1959). It is truly a geological map--the "Bande Sabloneuse" is shown along the east coast and the "Bande Marneuse" in shaded pattern. The "Bande Schisteuse ou Metallique," farther west, includes the whole Mississippi Valley. "Illinois" is located on the map and in it the symbols for "cristal," iron, lead, and other minerals, and nearby, coal and petroleum. The interesting legend is made up in part of the ancient alchemical symbols.

The final defeat of the French in North America and the peace of Versailles in 1763 opened the "Western Country" to legitimate and authorized exploration, travel and trade. For over a century this region had been either French territory or under *de facto* French control and the English on the Atlantic seaboard knew of it only vaguely through reports by the French or from oral reports of a few daring English frontiersmen who made quick and clandestine trips as far as the lower Ohio Valley. Immediately after 1763 several colonial Americans, army officers who had fought the French and their Indian allies, set out on exploratory and mapping expeditions in the newly acquired vast territory of the Mississippi-Great Lakes basins. The best known of these, Robert Rogers, Jonathan Carver, and Thomas Hutchins, produced books in the west fifteen years that informed English-Americans about the "Western Country" and included geological information. It is these books that provided knowledge of the geography and information about drainage, topography, and geological features. These, in turn, provided a basis for the accounts of more scientific travelers who, beginning with Volney in 1803, published organized accounts of the geology of the trans-Allegheny region.

LEWIS EVANS

Although Lewis Evans (1700-1756) in his travels reached only the eastern part of the Mississippi Basin, his great map of 1755 included territory as far west as Illinois and deserves an honored place in the earliest history of geology of the Middle West (White, 1951). Lewis Evans was the great colonial cartographer, surveyor, proto-geologist, engraver, and pamphleteer (Gipson, 1939). His map of the British colonies is almost a geological map, for he showed on it mountain ridges, the Allegheny front, the glaciated part of northern Ohio--"level rich land intermix'd with swamps and ponds"--location of deposits of petroleum, coal, clay, freestone, salt, and "elephant bones." Extending to the Mississippi, the map has an inset portion for Illinois and Indiana.

Lewis Evans' Map of the Middle British Colonies (1755) extended as far west as the Mississippi River and the accompanying text, An Analysis of the General Map, contains some descriptive material of the Mississippi Valley, most of which he had secured from others. His great map was reprinted with or without credit over a score of times through a period of more than 40 years. This map and accompanying text is one of the great treasures of early American geology. The Analysis has many pages of organized geologic and geomorphic description and some theoretical speculations of high merit on fossils, isostasy, erosion, peneplains, and Coastal Plain origin. Evans himself did not have personal knowledge of the country west of Pennsylvania, but he effectively synthesized the more important and accurate observations of others.

Evans' classification of physiographic provinces of the United States is the first such attempt; it is accurate and in broad outlines it is followed today. Though usually unacknowledged, his physiographic classification and descriptions provided the basis for classification by many later authors. His work was important and widely influential, and had he lived longer he might well have held a position in America comparable to that of the Englishman, William Smith, who is considered the "father of stratigraphy."

PETER KALM

Although Peter Kalm (1716-1779), who traveled in the United States from 1749 to 1751, only touched the edge of the Mississippi Valley, he did get beyond the mountains into the flat-lying Paleozoic strata and he knew that a very extensive plains region of horizontal beds extended far to the west. His descriptions of Niagara Falls, published in 1751 (in Bartram, 1751), was a very early one in English.

Kalm's Travels, published in Swedish in 1753, was published in English in 1770-1771. His book contains more than 150 excellent geological observations and includes perceptive speculations about the origins of rocks and their structures. Although his remarks are almost entirely on the eastern part of the continent, he deserves at least mention here. His descriptions, not available in English until 1771, were widely quoted in Europe and in America were the source of information to many writers beginning with the geographer Morse (1789).

MAJOR ROBERT ROGERS

Major Robert Rogers (1731-1795) of Rogers' Rangers fame, was born in New England and early achieved brilliant success in the Seven Year War and in campaign against Quebec. Immediately afterward he was sent to command the post at Detroit, from which he traveled even farther northwest. He soon went to England, where his book, A concise account of North America, was published in 1765. He returned to the Great Lakes region in 1766 and was commander at Mackinac when Jonathan Carver set out on his trip to the northwest. Rogers returned to England, where in London his career ranged from "checkered" to distinctly unhappy. The unfortunate circumstances of his later life are a sharp contrast to the brilliant military, geographic, and literary successes of his earlier career.

Rogers' book is based on his personal knowledge of the interior through his journeys along several routes in Canada, his travels in the Great Lakes and Mississippi basins, and his years of residence on the far frontier. In his book he describes "the face of the country" and includes considerable detail about lakes and rivers. His geological references are not collected in one section, but are mentioned in connection with other matters. Rogers knew about the oil seeps in Oil Creek, Pennsylvania, referring to them as:

> ...those remarkable springs, greatly esteemed by the Indians as a remedy for almost every disease; they are called the oil-springs, on account of an oleous substance that issues forth with the water, and rises upon the surface of it. The Indians use these springs for consumptions, asthmas, and various internal disorders, by drinking the water; and for rheumatick pains, strains, dislocations, &c. by bathing, with great success. (p. 177)

Rogers speaks of the territory that we now know as "Middle West" in a most favorable way--too optimistically indeed of the prospect of a great city at the mouth of the Ohio River; he could not know in 1765 that it would be the railroad nexus at the south end of Lake Michigan--Chicago--and not Cairo at the junctions of the rivers that would be one of the world's major cities.

> The country between the lakes and the junction of the Ohio and Mississipi, for several hundred miles, and all the country between Fort St. Joseph and the Green Bay, and between Detroit and the Illinois, and even much further north than Detroit, is level, the soil excellent, the climate healthy and agreeable, and the winters moderate and short. Its natural productions are numerous and valuable. It is sufficiently, but not too thickly timbered; what there is, is tall and fair, and fit for any common use. In short, no country in this quarter, if any in the world, is capable of larger or richer improvements than this.
>
> There is a good coal-mine near Fort Pitt, made use of by the garrison for fuel; and what is still more in commendation of this country, it is extremely well watered by springs and rivulets, and has an easy communication with the whole world from the mouth of the Mississippi, and with great part of the interior country of North America, by its several branches, the Wabach, Masauris, Yeahtanees, the Ohio, &c. and with the great lakes by way of Presque Isle, where a small expence would turn the waters of the lakes into the Ohio. At present the portage is but a small distance, and the land level. Indeed such is the situation of this country, that, at or near the junction of the Ohio and Mississipi, in my opinion, within a century or two, it will be the largest city in the world; for hither flow, and here center, the exports of all the country upon the Mississipi, above and upon the Ohio, the Misauris, the Wabach, the Tanesee, the great lakes, &c. &c. The imports to this country will be most easy and natural up the rivers St. Lawrence and Hudson, by way of Montreal and Albany, and up the Delaware to Philadelphia, and from thence over the mountains, the navigation up the Mississipi being difficult. (p. 199-200)

Rogers' <u>Concise account</u> was the first book in English to give information from personal knowledge of the western country. It is particularly important because, scanty as the geological information was, it did provide a basis for elaboration by later writers. His book has always been highly prized as an item of early Americana, but it also deserves a place in the very early dawn stage of history of geology in the Mississippi basin.

JONATHAN CARVER

Jonathan Carver was born in 1732 in Connecticut. Becoming an Army officer at the age of 18, he fought in the Indian wars and was a captain at the conclusion of the French and Indian War in 1763. He then set out on a long series of travels through the western and northwestern parts of the British dominions, reaching as far as the headwaters of the Mississippi in northern Minnesota in 1766-1768. In 1769 he went to England, as he says "to communicate the discoveries I had made, and to render them beneficial to the kingdom." He was bitterly and justly disappointed by the lack of recognition of his contributions to the knowledge of the far frontier and of the Indian tribes with whom he had lived. He assembled his notes and maps for publication in 1778 as Travels through the interior parts of North America. Although the book was successful, passing through three editions by 1781, Carver derived little profit from it, and the privations he endured (far worse in London than in the wilderness!) led to his early death in 1780. It is sad to realize that the present price of one copy of any of the editions of his book would have relieved for a long period the extreme hardships of his family in England and would have prevented Captain Carver's own death from actual destitution.

Carver's book describes the topography of the Great Lakes region, the Wisconsin River Valley, and the Upper Mississippi Valley. South of the Wisconsin River he saw the region of the driftless area which he described in the following terms:

> Whilst I staid here, I took view of some mountains that lie about fifteen miles to the southward, and abound in lead ore. I ascended one of the highest of these, and had an extensive view of the country. For many miles nothing was to be seen but lesser mountains, which appeared at a disstance like haycocks, they being free from trees. Only a few groves of hickery, and stunted oaks, covered some of the vallies. So plentiful is lead here, that I saw large quantities of it lying about the streets in the town belonging to the Saukies, and it seemed to be as good as the produce of other countries. (p. 47-48)

Carver was one of the first of the English to record the changes in the level of the Great Lakes which had long been known to the French. His observations particularly referred to Mackinac, where he had stayed at the fort, which at that time was under the command of Major Robert Rogers and no doubt the two men discussed this phenomenon:

> I had like to have omitted a very extraordinary circumstance relative to these Straights. Acording to observations made by the French, whilst they were in possession of the fort, although there is no diurnal flood or ebb to be perceived in these waters, yet, from an exact attention to their state, a periodical alteration in them has been discovered. It was observed that they arose by gradual, but almost imperceptible degrees till they had reached the height of about three feet. This was accomplished in seven years and a half; and in the same space they as gently decreased, till they had reached their former situation; so that in fifteen years they had completed this inexplicable revolution. At the time I was there the truth of these observations could not be confirmed by the English, as they had then been only a few years in possession of the fort; but they all agreed that some alteration in the limits of the Straights was apparent. All these lakes are so affected by the winds, as sometimes to have the appearance of a tide, according as they happen to blow; but this is only temporary and partial. (p. 146-147)

Carver described the features of various lakes and particularly those of Lake Erie, which discharges through the Niagara River into Lake Ontario. His description of Niagara Falls is disappointingly brief because he says it has been "so frequently described that I shall omit giving a particular description of them" (p. 69).

Carver's book deals at great length with the Indian tribes with whom he lived. His book includes a long chapter of organized observations "On the Beasts, Birds, Fishes, Reptiles, and Insects, which are found in the interior Parts of North America." Another long chapter is on "Trees, Shrubs, Roots, Herbs, Flowers &c." His "Appendix" deals with drainage and transportation and his estimate of the more favorable parts for future settlement and includes some specific recommendations keyed to the large map in his text. Inasmuch as he deals in such a well-organized way with Indians, animals, plants, and prospects for settlement, we regret that he did not include a similar chapter on geology. From casual comments throughout the book it is evident he could have included such a chapter, but perhaps the book was already getting too long.

Carver's book in the various editions was widely used and quoted by those who came after him. It is useful for the knowledge of land routes, water routes for travel; and some nuggets of geology may be mined from it by close attention to references primarily about other subjects.

THOMAS HUTCHINS

The little book, A Topographic Description, by Thomas Hutchins (1730-1789) was written in 1778, but it was based on personal knowledge he had gained during the preceding 25 or more years. The author was a military engineer, surveyor,and cartographer who had traveled as far as the Illinois country and whose book was a source of information for many years and was often printed with or without acknowledgement. Hutchins was not only a traveler in the central interior, but had also been a resident of the region in his younger days. He was the first to map accurately the Ohio River in 1766. His field notes on the sources of the river included many comments on the materials of the banks and channel and on the rocks along the shore (Bond, 1942). He accompanied Col. Bouquet's march agains the Indians into central Ohio in 1764 and made a very accurate map of the route (Williams, 1960). Earlier, in 1762, he had traveled across Ohio to Sandusky and thence to Detroit, Mackinac, and Green Bay, returning across Illinois to the Wabash Valley and then to Portsmouth, Ohio (Bond, 1942, p. 14). He thus had the most extensive personal knowledge of anyone to that time when he wrote his book of 1778.

Hutchins' book contains many geological observations. He noted coal, salt springs, iron ore, and lead deposits along the Ohio and Scioto Rivers, and the great vertebrate fossils at "Big Bone Springs," Kentucky:

> Most of the Hills on both sides of the Ohio are filled with excellent coal, and a coal mine was in the year 1760 opened opposite to Fort Pitt on the river Monongahela, for the use of that Garrison. Salt Springs, as well as Iron Ore, and rich Lead Mines, are found bordering upon the river Ohio. One of the latter, is opened on a branch of the Sioto River; and there the Indian natives supply themselves with a considerable part of the lead, which they use in their wars, and hunting.
>
> About 584 miles below Fort Pitt, and on the eastern side of the Ohio river, about three miles from it, at the head of a small Creek or Run, where are several large and miry Salt Springs, are found numbers of large bones, teeth, and tusks, commonly supposed to be those of Elephants:--but the celebrated Doctor Hunter of London, in his ingenious and curious Observations on these bones, &c. has supposed them to belong to some Carnivorous animal, larger than an ordinary Elephant. (p. 10-11)

Hutchins' reference to a silver mine near the present location of Lafayette, Indiana, was repeated by other writers for many years. Hutchins was not referring to an actual opening from which ore was produced, but rather to a "deposit" and perhaps only to a single specimen--the word "mine" having this meaning in earlier times. By the time he wrote, this use of the word was already obsolete. Perhaps the material he saw was marcasite, the light colored iron sulfide. At any rate, Hutchins was quite correct, as he usually was, about the salt-springs, coal, clay and limestone in the Wabash Valley:

> A silver mine has been discovered about 28 miles above Ouiatanon, on the northern side of the Wabash, and probably others may be found hereafter. The Wabash abounds with Salt Springs, and any quantity of salt may be made from them, in the manner now done at the Saline in the Illinois country: the hills are replenished with the best coal, and there is plenty of Lime and Free Stone, Blue, Yellow, and White Clay, for Glass works and Pottery. Two French settlements are established on the Wabash, called Post Vincient and Ouiatanon; the first is 150 miles, and the other 262 miles from its mouth. (p. 28)

Hutchins' intimate knowledge of the Ohio River and the details of the Mississippi upstream from the Ohio stemmed from his careful surveys which he recorded in a beautifully detailed map of parts of these valleys. Whereas Rogers, the line officer, thought the location of the junction of the Ohio and Mississippi rivers would be the site of an important city; Hutchins, the engineer, readily recognized that the location was subject to severe floodings and he knew, of course, that it would be unsuitable for a major city. Hutchins' description of the valleys, the origin of the flood plain material and the rock walls of the valleys are those of an engineer with a keen eye for topography:

> I now proceed to give a Description of that part of my Map called the Illinois country lying between the Missisippi westerly, the Illinois River northerly, the Wabash easterly, and the Ohio southerly.
>
> The land at the confluence, or Fork of the rivers Missisippi and Ohio, is above 20 feet higher than the common surface of these Rivers; yet so considerable are the Spring floods, that it is generally overflowed for about a week, as are the lands for several miles back in the country--The soil at the Fork is composed of Mud, Earth and Sand, accumulated from the Ohio and Missisippi

> Rivers. It is exceedingly fertile, and in its natural state yields Hemp, Pea-Vines, Grass, &c. and a great variety of trees, and in particular, the Aspen Tree of an unusual height and thickness.
>
> For 25 miles up the Missisippi (from the Ohio) the country is rich, level, and well timbered;--and then several gentle rising grounds appear, which gradually diminish at the distance of between four and five miles eastward from the River. From thence to the Kaskaskias River is 65 miles. The country is a mixture of hills and vallies; some of the former are rocky and steep;--but they, as well as the vallies, are shaded with fine Oaks, Hickory, Walnut, Ash, and Mulberry trees, &c. Some of the high grounds afford most pleasant situations for settlements. Their elevated and airy positions, together with the great luxuriance of the Soil, everywhere yielding plenty of Grass and useful plants, promise health and ample returns to industrious settlers.
>
> Many quarries of Lime, Freestone, and Marble, have been discovered in this part of the country....
>
> The high grounds, just mentioned, continue along the eastern side of the Kaskaskias River at a small distance from it, for the space of five miles and a half, to the Kaskaskias village; then they incline more towards that River, and run nearly parallel with the eastern bank of the Missisippi, at the distance of about three miles in some parts, and four miles in other parts from it. These are principally composed of Lime and Free stone, and are from 100 to 130 feet high, divided in several places by deep cavities, through which many small rivulets pass before they fall into the Missisippi. The sides of these hills, fronting this River, are in many places perpendicular--and appear like solid pieces of Stone Masonry, of various colours, figures and sizes. (p. 33-36)

The large map accompanying the report is significant geologically because it shows "petroleum" at several places and had many other notes on coal, clay, freestone, and topography. It shows a one-half mile outcrop ("mine") of coal on the Illinois River near the mouth of the Vermillion River. The Mississippi Valley part of the map, precisely the part of our present interest, is beautifully reproduced as Plate 29 in a recent atlas of the Illinois State Museum (Tucker, 1942).

JOHANN DAVID SCHÖPF

Another observer whose work dealt almost entirely with the eastern part of the United States, but who made some mention of the "western territory," was Dr. Johann Schöpf (1752-1800). During the Revolutionary War, Schöpf, a scientist, scholar, and academician, was attached as a physician to the Hessian troops. Immediately after the war he set out on travels which took him through the eastern and southern part of the country and as far west as Pittsburgh, which brought him into the Mississippi Valley. Besides being a keen and competent observer, he was a lucid, compact writer, and in the years following his travels he published several books and many articles about the scientific features of America.

Schöpf's book on geology and minerals, Beyträge zur mineralogischen Kenntniss des Östlichen Theils von Nordamerika und seiner Gebürge, 1787, never translated into English, is actually the first book on the geology of the United States. [1977 - Now happily reprinted and translated, with elaborate annotations by E. M. Spieker in Geology of Eastern North America, New York, Hafner Press, 1972, 384 pp.] It was followed the next year by a two-volume book on his travels in the United States (1788) and the two books form almost a single entity. The Reise, translated in 1911 as Travels in the Confederation, gives a day-to-day account of what he saw and the Beyträge gives an organized summary of the geology of his reports.

Schöpf was very well acquainted with the European literature on mineralogy and with the American reports of Kalm, Catesby, Evans, Pownall and others. From his own observations he confirmed Evans' division of the country into Coastal Plain, Piedmont, and the Mountains. He described the attitude and bedding of the strata at various places, and the relations of sand, clay and fossiliferous limestone. He described and named various fossils. Although he gave most of his attention to the eastern part of the country, with which he was most familiar, he did describe the "western country" from his own knowledge as far as Pittsburgh and the upper Ohio Valley. His description of the coal and associated shale and limestone at Pittsburgh is accurate and probably the earliest published description by a scientist of this, the world's greatest coal bed.

The concluding theoretical part of Schöpf's book is notably perceptive in its discussion of sedimentation in ancient seas, with their falling and rising levels, and of the very long time necessary for the extensive erosion of mountains and plateaus.

THOMAS JEFFERSON

Although Thomas Jefferson (1743-1806) was never in the Mississippi Valley, his Notes on the State of Virginia (1788) contains some material on the region. He wrote of the caverns in "Limestone-country," and of iron, lead, salt, and coal deposits in Ohio, Kentucky, and Illinois. "We hear of limestone on the Mississippi and Ohio," he said, and in the valleys between (but not on) the mountains. The mammoth fossils near the Ohio River and elsewhere greatly impressed him; he thought they were from elephant-like animals, but not really from elephants. In later years he assembled in the White House a collection of mammoth, mastodon, and megalonyx bones (Martin, 1952, p. 107). The invertebrate fossils in limestone and shale he recognized as marine, but as to why they were now above sea level, as much as 15,000 feet in the Andes, he said, "this great phaenomenon is as yet unsolved." Because his book enjoyed wide popularity, it is important for understanding what was geologically known and considered at the time.

JEDIDIAH MORSE

The state of knowledge in 1789 of the physiography and geology of the country is illustrated in American Geography by Jedidiah Morse (1760-1826). It contains lengthy sections on the "Face of the Country," both for the whole nation, and for separate states and territories. Morse was probably never in the Middle West and his descriptions are obviously derived from the work of others. However, he was a skillful synthesizer and either quoted directly (often without credit) the observations of Hutchins, Evans, Jefferson, Kalm, and others, or used them to make a connected description of his own, which was always more physiographic than stratigraphic. Some of his information was from sources which so far remain unidentified. An interesting study could be made of the successive editions of Morse to see the steady growth of information after 1789 and of his use of it over a period of more than 20 years. In his first edition Morse included a section (p. 403) on the rivers, mineral resources, topography, and rocks of "The Western Territory," (p. 458-462) in which in 1789, of course, no states had yet been established.

GILBERT IMLAY

Gilbert Imlay (1754-1828) was a native American, originally from New Jersey, who was an officer in the Revolutionary War. In 1783 he took up a large tract of land in Kentucky where he lived and which he used as a travel base for several years. He went to England, presumably before 1792, when the first edition of his book, Topographical Descriptions of the Western Territory, was published in London. He was involved in French political affairs in 1793 and in Paris formed the famous, but ungentlemanly alliance with Mary Wollstonecraft (later Mrs. Godwin, whose daughter Mary W. Godwin was the inamorata and later the wife of the poet Shelley). Very little is known of Imlay's career after 1795, and just how, where, and when he secured the voluminous materials for the greatly enlarged 1797 edition of his book is a puzzle.

Imlay's book, of which the first edition was published in 1792, is arranged as a series of descriptions rather than as a travel narrative. He deals particularly with Kentucky, but includes some descriptions of features of the states to the north. He states that "the western country is those parts which are watered by streams running into the Mississippi" (p. 32) and that the upper Ohio basin "abounds in coal" (p. 34). Imlay was impressed by the upland level above the Ohio River, which was not evident from the river: "You ascend a considerable distance from the shore of the Ohio, when you would suppose you had arrived at the summit of a mountain, you find yourself upon an extensive level" (p. 39). Imlay had seen the great fossils of the Big Bone Lick and includes a brief description:

> Between the mouths of Licking and Kentucky lies the Great Bone Lick, which is justly celebrated for the remarkable bones which are found there, and which gave name to the place. Several of those bones have been sent to Europe; but I believe no natural historian has been able to give any decided opinion to what class of animals they belonged. Buffon has called them the Mammouth; but I am at a loss to know from what authority, as we have no tradition either oral or written, that gives an account of any species of animals which were as large as those must have been, judging by the magnitude of the bones. Buffon says, that similar bones have been found both in Ireland (if I am not mistaken)and in some part of Asia. It appears somewhat extraordinary, at the first view, that we should discover manifest proofs of there having existed animals of which we can form no adequate idea, and which in size must have far exceeded anything now known upon earth; and those signs too, in climates where the elephant (the largest animal now in existence) is never found. (p. 43-44)

As Imlay continued his description of the Ohio Valley, he mentioned the rapids of the Ohio at Louisville, which were such an impediment to travel in the early days, and says, "they are occasioned by a ledge of rocks" (p. 47).

Imlay knew that the western country would eventually be divided into a number of states and he proposed that the division be based on topographic features which would give natural boundaries. He thought one of the major boundaries would be the divide between the Great Lakes and the Wabash and Illinois rivers (p. 75-76).

As did many travelers in the Ohio Valley, Imlay noted the presence of salt springs, which were so very important, as salt was one of the two absolute necessities of the settlers. (Ammunition--lead and powder--was the other.) Any local source would be of the greatest importance, as this would save the long and tedious transport by pack horse over the mountains from the east. He describes the earlier production of salt which was precipitated from the water of salt springs in household kettles; but he writes "salt is now manufactured in plenty and sold cheap" (p. 119). He describes in detail the digging of wells and the method of producing salt. He was impressed by other mineral resources as his description will show:

> We have various other minerals, such as iron (which is the most useful), copper, lead, sulphur, nitre, &c. &c. Iron ore is found in great plenty upon the northern branches of Licking creek, and likewise upon the waters of Green river. A lead mine has been worked many years with considerable profit, which lies in the county of Montgomery, upon the waters of the Great Kanhaway. There is another between the Cumberland and Tenasee rivers which is said to be very valuable, and its ore is more pure than any other which has been discovered in America. But the lead mine on the Mississippi must prove inexhaustible. It extends from the mouth of Rock river more than 100 miles upwards. Besides these there are several others, some of which lie on the Spanish side of the Mississippi, and have been used for years past. Copper mines have been discovered in several places, but the mine on the Wabash is, perhaps, the richest vein of native copper in the bowels of the whole earth; and no doubt will render all the others of little or no value. Sulphur is found in several places in abundance; and nitre is made from earth which is collected from caves and other places to which the wet has not penetrated. The making of this salt, in this country, is so common, that many of the settlers manufacture their own gunpowder. This earth is discovered

in greater plenty on the waters of Green river, than it is in any other part of Kentucky. But, perhaps still farther southward, it will be found in greater plenty. However, it is so common in every part of the country that it might be made a considerable article for exportation. I have heard of black lead mines upon the head waters of the Kentucky, but I have not been able to procure any certain information respecting them. But I should conceive that there can be little doubt that, when the country, and particularly the mountainous parts of it, are well explored, all of the useful minerals will be found in abundance.

I have already mentioned the coal mines in the upper parts of the Ohio country; besides which there are great quantities of coal upon the upper branches of the Mississippi. It is particularly favourable that this mineral lies at the heads of our larger rivers, as it can be sent down with the greatest facility; and it is very certain that the great body of it which the Ohio country alone contains, is equal to answer all the purposes for which it may be wanted throughout this extensive empire.

Though the champaign part of this country has no stone on its surface, yet every where limestone is found from 6 to 15 feet below it. Most of the bottoms of our rivulets and streams are paved with this stone. It is very easily calcined, when it becomes excellent lime. It is also convenient for building, by reason of its peculiar smoothness, and the ease with which it may be worked into any form. Besides this stone, which is the most common, every other kind of stone is found which is either useful or ornamental, such as flint, grindstone, and millstones, of a very good quality, which have been reckoned equal to French burrs. There is the greatest plenty of marble upon the banks of the Kentucky, particularly at Leesburg. I have not seen any which has been polished; but judges in that business give us the most flattering ideas of its quality.

Clay is very common in every part of this country which is proper for bricks; and there is a superior kind on the Beech fork of Salt river, which no doubt might be manufactured into good porcelain. Carver has mentioned a clay of this sort which he saw above St. Anthony's falls. Marle, chalk, gypsum, and ochres, are found in various parts.

> Mr. Jefferson has described the medicinal, inflammable, bituminous, and other springs, very accurately; and as there have been no discoveries or light thrown upon the subject since he wrote, I shall refer you to his book for a particular account of them. Indeed, his account of the natural history of this country is generally to be depended upon, so that it is scarcely possible to make any improvement upon it, until farther discoveries shall have arisen: I therefore confine myself to such objects as he has not taken notice of, and to such as have presented themselves since he wrote, occasionally making some strictures and animadversions upon his opinions and information. (p. 120-125)

We see that from his own observations that Imlay realized that there were important mineral resources of metals, fuel, salt and limestone. He realized that the extensive limestone area began about Maysville, Kentucky, and continued down the Ohio for a long distance. He knew that this stone was arranged in horizontal beds. It is interesting to note that Imlay, as well as other early travelers such as Hutchins, took particular note of the clay resources and seemed to realize that some of it was very refractory, sufficiently so that it could be used to make glass pots.

Although it is the first edition with which we are most concerned here, because it shows us what was known by 1792 and the information available at that date, it should be noted that the book passed through two later editions. In the highly prized third edition (1797) Imlay added many extensive notes to his earlier text and included reprinting of several original narratives and the entire works of Thomas Hutchins (already described) and John Filson, thus giving "the most complete information on the trans-Allegheny region available at the end of the 18th century" (Howes, 1962, p. 288). The reprinting of Filson's _Discovery, settlement, and present state of Kentucky_, first published in 1784, is especially important. Filson described the rivers and topography of the country. Under "Curiosities" he noted caves in limestone; "bitumen" used in lamps; salt deposits; and the large vertebrate fossils at Big Bone Lick, the origin of which he discussed in some detail (Imlay, 1797, p. 322-326).

Gilbert Imlay was an acute observer and had the ability to recognize the pertinence of the reports of others. His own writings and those of the others he paraphrased or reprinted, formed an important body of knowledge for those seeking information about the new "western country," either for settlement, travel, or scientific purpose.

C. F. VOLNEY

One of the most important books containing geological information about America to appear at the beginning of the nineteenth century was View of the Climate and Soil of the United States by Constantine F. Volney. This was a translation of 1804 by Charles B. Brown, the noted novelist, of Volney's book originally published in Paris in 1803 as Tableau du climat et du sol des États Unis. (Another translation by an unknown Englishman appeared in London at the same time.) The book is particularly important in the history of geology of the Mississippi Valley because such a large part is concerned with the country west of the Appalachian mountains. It is geologically notable for containing the first descriptions of any fossils from the interior, by the well-known J. B. Lamarck. It is likewise important for its climatologic and geographic analyses (Brown, 1938, p. 213-217).

Volney (1957-1820) was a Frenchman who had traveled extensively in the Mediterranean and the near Eastern regions and had written a book (1787) about his travels which included considerable geology. Narrowly escaping the guillotine in the French Revolution, he came to America in 1795 where he remained until 1798. He had a great deal of personal contact and later correspondence with Thomas Jefferson (Chinard, 1923) and was acquainted with all the important scientific and literary figures of the eastern region. His headquarters was generally in Philadelphia (Brown, 1938), from which he made journeys in New England, in the South, and one particularly extensive one to the upper Ohio Valley and thence to Cincinnati, Louisville, and to Vincennes on the Wabash. His return was by way of Cincinnati, Detroit, Niagara Falls, the Mohawk Valley, Albany, New York, and Philadelphia.

Volney's interest and knowledge of geology was developed long before his American experience. More than 10 years before he came to America he had acutely observed the rocks, fossils, stratification and structure in the mountains of Syria. He had written perceptively on the Nile delta and its deposits (1787). Volney's comparison of American physical features was generally not with western Europe, but with the "naked plains of Egypt, Asia, and the coast of the Mediterranean." He was especially impressed with the vast forests, and likewise with the cessation of the forest at the margin of the western prairies, which he repeatedly compared to the steppes.

He freely credited much of his geological information on America to Mitchill (1798) and others in the United States. He had a high opinion of the work of Hutchins (1778) and was well acquainted with the descriptions of Lewis Evans (1755), Imlay (1792), and Morse (1789). He was familiar with the report of Alexander Mackenzie (1802) and indeed furnished some notes to the 1802 edition of that book. Volney was personally acquainted with William Maclure and we may assume he discussed American geology with him (Chinard, 1923, p. 121, note).

Volney's book contains the first extensive, organized account of the physical features of the trans-Allegheny region, which he describes as the:

> ...region, situate westward of the Allegheny, [which] may be termed the valley of the Mississippi, because all the streams which traverse it are ultimately lost in that river, and, of consequence, must flow upon a plane, every where, and upon the whole, inclining towards its channel. This valley, or rather slope of a valley, is bounded eastward by the Allegheny, northward by the Lakes Michigan, Erie, and Ontario, and southward by the Floridas....
>
> The valley of the Mississippi comprehends western Georgia, Tennessee, Kentucky, the Mississippi and North-Ohio territories, and some western districts of Virginia, Pennsylvania, and New York. The people of the maritime provinces are accustomed to distinguish this space by the names of the back country, the back woods, the wilderness, and, more fancifully, the western waters (p. 17: all quotations are from Brown's 1804 translation).........
>
> The ... district ... bounded southward by the Ohio, by the great lakes on the north, and on the east and west by the Allegheny Mountain and the Mississippi respectively ... [is], called the Northwestern territory, in consequence of its slender population [it] has not yet been erected into a state. The surface is either plain, or composed of gentle inequalities and undulations. It would be difficult to meet with a hill five hundred feet high; and the western parts of it, between the Wabash and the Mississippi, are little else than immense flats or natural meadows. Regular as this surface is, it still inclines in such a manner as to give opposite directions to the course of great and numerous rivers; some of which flow, by the Mississippi, into the Gulph of Mexico, others into the north Atlantic, by the St. Laurence, and others, lower down, into the same great reservoir, by the Mohawk, Hudson, and Susquehannah. Hence it appears, that the Allegheny, from which the latter streams originate, forms a kind of rampart to this vast undulating surface, which is nearly on a level with its summit. The surface, in general, is so slightly inclined, that the rivers descend slowly and circuitously, and frequently stagnate into swamps....

> These rivers, different in this respect from those of Kentucky, flow over an earthen bed; and this arises not only from the general equality of surface, and consequent tardiness of the current, but from the argillaceous or clayey nature of the soil, which their waters cannot desjoint or penetrate. This circumstance will hereafter be favourable to commerce and agriculture. (p. 23, 24)

Volney knew very well the distribution of various rocks and something about their structure. He was even more impressed than those observers before him with the great extent of flat-lying strata, which begin at the western margin of the Appalachian Mountains and continue westward and north-westward for thousands of miles. He was particularly aware of the vast extent of the limestone strata and was impressed by the fossils they contained, as shown by this section of his report on "The Calcareous or Lime Stone Region":

> The third or calcareous region comprises all the western country, beyond the mountains... The portion that has fallen under my own view, and which lies between the Tenessee and the St. Laurence, and between the mountains and the Mississippi, is an immense rock or bed of lime stone, disposed horizontally in lamina or layers, of one or more inches in thickness, of a close and firm texture, and generally of a grey colour.... This stratum is overlayed sometimes by clay, and sometimes by gravel. Above all, at the surface, is a mass of black mould, which is more abundant and of greater depth, sometimes fifteen feet, in the vales and bottoms, and shallower and more scanty in the slopes and heights, where it sometimes dwindles to a depth of six or eight inches. This circumstance, as well as the foliated form of the stone, bears testimony to the ancient influence of the waters of the ocean.
>
> In the country round Pittsburg, on the Ohio, in the district of Green Briar, on the Kenhawah, and throughout Kentucky, an examination always leads to the grand calcareous foundation. I have seen it naked in the channels of all the brooks and rivers from the Kenhawah to the rapids of Ohio, near Louisville. On the road from Cincinnati to Lake Erie, it is found to be the flooring of the Miami and Clay [Auglaise] Rivers. Lake Erie is probably bottomed upon a dark-coloured schist, but specimens are continually found of calcareous substances. A lime stone rock forms the cataract of Niagara, extends itself into the Genessee country, and accompanies the river to Quebec....

The breaches made in this stratum or rock give rise to the funnels or pits before-mentioned, which swallow up the rains and rivulets. I met with curious instances of this nature near Green Briar, in Virginia, and at Sinking Spring, in Genessee, where a spring appears at the bottom of a pit, and, after a course of only six feet, sinks again into the earth....

From Louisville to the White River, where it ends abruptly, the naked lime stone rock may be seen through the limpid course of all the streams....

At Louisville, the first layer or stratum, on the lofty bank of the river, is a black earth, three feet thick. Under this is a mass of thin sand, from fourteen to fifteen feet thick, without any marine exuviae; but below this is another layer of sand, six or ten inches thick, which contains these remains: from thence to the bottom of the river is a stratum of large, heavy gravel. The whole bank is twenty-five feet high....

At the spot above-mentioned, near Louisville, the rock is three feet thick, and water is found ten feet below the surface. At other places, the thickness of the stratum is greater. The rocks which form the rapids of the Ohio, below Louisville, belong to this great calcareous bed. In these lower masses, at their surface, are found a great number of petrifactions, ...

On my return to Paris, I submitted these petrified shells to the examination of one of our most skillful naturalists, M. Lamark; and I cannot do better than to insert, in this place, his acount of them, in his own words:

"I have carefully examined the fossil specimens with which you have entrusted me, and which you collected in North America.

I cannot hesitate to refer them to the *terrebratula* genus of fossils, in irregular masses. This is a new genus, explained in my system of *Animals without Spine*. These almost all of them belong to that class which are grooved or fluted longitudinally from top to bottom, like those which Linnaeus has denominated *anomia dorsata*. In these specimens, we see nothing but the interior mass, the stony or concrete substance with which their cavities have been gradually filled up, during their long abode in the bowels of the earth. We nevertheless still find on many of them small whitish fragments of the primitive shell itself.

In the specimen from Cincinnati, we may distinctly trace three kinds of fossil shells: first, a species of terrebratula with large grooves or flutings, and which resembles the figure given in the New Encyclopædia, pl. 241, fol. 3; secondly, a species not grooved or fluted, but jagged or notched; and thirdly, a bivalve shell, with a few jaggs, which I cannot assign to its proper genus, not being able to examine the joint....

After carefully considering these specimens, it is clear to me, that the districts of North America which furnished them have formerly been covered by the sea, or at least they point out that portion of the surface which has once been the _bottom_, and not the _banks_ of the sea; for they all belong to the class of pelagia (see my Hydrogeology, pp. 64, 70, 71), which, like the gryphytes, ammonites, orthoceratistes, belemnites, encrinites, &c. are found only at the greatest depths of the sea, and never the shore. The greatest part of the shells are only known in a fossil state. On the whole, your enquiries have ascertained the mineralogical state of North America, as far as can be conveniently done, and justify us in concluding that, among its mineral substances, none of the _littoral_ shells are to be found."

The western stratum is nearly horizontal, and forms the country into a sort of immense flat table. In the eastern district, ...the strata is generally broken and irregular. When found regularly inclined, the inclination is most commonly from 40 to 50 degrees....

From these facts we are led to conclude, that the Atlantic coast has been shaken and overturned by earthquakes, to which we shall herafter show it has been subject, while all the country westward of the Allegheny has been undisturbed. (p. 46-54)

Volney recognized that all this limestone must have been deposited in great bodies of water, but he did not guess how many episodes of appearance, disappearance and reappearance of inland seas were required to explain the whole sequence of these strata and their content of fossils, some of which the famous Lamarck identified for him after his return to France. A recent analysis of Lamarck's communication to Volney is presented by Carozzi (Lamarck, 1964, note, p. 63).

Volney was the first to systematically collect geological specimens from the "western country" and probably his collection, arranged and augmented in Philadelphia by gifts, was the best synoptic collection of American geological specimens made to that time. He, himself, was competent to identify ordinary minerals and rocks and had the advice of de la Metherie and Lamarck after he returned to France. His specimens were placed in the cabinet of de la Metherie in Paris (p. 35).

Volney was the first to illustrate a geological report on America by geologic sections. His diagram of Niagara Falls, while not showing the limestone (dolomite) caprock and underlying shale (which he discussed clearly in the text), is a very early, and perhaps first attempt, at a section of the Falls and of the valley downstream. His description (p. 68) of the terraces at Cincinnati and their composition of gravel is accompanied by a cross-section and measured columnar sections. These are noteworthy both for their accurate description and for the explanation of their origin.

Geological divisions are only partially shown on the maps in any of the translations, although Volney showed these divisions on his map in the Paris edition of 1803. In a few, very rare, large paper copies the map is colored geologically--the calcareous regions in green, the granitic region in red, and the sandstone region in brown. This is the first colored geological map of the United States and preceded by six years the map of Maclure which is usually regarded as the first geological map of the United States (Wells, 1959). Volney gracefully acknowledged his debt to the earlier classifications of Evans, Mitchill, and the observations of Kalm (1770-1771) and others. His own observations and the expert identification of the many specimens he collected confirmed, in a broad way, the earlier divisions, but his own analysis, being more precise, was a distinct step forward in depicting broad geological features of the United States, especially of the trans-Allegheny part.

Volney's work is straight-forward description and an analysis of origin based on the immediate facts at hand rather than on some preconceived "theory of the earth." He did not use the terms "Primary," "Secondary," and "Transition." He did not call upon the Flood to explain anything he saw. It is not clear just what his concept of time may have been, but as an intimate of Lamarck, who assumed many hundreds of millions of years for geological time (Lamarck, 1802, English translation, p. 49), it is likely that Volney was at least hospitable to such views of long extent of geological time. If he was a believer in a universal ocean, there is no hint of such belief in his writings. Indeed, there is just the opposite impression, as he speaks of the partial extent of the seas over the coastal plain and the one-time extent of seas, large to be sure, in the interior, but which had definite boundaries some place or other, the eastern boundary of which he thought he had seen.

The interpretation of depositon of the flat-lying strata west of the Appalachian Mountains in a large lake recurs over and over again in the early writings (Merrill, 1924, p. 50-53) and Volney devotes much space to it (p. 60-70). This concept persisted far into the nineteenth century. It was supposed that one or more large lakes had drained through gaps in the mountains which formed the eastern margin. The idea goes back to at least 1743 in an observation of Lewis Evans, who got the idea from even earlier informants (White 1959, p. 153). The presence of marine fossils in the limestone convinced Volney that the lakes were either remnants of inland seas or were actually arms of the sea. The origin, development, and decline of this theory of inland lakes breaking through mountain barriers would form a whole study in itself. It is easy to see how this idea occurred to those who had seen the Great Lakes and realized that they must once have been much more extensive.

The descriptions and concepts of the geology of the western interior presented by C. F. Volney--politician, linguist, philosopher, geographer, and proto-geologist--were widely disseminated in continental Europe, in Great Britain, and in the United States. They formed a general basis for compilations and for further classification of importance equal to those for the eastern part of North America by Lewis Evans just 50 years earlier.

CONCLUSION

Beginning with Volney, the history of geology in the "western country" becomes more and more that of organized scientific description and analysis, in contrast to the "observations" of those who preceded him. It therefore marks a suitable place to pause in the study of the history of geology of the central interior. My next studies will deal with the organized reports of William Maclure in 1809 and 1817, of Daniel Drake in 1815, of John Bradbury in 1817, and of David Thomas in 1819.

In the 1820's much more extensive reports were published by professional scientists, some of whom were professional (or almost professional) geologists--Schoolcraft, Featherstonhaugh, Nuttall, James, Keating, Long, and others. (For references see White and Slanker, 1962). In the 1830's began the state geological surveys and with the reports at the end of this decade by Mather in Ohio and Owen in Indiana we enter the era of "modern" geology in the central interior.

There was as much geological knowledge of the west as of the east at the beginning of the 19th century--indeed there was even <u>more</u> understanding of western geology! In the west there was no use of the terms about Primary, Secondary, and Transition rocks--the rocks were marine sediments; the valleys had been excavated by streams; no "convulsions" had disturbed the strata; no Deluge was required to explain the features. Admittedly, the eastern geology was more complicated for the origin of the igneous, metamorphic and folded sedimentary rock. The eastern view was of deposition in a "primeval ocean"--a far different ocean from the actualistic seas in which were deposited the western strata with the well-preserved fossils that Lamarck so clearly recognized.

As my studies progress, it becomes more and more apparent that geology done by men was indeed subject to "environmental control." Just as the cratonic geology of the interior was different from the eugeosynclinal-miogeosynclinal geology of the east, so the geologists of the two areas under discussion were different and the geological ideas and theories they present were different. Perhaps more open spaces led to more open and independent minds!

BIBLIOGRAPHY

Banta, R. E. The Ohio. New York: Rinehart and Company, Inc., 1949. 542 p.

Bartram, John. Observations on the inhabitants, climate, soil, rivers, productions, animals, and other matters worthy of notice. Made by Mr. John Bartram, in his travels from Pensilvania to Onandago.... To which is annex'd a curious account of the cataracts at Niagara by Mr. Peter Kalm, a Swedish gentleman who travelled there. London: J. Whiston and B. White, 1751. 94 p., 1 pl.

Bond, B. W. (editor). The courses of the Ohio River taken by Lt. T. Hutchins Anno 1766 and two accompanying Maps. Cincinnati: Historical and Philosophical Society of Ohio, 1942. 85 p., 11 figs., 3 maps.

Brown, R. B. Materials bearing upon the geography of the Atlantic Seaboard, 1790-1810. Annals Assoc. Am. Geographers. Vol. 28, pp. 201-231, 9 figs., 1938.

Carver, John. Travels through the interior parts of North America, in the years 1766, 1767, and 1768. London: J. Walter, 1778. 544 p., 4 pl. 2 maps. (Many later editions.)

Chinard, Gilbert. Volney et l'Amérique d'après des documents inédits et sa correspondence avec Jefferson. Baltimore: John Hopkins Press, 1923. John Hopkins Studies in Romance Literature and Languages, vol. 1, 207 p., port.

Evans, Lewis. Geographical, historical, political, philosophical and mechanical essays. The first, containing an analysis of a general map of the middle British colonies in America.... Philadelphia: B. Franklin and D. Hall, 1755. 32 p. The large map was published separately. (Facsimile of text and map in Gipson, 1939.)

Gipson, L. H. Lewis Evans, to which is added...facsimile of his... Analysis of a General Map...Also facsimile of Evans' Maps. Philadelphia: Historical Society of Pennsylvania, 1939. 219 p., 6 maps.

Guettard, Jean-Etienne. Mémoire dan lequel on compare le Canada à la Suisse, par rapport à ses mineraux, Histoire de L'Academie Royale des Sciences. Paris: 1752. p. 189-220, 323-360, 524-538.

Howes, Wright. U.S. iana (1650-1950). A selective bibliography. New York: R. R. Bowker Company for the Newberry Library, 1962. 652 p.

Hutchins, Thomas. A topographical description of Virginia, Pennsylvania, Maryland, and North Carolina.... London: J. Almon., 1778, ii. 67 p., 2 pl. (The map was issued separately.)

Imlay, Gilbert. A topographical description of the western territory of North America.... London: J. Debrett, 1792, xv, 247 p. (Third edition, 1797, xii, 598, xxvi p., 4 maps.)

Jefferson, Thomas. Notes on the State of Virginia. Philadelphia: Pritchard and Hall, 1788. 244 p. (First French edition 1786; first London 1787.)

Kalm, Peter. Travels into North America. Warrington - London: 1770-1771. 3 vols., 24, 400, 352; 3, 310, 14, map, 6 pl., translated by J. R. Forster; another ed. 1772; 1938. (Original Swedish 1753-1761; German and Dutch editions, partial French editions.)

Lamarck, J. B. Hydrogéologie, or recherches sur l'influence qu'ont les eaux sur la surface du globe terrestre.... Paris: Chez l'Auteur, An X (1802), 268 p.

__________. Hydrogeology by J. B. Lamarck. translated by A. V. Carozzi. Urbana: University of Illinois Press, 1964. vi, 152 p. (Many extensive notes.)

Mackenzie, Alexander. Voyages from Montreal, on the River St. Laurence, to the Frozen and Pacific oceans; in the years 1789 and 1793.... With original notes by Bougainville, and Volney.... London: T. Cadell, etc., 1802. 2 vols. xiv, 290; 332 p. (The first edition of 1801 did not have the "Notes.")

Martin, E. T. Thomas Jefferson; Scientist. New York: Henry Schuman, 1952. x, 289, 15 pl.

Merrill, G. P. The first one hundred years of American geology. New Haven: Yale University Press, 1924. xxi, 773 p., 130 figs., map; Hafner, 1964.

Mitchill, S. L. A sketch of the mineralogical and geological history of the State of New York.... Medical Repository. vol. 1, 1798, p. 293-314, 445-452; vol. 3, 1800, p. 325-335; vol. 5, 1801, p. 212-214.

Morse, Jedidiah. The American geography; or, a view of the present situation of the United States of America.... Elizabethtown: Shepard Kollock, 1789, 534 p., 2 maps. (Also many later editions.)

Rogers, Robert. A concise account of North America:...situation, extent, climate, soil...also of the interior or westerly parts of the country.... London: J. Millan, 1765. vii, 264 p.

Schöpf, Johann David. Beyträge zur mineralogischen Kenntniss des Östlichen Theils von Nordamerika und seiner Gebürge. Erlangen: J. J. Palm, 1787. 12, 194, 2 p.

__________. Reise durch einige der mittlern und südlichen Vereinigten nordamerikanischen Staaten....1783 und 1784. Erlangen: J. J. Palm, 1788. 2 vol. 24, 644, 2, 40, 352 p., map.

__________. Travels in the Confederation, 1783-1784. Translated and edited by A. J. Morrison. Philadelphia: William J. Campbell, 1911. 2 vol. x, 426, 344.

Tucker. S. J. Indian villages of the Illinois Country, Part I. Springfield: Illinois State Museum. 1942, xiii, 18, 54 pl.

Volney, C. F. Tableau du climate et du sol des États Unis d'Amérique. Suivi d'éclaircissemens sur la Florida, sur la colonie Française du Scioto, sur quelques colonies Canadiennes et sur les sauvages. Enriche de quatre planches gravées, dont deux géographiques et une coupe figuree de la ĉhute de Niagara. Paris: Coucier, Dentu, An XII 1803. 2 vol. xvi, 300, 532 p., 2 folding maps, 2 folding pl. A second state has vol. 2, 524, 8, 4 p. Large paper copies ("quarto") are known. There are also later editions, including English, German and Italian translations.

__________. View of the climate and soil of the United States of America: to which are annexed some accounts of Florids, the French colony on the Scioto, certain Canadian colonies, and the Savages or Natives: translated from the French of C. F. Volney. London: J. Johnson, 1804. xxiv, vi, 503 p., 2 folding maps and 2 folding p. (Translator unknown.)

__________. A view of the soil and climate of the United States of America...with supplementary remarks upon Florida; on the French Colonies, on the Mississippi and Ohio, and in Canada; and on the aboriginal tribes of America. Translated with occasional remarks by C. B. Brown. Philadelphia: J. Conrad and Co., 1804. xxvii, 446 p., 2 maps, 2 pl.

__________. Voyage en Syrie et en Egypte, pendent les années 1783, 1784, et 1785.... Paris: Desenne 1787. 2 vols. English translation 1787. (Many later editions.)

Wells, J. W. Notes on the earliest geological maps of the United States, 1756-1832. Jour. Washington Academy of Sciences, vol. 49, 1959. p. 198-204.

White, G. W. Lewis Evans' contributions to early American geology--1743-1755. Ill. Acad. Science Trans., vol. 44, p. 152-158, 1951.

__________. Early American geology. Scientific Monthly, vol. 76, p. 134-141, 1953.

__________, and Slanker, B. O. Early geology in the Mississippi Valley: an exhibition of selected works held in the University of Illinois Library, November, 1962. Urbana: University of Illinois Library 1962. 26 p.

Williams, E. G. (editor). The orderly book of Colonel Henry Boquet's Expedition against the Indians. 1764. Pittsburgh: Mayer Press, 1960. 74 p., maps.

BIBLIOGRAPHICAL ESSAYS

EARLY GEOLOGY IN THE MISSISSIPPI VALLEY

An Exhibiton of Selected Works
Held in the University of Illinois Library at Urbana
November, 1962

by
George W. White
Geology Department

and
Barbara O. Slanker
University Library

University of Illinois
.Urbana, 1962

Foreword

A tradition that scientists, as a group, are unconcerned with the historical aspects of their disciplines, and generally inclined to leave to humanists and social scientists explorations of the past, is not borne out at the University of Illinois. A History of Science Society, established in 1954, has enlisted the support and active participation of faculty members from every scientific division in the institution, as well as the enthusiastic interest of scholars in other fields. In addition, formal courses in the history of their subjects are currently offered by Agriculture, Biology, Chemistry, and Engineering, and as sections of courses in other areas of science and technology.

One of the moving forces behind this lively attention to the history of science is Professor George W. White, Head of the Department of Geology, an avid collector himself, who, since joining the faculty fifteen years ago, has been instrumental in developing an outstanding library collection relating to the historical phases of science as a whole, with particular reference to the history of geology. The interest is natural — no scientists are more conscious of the past than are geologists. A strong case can be made out, in fact, for requiring every student majoring in the field of history to have as a prerequisite courses in paleontology, for proper perspective upon the relatively recent events of the last few thousand years.

The exhibition on early geological writings for the Mississippi Valley, prepared by Professor White and Mrs. Slanker and recorded in the present catalog, is a sampling of the riches available in the Illinois collection. The detailed annotations reveal the concern for geological matters of a variety of commentators, American and foreign, dating back to the earliest days of exploration and discovery.

Robert B. Downs
Dean of Library Administration

Introduction

The growth and development of geology in America can be traced through the writings of observant men, scientists, proto-geologists, and professional geologists. One of the earliest works to include geological description was A Briefe and True Report of the Newfound Land of Virginia, by Thomas Hariot, which was published in 1588. Reports of the earliest explorers often contained geological notes, and early descriptions of the new lands included sections on soils, minerals, and the "face of the country."

Besides laying the foundation for later developments in the science, the early books served several other important purposes. People east of the Allegheny Mountains consulted them to learn about the topography of the land they would pioneer. Information about the mineral resources, especially fuel, salt, iron, and lead, and the location of springs and permanent streams was vital to them. People west of the Allegheny Mountains read the books to find what lay further west. Armed with a knowledge of the country, they were stimulated to move farther west into the less forested areas of the great Mississippi Valley and its tributaries.

Arm chair travelers from all parts of the world read the books for information about the West. Readers in Europe sometimes read the American edition, but sometimes the London edition or a translation. The geological observations were absorbed, copied, and recopied, and gradually entered into the body of common knowledge.

Following the early books are reports by scientists and scientifically inclined travelers. There are also truly professional reports by scientists in government service. The states were far ahead of the federal government in supporting geological surveys, and their reports began appearing about 1830. This group of reports forms the beginning of the vast body of technical knowledge that accumulated over the following century and a quarter.

Limitations of space make it necessary to show only representatives of the University of Illinois Library's outstanding collection of books dealing with early American geology. The works of Morton, Josselyn and Captain John Smith, as well as those of Imlay, Carver, and Pike could have been used, but lack of space precluded their inclusion. Michaux, Brackenridge, S. L. Mitchill, Amos Eaton, and Constantine Rafinesque are omitted for the same reason. Though valuable and voluminous, none of the works of the French explorers, missionaries, and writers who came into the Mississippi Valley from Canada is included.

In studying the items for the exhibit, a contrast between the eastern "authorities" and the western workers became apparent. The latter group was not so apt to be tinged with Wernerian Neptunism, or if they were, they tended to modify and even discard the hampering dogmas. The westerners described the rocks as they were and made their own speculations about their origin instead of looking for German rocks. The Noachian deluge and Usher chronology exerted less control over their thought. It almost seems that the open spaces both attracted and led to the development of more open minds.

The books selected for this exhibition in the University of Illinois Library have been limited primarily to publications in book form in English that dealt with the Mississippi Valley or some part of it. Though the upper valley of the Mississippi and the Ohio Valley, really the Middle West, have been emphasized, the Allegheny Plateau region has been included. With one or two exceptions, the earliest edition of a title is shown, because it was the edition which first made the author's ideas and observations known to his contemporaries. Generally it was the edition which was quoted, with or without credit.

A quotation from the travel diary of David Thomas, a Quaker engineer of the early nineteenth century, concludes the introduction. A century and a half ago Thomas made the following prophecy of the prairies:

> In traversing such delightful regions, the mind acquires a degree of cheerfulness that rarely attends it in the deep gloom of the forest . . . The dark days of (this) country are past. In fancy must (we) view the current of population breaking from the mountains, full, broad, resistless; and the vast and long deserted plains of the Mississippi, fill with life, with intellect, and with elegance.

1. Hariot, Thomas, 1560-1621.

A Brief and True Report of the New Found Land of Virginia. A Facsimile Reproduction of the 1588 Quarto. Ann Arbor, 1931.

A Briefe and True Report of the New Found Land of Virginia... (In Hakluyt, Richard. *The Principal Navigations...* London, 1598-1600.)

Thomas Hariot was the first scientist to visit America and record his observations of the new country. Under the patronage of Sir Walter Raleigh he came to Virginia in 1585 to see the country. As an accompanying staff officer of Raleigh's expedition, Hariot served as consulting scientist and recorder. The reports he wrote included material on natural history and geology. Though his connection with the Mississippi Valley is exceedingly tenuous, his report is included because it was the first scientific report on America in English.

Hariot recognized the Coastal Plain and the Piedmont and knew that there were mountains still farther to the west. He described some mineral deposits and their assays by Thomas Vaughan, the "mineral man" and assayer who accompanied the expedition. Hariot clearly described the mineral resources of the Coastal Plain, and from other sources it is apparent that he recognized fossils in the rocks.

Since only five copies of this earliest item of scientific Americana are known to exist and just two of them are in the United States, a facsimile of the report is shown. The report was reprinted by Theodore DeBry in 1590 in Latin, English, and German and engravings of the water colors of John White, a member of the expedition, were added.

The great compiler of voyages and travels, Richard Hakluyt, included Hariot's report in the third volume of his Voyages. This reprint in black letter (the original is in Roman type) is almost exactly that of the original except for some additional introductory material. It is probable that most readers derived their knowledge of Hariot's report from this Hakluyt edition.

2. Evans, Lewis, 1700-1756.

Geographical, Historical, Political, Philosophical and Mechanical Essays. The First, Containing An Analysis of A General Map of the Middle British Colonies in America... Philadelphia: Printed by B. Franklin and D. Hall, 1755.

3. Pownall, Thomas, 1722-1805

A Topographic Description of Such Parts of North

America as Are Contained in the (Annexed) Map of the Middle British Colonies &c. in North America. London, J. Almon, 1776.

4. Stevens, Henry Newton, 1855-1930.
Lewis Evans, His Map of the Middle British Colonies in America; A Comparative Account of Ten Different Editions Published Between 1755 *and* 1807. London, H. Stevens, Son, and Stiles, 1905.

Lewis Evans was the great colonial cartographer, surveyor, proto-geologist, engraver, and pamphleteer. His map of the British colonies is almost a geological map, for he showed on it mountain ridges, the Allegheny front, the glaciated part of northern Ohio — "level rich land intermix'd with swamps & ponds" — the location of deposits of petroleum, coal, clay, freestone, salt, and "elephant bones." Extending to the Mississippi, the map has an inset portion for Illinois and Indiana.

The Analysis has many pages of organized geologic and geomorphic description and some theoretical speculations of high merit on fossils, isostasy, erosion, peneplains, and Coastal Plain origin. Evans himself did not have personal knowledge of the country west of Pennsylvania, but he effectively synthesized the more important and accurate observations of others.

Evans' classification of physiographic provinces of the United States is the first such attempt; it is accurate and in broad outlines it is followed today. Though usually unacknowledged, his physiographic classification and descriptions provided the basis for classification by many later authors. His work was important and widely influential, and had he lived longer he might well have held a position in America comparable to that of the Englishman William Smith who is considered the father of stratigraphy.

Thomas Pownall, colonial governor and friend and supporter of Evans, had encouraged his 1755 publication and possibly may have collaborated with him. In 1776 when Pownall reprinted the map, he revised and expanded the text to include even more physiographic material on the western country, especially from the surveys and observations of Gordon, Gist, and Hutchins during the 1760's.

The Evans map was reprinted with and without credit at least twenty-six times over more than forty years. A copy of the original 1755 map is exhibited along with the Pownall reprint. The Illinois collection contains eight other editions of the map which are not shown. These maps formed the basis for Henry Stevens' 1905 monograph on the ten various forms of the map; his later studies raised the number of editions to twenty-six.

The three books and ten maps are contained in a sumptuous case of green niger lined with red niger and chamois. The three books, all large paper copies, are bound in matching green niger by W. Pratt.

5. Hutchins, Thomas, 1730-1789.

A Topographical Description of Virginia, Pennsylvania, Maryland, and North Carolina, Comprehending the Rivers Ohio, Kenhawa, Sioto, Cherokee, Wabash, Illinois, Mississippi... London, Printed for the Author, and Sold by J. Almon, 1778.

The maps and notes on the geography of Ohio, Indiana, and Illinois by Thomas Hutchins, a military engineer and surveyor, were the first written from personal observation. The notes contained geological observations, and they formed a valuable source of information for many years.

In the notes Hutchins described coal, salt springs, iron ore, and lead deposits along the Ohio and Scioto Rivers, and the great fossils at "Big Bone Springs," Kentucky. In some detail he discussed the minerals of the Muskingum and Wabash valleys and gave special attention to the clay deposits which he said would be useful for pottery and glass pots. He described the Lake Erie shore and also recognized that the strata of the region west of the mountains are essentially horizontal.

The map is significant because it shows "petroleum" at several places and has many other notes on coal, clay, freestone, and topography. Also, it shows a one-half mile coal deposit ("mine") on the Illinois River near the mouth of the Vermillion River.

The Illinois copy of this book is an exceptional one, bound by F. Bedford for H. Stevens, and the map is a particularly fine example. Two of the four sections of the map, including that part showing the "Illinois Country," are displayed. The book and map are actually seperate publications, but they are brought together here in a beautifully simple case by Sangorski and Sutliffe.

6. Beverley, Robert, 1673-1722.

The History of the Present State of Virginia, in Four Parts... London, Printed for R. Parker, 1705.

Robert Beverley, Virginia planter and lawyer, was the first native born American to write a book about America which included much about natural history and some worthwhile parts on geology. He wrote of streams, springs, clays, ores, and other mineral resources. He understood the relation of soil to vegetation, and the physiographic divisions, particularly of the Coastal Plain and Piedmont.

The bounds of colonial Virginia extended to the Mississippi River, and thus included much territory in the Mississippi basin. Though Beverley had never crossed the mountains, he referred to the observations of Captain Henry Batt, who penetrated to the Allegheny Plateau and reached the Mississippi basin, "a fine level Country again, and discovered a Rivulet that descended backwards. Down that stream they travell'd several days" and came to country where the Indians made salt. These are slender geological observations of the Mississippi valley, but perhaps they are enough to allow inclusion of this important book.

The book includes fourteen page size engravings of Virginia, mostly of Indians, that are copies of the DeBry engravings from the John White paintings of 1585. Beverley's explanations of the charming little pictures are detailed and informative, and impressions in this copy are particularly crisp ones. The Illinois copy of the book is bound in full brown gold-tooled morocco by the Club Bindery, 1900.

7. Guettard, Jean-Etienne, 1715-1786.

"Memoire Dans Lequel On Compare le Canada a la Suisse, par Rapport a Ses Mineraux," *Histoire de l'Academie Royale des Sciences* (Paris) Annee 1752. (1756), 189-220, 323-360, 524-538.

This curiously titled work is by the great French naturalist, Jean-Etienne Guettard who was never in North America. In compiling the map Guettard consulted many earlier publications and through extensive personal correspondence secured much other information not yet published. According to Wells, the map is the "oldest geological map of the continent as well as one of the oldest geological maps in general."

Guettard described minerals and mineral resources and the arrangement in belts of various kinds of rocks. He described and illustrated both Paleozoic brachiopods and Pleistocene mastodon teeth, the latter probably from Big Bone Lick in the Ohio Valley. Also, he suggested some correlation of the rocks of North America with those of Europe.

The volume is opened to show the map of North America. It is truly a geological map — note the "Bande Sabloneuse" along the east coast and the "Bande Marneuse" in shaded pattern. The "Bande Schisteuse ou Metallique," farther west, includes the whole Mississippi Valley. Note the location of "Illinois" on the map and in it the symbols for "cristal," iron, lead, and other minerals, and nearby, coal and petroleum. The interesting legend is made up in part of ancient alchemical symbols.

8. Jefferson, Thomas, 1743-1826.

Notes on the State of Virginia. Philadelphia: Printed and sold by Prichard and Hall, in Market Street, between Front and Second Streets. M.DCC.LXXXVIII.

Notes on Virginia, by Thomas Jefferson, is a synthesis of his personal observations of the eastern part of the country and his wide knowledge of written and oral accounts of the western areas of the state of Virginia which then extended to the Mississippi River. His descriptions and speculations of the origins of the water gap at Harpers Ferry, of the Delaware Water Gap and the nearby Wind Gap, and of the Natural Bridge were widely read and quoted for many years. He wrote of the caverns in "limestone - country," and of iron, lead, salt, and coal deposits in Ohio, Kentucky, and Illinois. "We hear of lime-stone on the Mississippi and Ohio," he said, and in the valleys between (but not on) the mountains.

The mammoth fossils near the Ohio River and elsewhere greatly impressed him; he thought they were from elephant-like animals, but not really from elephants. In later years he assembled a collection of mammoth, mastodon, and megalonyx bones in the White House. The invertebrate fossils in limestone and shale he recognized as marine, but as to why they were now above sea level, as much as 15,000 feet in the Andes, he said "this great phaenomenon is as yet unsolved." Since his book enjoyed wide popularity, it is important for understanding what was geologically known and considered at the time.

The copy shown here is the first American edition. The Notes were written in 1781-82, printed in France in 1784, and first printed in America in 1788. In the many editions which followed some changes and additions were made.

9. Schoepf, Johann David, 1752-1800.

Beytraege zur Mineralogischen Kenntniss des Ostlichen Theils von Nordamerika und Seiner Gebuerge. Erlangen, J. J. Palm, 1787.

During the Revolutionary War, Johann David Schoepf, a scientist, scholar, and academician, was attached as a physician to the Hessian troops. Immediately after the war Dr. Schoepf set out on travels which took him through the eastern and southern part of the country and as far west as Pittsburgh, which brought him into the Mississippi Valley. Besides being a keen and competent observer, he was a lucid, compact writer, and in the years following his travels he published several books and many articles about the scientific features of America.

Schoepf's book on geology and minerals is actually the first book on the geology of the U.S. It was followed the next year by a two-volume book on his travels in the United States and the two books form almost a single entity. The Travels gave a day-to-day account of what he saw, and the Beytraege gave an organized summary of the geology of the country, both from his own observations and from carefully considered reports.

Schoepf was very well acquainted with the European literature on mineralogy and with the American reports of Kalm, Stork, Catesby, Evans, Pownall and others. From his own observations he confirmed Evans' division of the country, into Coastal Plain, Piedmont, and the Mountains. He described the attitude and bedding of the strata at various places, as well as the relations of sand, clay and fossiliferous limestone. He described and named various fossils. Although he gave most of his attention to the eastern part of the country, with which he was most familiar, he did describe the "western country" from his own knowledge as far as Pittsburgh and the upper Ohio Valley. His description of the coal and associated shale and limestone at Pittsburgh is accurate and probably the earliest published description of this, the world's greatest coal bed, by a scientist.

The concluding theoretical part of Schoepf's book is amazingly modern in its discussion of sedimentation in ancient seas, with their falling and rising levels, and of the very long time necessary for the extensive erosion of mountains and plateaus.

The fine copy displayed has been loaned for this exhibit by the Edward Orton Memorial Library of Geology of the Ohio State University.

10. Volney, Constantin Francois Chasseboeuf, comte de, 1757-1820.

A View of the Soil and Climate of the United States of America . . . Translated, with Occasional Remarks, by C. B. Brown. Philadelphia, Pub. by J. Conrad & Co., 1804.

This book by Count Volney is not a journal of travels, but an organized description of the physical features of the country, much of which he saw himself as he travelled on foot as far west as the Wabash Valley from 1795 to 1798. A friend and associate of Dr. S. L. Mitchill, Volney derived some of his ideas from him, and he drew on Lewis Evans for some of his description of the Coastal Plain, the Piedmont and Appalachian Mountains. From his own knowledge he described and illustrated the terraces of the Ohio at Cincinnati and said they had been formed by the river flowing at once higher levels. His columnar sections of the terrace materials are the first geological columns of American Pleistocene deposits.

He observed that the rocks of the Allegheny Plateau were mainly sandstone, whereas the lower land to the east was largely limestone, which extended to or beyond the Mississippi River. The fossils he collected from rocks near Cincinnati were the ones which the great Lamarck reported in detail and said were deposited on the "bottom, and not the banks of the sea." Volney thought that the coal beds of the Mississippi Valley were consolidated deposits on ancient deltas deposited in just the same way as organic matter is now being laid down on the Mississippi delta, a remarkably modern conclusion. A long and valuable section dealt with Niagara Falls and its retreat.

Volney's book was published in Paris in 1803, and two English translations appeared the next year, one in London and another in Philadelphia. The Philadelphia edition which contains interesting notes by Charles Brockden Brown, the well-known early American novelist, is displayed. Because Volney's descriptions of America were read extensively, his work formed the basis for much that was written by others in the following years.

11. Mease, James, 1771-1846.

A Geological Account of the United States... Philadelphia, Published by Birch & Small, R. Carr, Printer, 1807.

The Account, by James Mease, was the first book with the word "geology" in the title to be published in the United States. It deals with a variety of subjects — climate, soils, lakes and rivers, plants and animals, as well as what is considered geology today. Some of the material was derived from Volney, some second or third hand from Morse, and some from other sources.

Dr. Mease, a physician of Philadelphia, may have traveled west of the mountains, but probably his journeys did not extend very far into the Ohio and Mississippi Valleys. He did recognize that the rocks of the Mississippi Valley were essentially horizontal strata that included a great deal of limestone. Though he added very little original material to American geology, his summaries of known information reached a wide audience, and therefore his book with all its insufficiencies is signifcant.

The copy exhibited is typical of the edition, with its closely trimmed 12mo signatures and its undistinguished binding.

12. Ashe, Thomas, 1770-1835.

Memoirs of Mammoth, and Various Other Extraordinary and Stupendous Bones, of Incognita, or Non-descript Animals... Liverpool, Printed by G. F. Harris, 1806.

Thomas Ashe was an English imposter, charlatan, and swindler who disliked America and Americans, especially those of the frontier. In his book of travels in the Ohio and Mississippi valleys, he frequently described things, he had not seen. It is likely, however, that he did see the great deposits of bones of fossil elephants and other great animals at Big Bone Lick, Kentucky. Ashe met Dr. Goforth of Cincinnati, who had a great collection of bones from the deposit, and persuaded the trusting physician to let him take the bones to England for mutual profit. The bones were displayed with great profit to Ashe, but none to Dr. Goforth.

There were many publications in the late eighteenth and early nineteenth century on fossil elephants and other large animals. The 1802 and 1803 books by Rembrandt Peale on the mammoth skeleton from the Hudson Valley are better known than this one by Ashe. The book by Ashe is displayed, however, because it describes material from the Mississippi basin.

13. Maclure, William, 1763-1840.

"Observations on the Geology of the United States, Explanatory of a Geological Map," *American Philosophical Society Transactions,* VI (1809), 411-428.

Except for Volney's scattered remarks on American geology and Dr. Mease's well-meaning but limited effort, this paper by William Maclure is the first summary of the geology of the United States and the first so titled. It is particularly important for the inclusion of the first geological map of the United States printed in America.

The paper appeared in the Transactions of the American Philosophical Society, the oldest scientific society in America. The Transactions contain scores of papers on American geology, including ones on the western country published as early as 1786.

Using Wernerian terminology, Maclure described the rocks of the country as Primary, Transition, Secondary, and Alluvial. He stated that the vast area west of the mountains was made of Secondary rocks, except in the wide alluvial valley of the Mississippi. The Secondary rocks he described as horizontal strata of limestone, sandstone, and coal.

The book is opened to show the crisp, bright, hand-colored map, one of the important features. This map provided the basis for geological knowledge of the country for more than twenty-five years. It perpetuated Wernerian terminology and attitudes in American geology long after they had lapsed into deserved oblivion in Europe. With all its faults and misfortunate theory, it is one of the most important single publications in early American geology.

14. Maclure, William, 1763-1840.

Observations on the Geology of the United States of America... Philadelphia, Printed for the Author by Abraham Small, 1817.

On January 20, 1817, Maclure presented a revision of his earlier report on geology of the United States before the American Philosophical Society. The revised report was published in two forms, as a book in 1817 and as an article in the Transactions (New Series) early in 1818. More European comparisons were introduced in the revised report and strict Wernerianism was somewhat questioned, but the classification and descriptions were actually little changed from those of 1809. A long section on weathering and soil formation from the various rocks was added. A discussion of the economic prospects of the country deduced from the geological resources and physiographic controls was appended. The geological map was changed only in minor ways, although a different base was used.

Maclure's ideas became known mainly through the publication of his report in book form. Forty years ago the work was unchallenged as the most important early book in American geology and Maclure was considered as the father of American geology. Today some geologists are apt to emphasize his errors and insufficiences rather than his truly great contributions. Perhaps in future assessments when his other publications are considered, Maclure will be restored to his former high and paternal place.

The Illinois copy, which is in the original boards, is a presentation copy signed by the author. The map displayed is in the second state on a revised geographical base, which shows the Arkansas Territory, established when the map was going through the press.

15. Forsyth, Gideon C.

"Geological, Topographical, and Medical Information Concerning the Eastern Part of the State of Ohio," *Medical Repository* XII, no. 4 (1809), 350-358.

Although the paper shown is of some geologic importance, it is displayed primarily to call attention to the journal in which it appeared, the Medical Repository. According to Meisel, this publication is "the first strictly scientific periodical in the United States." It was founded in 1797 and appeared until 1823.

While the journal's emphasis was on medical papers, it contained much on general natural history and geology. Of even greater importance are the editorials, notes, and reviews of the books and domestic and foreign articles on geology and the western expeditions of the day. Many of the notes were by the journal's founder, Dr. Samuel L. Mitchill (1764-1861), the scientifically influential, but wordy genius and polyhistor, who was equalled only by Thomas Jefferson in intellectual capacity, but who fell far short of Jefferson's intellectual discipline.

The paper by Dr. Forsyth, a physician of Wheeling, West Virginia, asserted that the Ohio River and its tributary streams gradually wore away their valleys and left the hills. He understood the origin of the terraces of the Ohio River and the cause of the fluctuation of the level of the ground water in the wells and in the terraces. He listed mineral resources, including salt springs which are of no consequence now, but were of utmost importance to the pioneer. Forsyth recognized the spring water is harder than river water, for he says "the former does not wash well."

16. Cutler, Jervis, 1768-1844.

A Topographical Description of the State of Ohio, Indiana Territory, and Louisiana. Comprehending the Ohio and Mississippi Rivers ... Boston, Pub. by C. Williams, J. Belcher, Printer, 1812.

One of the first settlers of Marietta was Jervis Cutler, surveyor, traveler, farmer, soldier, and later engraver. He had a keen eye for topography and seems to have understood the relation of topography to rock character and structure. The valleys of the Ohio, the lower Mississippi and the Red River were familiar to him, and he described in detail these valleys and the character of the uplands. In his book, he discussed oil springs, coal deposits, sandstone for building, iron ore, salt springs and furnaces, mineral springs, and the great fossil deposits of Big Bone Lick and other localities. His description of the lead mines of Missouri is important and may be the first of the geology of these deposits.

This unassuming book is of importance for its first-hand descriptions of natural features and geology and for its five plates of animals, Indians, and view of Cincinnati drawn and engraved by the author. The Illinois copy of this much read and often incomplete book is complete with all the plates and blanks.

17. Drake, Daniel, 1785-1852.

Natural and Statistical View, or Picture of Cincinnati and the Miami Country, Illustrated by Maps... Cincinnati, Printed by Looker and Wallace, 1815.

Daniel Drake was an outstanding physician in Cincinnati, an organizer of medical schools, a highly competent scientist, and a pungent writer. His contributions to early geology deserve to be more widely known and appreciated; his geological observations were accurate and his speculations and understanding of geological processes were much advanced for his time.

Natural and Statistical View, or Picture of Cincinnati was written after Drake had traveled extensively in Ohio, Kentucky, and Indiana. The section of the book concerning geology is the first organized treatise on the geology of western Ohio and eastern Indiana. It deals with the limestone bedrock and its fossils, alluvial deposits of the valleys and their terraces, the elephant fossils of Big Bone Lick (including an account of Ashe's swindle of Dr. Goforth), and the glacial drift and erratics of granite from north of the Great Lakes. Drake did not realize that the erratics had been transported by an ice sheet, but he speculated that they may have been brought to their present position in "cakes of ice." This is the first hint in America of any connection between erratics and ice. The appendix constitutes a valuable and accurate record of earthquake history, especially the time and severity of the New Madrid earthquakes of 1811-15 as recorded in Cincinnati.

Although the copy shown has been rebacked, it is quite crisp and the maps are notably fine.

18. Cleaveland, Parker, 1780-1858.

An Elementary Treatise on Mineralogy and Geology

Being an Introduction to the Study of These Sciences ... Boston, Published by Cummings and Hilliard, 1816.

Written by Parker Cleaveland, professor at Bowdoin, this first American book on mineralogy is noteworthy for its completeness; it compares favorably with any European work of the time. The book is divided into three sections, the first of which is a lengthy introduction explaining mineral properties. The second and longest section treats classification, crystallography, and the description and occurrence of minerals. A colored map derived from Maclure's map of 1809 forms the basis for the concluding section on the geology of the United States. On the map the rocks from the Appalachians to beyond the Mississippi are shown as "Secondary," mainly horizontally bedded sandstone and limestone, but the text makes it clear that coal, salt, clays and iron ores also occur.

The book was in such great demand that the first edition was soon sold out. In 1822 an extensively revised second edition was published in which was incorporated expanded information of great value about the mineral localities in the western country, especially Ohio, Illinois, and Missouri. The localities, descriptions and references to mineral deposits in the Mississippi Valley are significant additions to the later edition. The inclusion of ths material is an indication of the rapidly expanding knowledge of western geology and mineralogy.

A copy of the 1816 edition which was once owned by the mineralogist Benjamin Akerly is shown.

19. Brown, Samuel R., 1775-1817.

The Western Gazetteer; or Emigrant's Directory, Containing a Geographical Description of the Western States and Territories... Auburn, New York, Printed by H. C. Southwick, 1817.

Although Samuel R. Brown is almost unknown to historians of American science, his geological writings are of considerable importance. From personal observation, letters, and published sources, he described the western country. The sections on Ohio, Indiana, and Illinois are especially detailed and include many pages of perceptive geological references. The section on the "Face of the Country" which describes the "six different kinds of land in Illinois" is the first attempt to classify Illinois into physiographic regions. Repeated references to petroleum appear in the book.

The very important description of the raised beaches of Lake Erie is based in part on the geological chapter in his earlier book, largely unknown to geologists, Views of the Campaign of the Northwestern Army .... (Burlington, 1814). It is curious that Brown's valuable references to natural history and geology have escaped the notice of historians of biology and geology.

Although several American, Irish, and English editions of the book were published, the first American edition is displayed.

20. Thomas, David, 1776-1859.

Travels Through the Western Country in the Summer of 1816... Auburn, New York, Printed by David Rumsey, 1819.

The Quaker practical engineer, David Thomas, travelled from his home in Cayuga County, New York, to the Wabash Valley by way of Erie, Pittsburgh, and across Ohio to Cincinnati, northern Kentucky and Indiana to Vincennes and Terre Haute. The unusual route was arranged to take him from one Society of Friends to another. The first part of his pleasant and enjoyable book is a journal and the second part, notes and discussion.

The composition, structure, and continuity of the rock units were described so completely that it is possible to identify the very formations Thomas passed over. Repeatedly he noted drumlins, eskers, kettleholes, and various kinds of erratics, which he was sure came from the north. Along with Drake he considered ice a transporting agent, but he preferred a great flood origin for these deposits.

In his book, Thomas advanced proof that valleys were carved by the streams which flowed in them and that the Knobstone Escarpment of Indiana was an erosional feature. His description of Kentucky and Indiana caves is the most complete to his time. Drawing on his excellent knowledge of mineralogy and chemistry, he proposed a theory of cave origin by solution of limestone bedrock.

The many references, quotations, and long critical notes on other travel accounts and textbook theories make the book particularly valuable. Though the book is little known now, it was widely used at the time by compilers of gazetteers and through these media his ideas reached a wide audience.

Thomas' book must have been popular in its day, for most copies show evidence of intensive use. It was produced by a local printer from undistinguished type on poor paper with narrow margins and irregular folding. The map, however, is very well drawn and printed.

21. Nuttall, Thomas, 1786-1859.

A Journal of Travels into the Arkansa Territory, During the Year 1819... Philadelphia, Printed and Published by T. H. Palmer, 1821.

Though Thomas Nuttall was primarily a botanist, his book on his travels to Arkansas contains many acute geological observations and speculations. He compared the limestones of the interior and their fossils with those of Derbyshire, his early home. He commented on the continuity of the coal and other rocks along the Ohio Valley from Pittsburgh to Cairo and noticed the erratics in the outwash gravels along the valley. Nuttall's discussion of the rocks and structures of the Ouachita Mountains and his statement that he "cannot perceive any difference" between them and the Allegheny Mountains of Pennsylvania and Virginia is the first comparison of these two mountain systems. Farther west he saw badlands, and understood in what materials and under what conditions they were formed.

Recently it has been discovered (J. E. Graustein, "Nuttall's Travels into the Old Northwest," Chronica Botanica, XIV (1950), 1-88) that in a plant collecting journey in 1810-11 Nuttall recorded in his diary interesting and valuable geological notes on erratics, sedimentary rocks and oil springs in northwestern Pennsylvania and on raised beaches of Lake Erie in Ohio. It is greatly to be regretted that these observations, together with those of the trip up the Missouri, were not published by Nuttall.

22. Bradbury, John, 1768-1823.

Travels in the Interior of America, in the Years 1809, 1810, *and* 1811... Liverpool, Printed for the Author by Smith and Galway and Published by Sherwood, Neely and Jones, London, 1817.

After coming to St. Louis from the Atlantic coast, the Scotch-English botanist John Bradbury spent two years traveling throughout the Mississippi Valley. In addition to covering the whole of the central Mississippi Valley, his travels extended up the Missouri River almost to Montana. The observations he made during his travels are clear and factual, not twisted and compressed to fit preconceived theories or systems.

The first half of Bradbury's book is an exciting and apparently unexaggerated travel narrative which includes some hair-raising material on Indians. Though the geology is woven into the narrative, it is also assembled and organized in the last half of the book.

Bradbury described the essentially flat-lying strata from the Alleghenies to the Rockies and knew of their fossil content and continuity. Clearly he recognized the erosional origin of topography and understood that differences in form — badlands, smooth slopes, wide or narrow flood plains — are due to differences in material, slope, climate, and vegetation. Since he thought that the slow action of known forces accounted for the present condition of the earth, he was one of the very early uniformitarians in America.

When the great earthquake series began at New Madrid December 16, 1811, Bradbury and his boat crew were on their way from St. Louis to New Orleans and were only a few miles away. Bradbury was the scientist closest to the center of the tremors and his vivid and trustworthy account itself entitles him to scientific recognition. Unfortunately Bradbury's important work is little known to historians of American natural history and totally unknown to historians of geology; a full study of the man and his scientific contribution is long overdue.

The 1817 edition of the book displayed was published without a map or other illustration; an edition published two years later included a map.

23. *American Journal of Science.* 1818—

In 1818 Benjamin Silliman (1789-1864) of Yale published the first volume of the American Journal of Science which is the oldest continuously published scientific journal in America. From its outset the Journal emphasized geology and mineralogy and devoted impressively generous space to papers on geology of the area west of the mountains.

In the early part of the last century, the Journal was an important means of disseminating geological information to scientists working in the west and to persons interested in the geology of the new territory. Through the articles, reviews, notes, and discussions that appeared in the Journal it is possible to trace the development of geology in the West.

In the very first number of the first volume of the Journal on page fifty-two is a communication by Joseph Baldwin of Shawnee Town on " . . . some interesting specimens of fluor spar. They are found not far from Shawnee Town (Illinois) on the banks of the Ohio . . . " The second number of the volume contained a long paper by Caleb Atwater of Ohio "On the Prairies and Barrens of the West," in which he suggested that the prairies may be parts of an old lake bottom, perhaps a once much more extensive Lake Erie.

The financial problems of the day soon oppressed the editor and many of his western subscribers. In volume four the editor tells of the wide distribution of the Journal in the U.S. and overseas and that " . . . it is not without patronage beyond the Alleghenies, but the state of the currency has made it necessary to relinquish an extensive subscription in these regions." Nevertheless, the Journal continued to print many articles on the geology of the West.

The Library has a complete set of the Journal. The early volumes were acquired as acquisition No. 681 when the University was still known as the Illinois Industrial University.

24. James, Edwin, 1797-1861.

An Account of an Expedition from Pittsburgh to the Rocky Mountains . . . London, Printed for Longman, Hurst, Rees, Orme, and Brown, 1823.

Edwin James was geologist of the 1819 expedition to the Rocky Mountains led by Major Stephen H. Long, and it was James who wrote the major part of the report of the expedition. Major Long himself is noteworthy in the history of American science, because he assembled an able scientific staff and encouraged them to record and publish promptly their observations, to which he added his valuable summaries. The wealth of immediately reported geologic information from the 1819 and the 1823 expeditions is in marked contrast to the paucity of such material in the reports of other great explorers, such as Zebulon Pike and Lewis and Clark.

James' report supplied information about the geology of the Great Plains and the Rocky Mountains used by succeeding writers for many years. In the chapter on geology James dealt at length with the rocks and structure of the Rocky Mountains, especially with the upturned sedimentary rocks along the eastern front of the mountains. His concept of the vast extent and continuity of the various strata from the Rockies far eastward is much better realized from his text than from the sections he included in his report.

Additonal contributions were made by other members of the expedition. The chapter by Major Long entitled "General description of the country . . . " is one of the first physiographic descriptions of the Great Plains.

First published in Philadelphia in 1823, the report contained a map which showed some geological boundaries, and thus is the first geological map of the trans-Mississippi region. The three-volume London edition of the report also appeared in 1823. The second volume of the London edition is displayed to show the colored plate of the canyon of the South Platte recorded by S. Seymour, senior artist of the expedition. The plate is not included in the American edition.

25. Keating, William Hypolitus, 1799-1840.

Narrative of an Expedition to the Source of St. Peter's River, Lake Winnepeek, Lake of the Woods, &c., Performed in the Year 1823 . . . London, G. B. Whittaker, 1825.

William H. Keating served as geologist, chemist, mineralogist, and historian of Major Stephen H. Long's expedition to the Upper Mississippi Valley, the Red River, Lake Winnipeg and Lake Superior. There was scarcely any part of the route from Philadelphia to Columbus to Chicago and thence to Winnipeg and Lake Superior to which Keating did not contribute important new geological observations. His Narrative of an Expedition is one of the important geological books of the period.

The observations Keating made of Ohio geology are especially important. Having followed erratic boulders from Zanesville, Ohio, to Canada, Keating recognized their source and speculated on the method of their transportation, but he did not realize their glacial origin. He understood that the Precambrian rocks of the Canadian shield extended southward underneath a covering of almost horizontal sedimentary rocks which contained fossils.

Narrative of an Expedition was first published in Philadelphia in

1824. An edition was published in London in 1825 to meet the popular demand for information about the American West. The first volume of the London edition is displayed.

26. Schoolcraft, Henry Rowe, 1793-1864.
A View of the Lead Mines of Missouri . . . New York, C. Wiley & Co., 1819.

Schoolcraft was an inveterate traveler through the whole Mississippi Valley and the Great Lakes region. He wrote several books and many articles that included geological and mineralogical observations of merit and importance. His Lead Mines of Missouri was the first American work on economic geology. Many aspects of geology are considered — a history of the lead mines, mine descriptions, smelting, production figures, and end uses. The book also contains an extensive "catalogue of the minerals of the Western country" with notes on the general geology of all the western states.

27. Schoolcraft, Henry Rowe, 1793-1864.
Narrative Journal of Travels from Detroit Northwest Through the Great Chain of American Lakes to the Sources of the Mississippi River in the Year 1820. Albany, Published by E. & E. Hosford, 1821.

Schoolcraft's account of an expedition through the Great Lakes and much of the upper Mississippi Valley contains many passages on geology and mineral resources. It includes some description of the geology of the basin of Lake Michigan and of the "drift clay" as well as the limestone along the shore. Schoolcraft was able to trace "the continuance of the calcareous, horizontal series of the formations of the Fox Valley, and of the islands of Green Bay, quite around those northern waters to the closing up of the surveys at Point St. Ignace and Michilimackinac." This is one of the earliest reports of actual tracing and correlation of stratigraphic units in the Mississippi Valley, and it marks the beginning of the kind of observations which formed the basis for the rapidly developing geological knowledge of the next few years.

28. Schoolcraft, Henry Rowe, 1793-1864.
A Memoir on the Geological Position of a Fossil Tree, Discovered in the Secondary Rocks of the River Des Plaines . . . Albany, Printed by E. and E. Hosford, 1822.

"A petrified tree, which is found in the bed of the River Des Plaines about four rods above its junction with the Kankakee," is the subject of this report by Schoolcraft. More than forty-one feet in length, the diameter of the tree ranged from one and one half to two and one half feet. Apparently the tree suffered some decay before burial in the sand, making it impossible to tell which of the large Pennsylvanian trees it might have been. Schoolcraft erroneously identified it as a black walnut, a tree which did not appear until about 170 million years later.

The rock in which the tree was embedded Schoolcraft identified as "newer Floetz sand stone," which he might also have called "Secondary," a term then used for the horizontal Paleozoic strata. Schoolcraft noted that the sandstone where the tree was found had a few other fossils. He observed that the material making up the surface of the adjacent prairie contained granite and other far-traveled rocks which he knew to be common in the northern states to the east.

This pamphlet is displayed not because of its scientific importance, but because it is one of the earliest separate publications on an American fossil plant, and because it has Illinois interest. The description was published also in essentially the same form in a paper in the American Journal of Science, v. 4, 1822.

29. Schoolcraft, Henry Rowe, 1793-1864.

Travels in the Central Portions of the Mississippi Valley: Comprising Observations on Its Mineral Geography... New York, Collins and Hannay, 1825.

In this book, Schoolcraft tells of his travels from Detroit to Toledo and Ft. Wayne and down the Wabash and Ohio to Shawneetown, Illinois, and thence overland to St. Louis and up the Illinois River to Chicago. His extensive observations on minerals, fossils, physiography, and sedimentary rock units are scattered through the book and if collected and summarized would have been a respectable scientific contribution.

The copy exhibited is opened to the map which shows the route of the travels.

30. Schoolcraft, Henry Rowe, 1793-1864.

Narrative of an Expedition Through the Upper Mississippi to Itasca Lake, the Actual Source of This River... New York, Harper, 1834.

The report of the expedition which reached the actual source of the Mississippi River forms the most important part of this book, but a great deal of incidental information is included. Numerous notes on geology are contained in the work. Schoolcraft recognized the passage from Paleozoic sedimentary rocks to the Precambrian crystalline rocks, and he mentioned the presence of glacial erratics though he did not understand their origin. Reports on copper deposits in Michigan, fossils, and lead mines and geology of the Galena, Illinois, region are mentioned.

Narrative of an Expedition and the Narrative Journal of Travels originally published in 1821 were reprinted in one volume in 1855 with some rearrangement and the addition of correspondence and numerous appendices, some of great geological interest and value.

31. Shepard, C. U., 1804-1886.

"Geology of Upper Illinois," *American Journal of Science,* XXXIV (1834), 134-161.

A mineralogist and chemist, Dr. C. U. Shepard was concerned primarily with mineralogy and eastern geology. Paleontology had some interest for him, and his writing contains expert and perceptive comments on it. If Dr. Shepard could have given more attention to Illinois and its geology, he could have contributed much to the advancement of knowledge of the state, its geology, and its mineral resources.

Shepard's penetrating remarks on northern Illinois were published in a decade in which Illinois geology was not being advanced very rapidly. The remarks were made as a result of a visit to the Michigan and Illinois Canal, possibly as a consultant, because he discussed the engineering geology of the canal's construction. This is the first publication in which the geology of Chicago is described, and it is noteworthy that the engineering soil conditions were recognized at this early time. Shepard described the succession of raised beaches of Lake Michigan, speculating on the length of time required for their formation; the "Magnesian Limestone" near Chicago, illustrating its characteristic fossils; and the northern margin of coal bearing strata, including some of their fossils.

The petrographic description and analysis of the coal near Peru contained in the article may be the first for an Illinois coal. Shepard also recognized the superior clays near Peru as well as the "extensive beds of pure, white sand" (St. Peter), and for both he predicted a promising economic future, a prediction which has been brilliantly fulfilled. The paper shows the influence of geology on feasability and cost of construction projects. In turn, the new canal excavations furnished information for pure science just as they had earlier in Great Britain to William Smith.

The paper is opened to show the map of the rocks near Peru and La Salle and the section which is the first to show the disturbed strata of the La Salle anticline.

32. Featherstonhaugh, George W., 1780-1866.

Geological Report of an Examination Made in 1834, *of the Elevated Country Between the Missouri and the Red Rivers.* Washington, 1835.

A respected English geologist, George W. Featherstonhaugh (the name is said to have been pronounced "Fernshaw" or "Fanshaw") was a fellow of the Royal Society and the Geological Society of London. He knew of the investigations in the puzzling lower Paleozoic rocks by Sedgwick and Murchison, and he used some of their findings in advance of complete publication. Appointed "U.S. Geologist" by the War Department, he was the first to bear the title. Though he was one of the best equipped men of his time to examine American rocks, his writings were hurried and require careful reading to extract the really valuable contributions.

In his report on the Missouri and Arkansas country he compared the rocks and fossils with those of Great Britain. He recognized facies fossils as well as index fossils. The vein and disseminated lead deposits of Missouri he described accurately, and his theory of origin is quite modern. Wisely he pointed out that proper theory would lead to great extension of known deposits and to increased production. He proposed that valleys in limestone between parallel ridges of sandstone in both the Appalachian and Ouachita mountains were stream eroded from a once level plain. His descriptions of Hot Springs and Magnet Cove in Arkansas are excellent.

The long section on shape, origin and modification of river channels, especially of the Arkansas River, is of special importance. The suggestion he made that the Army Engineers study sediment load, transport, and deposition under varying conditions has brought brilliant results a century later. Had the report been properly organized, illustrated and edited, it would have been a major contribution to American geology. Though it is important, it requires the laborious effort of a sympathetic reader to extract from it the remarkably prescient material. If Featherstonhaugh had been fortunate enough to have had a Playfair to interpret him to the world, his name would be much better known.

The report appeared in two different forms in the same year, a House document and a Senate document; a copy of the House document is displayed. The House document also appeared with a variant title page which omitted the Congressional references. The accompanying very long colored section of only moderate quality is too large to be displayed.

33. Featherstonhaugh, George W., 1780-1866.

Report of a Geological Reconnaissance Made in 1835, *From the Seat of Government, by Way of Green Bay and the Wisconsin Territory, to the Coteau de Prairie* ... Washington, 1836.

Featherstonhaugh's second trip as "U.S. Government Geologist" was from Washington to northeastern South Dakota. He described the geology of the area from the Blue Ridge and the folded Appalachians to Pittsburgh, thence by way of Ravenna to Cleveland, noting the erratic boulders and near Lake Erie the raised beaches of earlier lake stages. Passing through Detroit, Mackinac, and Green Bay, he ascended the Fox River to the portage to the Wisconsin River and thence traveled downstream to Prairie du Chien. On his way up the Mississippi and

Minnesota Rivers he described the rocks and had much to say about the appearance of the granite beyond and below the Paleozoic rocks. He ascended the Coteau and noted the little lakes on its surface. The report seems to be a rapid and unrevised compilation from much more extensive notes by a tired man who had to meet a deadline.

The summary of stratigraphic geology in America, the best to that time, is the most important feature of the report. Featherstonhaugh was probably the first in America to use the terms Cambrian and Silurian in a geologic column, and his is a very early American use of the terms Cretaceous, Eocene, Miocene, and Pliocene. His correlation chart of Carboniferous and Old Red (Devonian) rocks of America, Belgium, Ireland, and England is remarkably perceptive. Cretaceous rocks of the upper Missouri basin he identified on the basis of index fossils.

In discussing the rise and diminution of trilobites and other Paleozoic animals and of Mesozoic reptiles, as well as Carboniferous and Mesozoic plants, his ideas were unmistakably evolutionary. He understood the ecological implications of some of the animal and plant fossils, but he extended the so-called Carboniferous limestone too widely, including in it much that was actually Devonian and pre-Devonian. If Featherstonhaugh's stratigraphic summary had appeared in another form and in a separate publication, rather than a government document, it would have had wider influence.

There are apparently two forms of publication of this report, an earlier one without a map here displayed, and another the same year with a map. A much longer narrative form of the report was published in two volumes in England in 1847, as A Canoe Voyage up the Minnay Sotor, which contained two very useful maps and much additional material.

34. Nicollet, Joseph Nicolas, 1786-1843.

Report Intended to Illustrate a Map of the Hydrographical Basin of the Upper Mississippi River. Washington, Blair and Rives, 1843.

The French engineer J. N. Nicollet was just able to complete his great map and a compact topographic and geologic explanation before his death, but unfortunately he did not complete a proposed book on physiography. The map is exceedingly detailed, and its accuracy far surpassed any previous map of the West. His excellent work deserves to be better known.

In his geologic descriptions Nicollet wrote of the Paleozoic strata and reported on the Cretaceous rocks farther west. Accurately he described "erratic deposites", and he wrote about the character and distribution of "erratic blocks." Though he is one of the earliest geologists to use the term "erratics," he did not speculate on their origin. He described the White River badlands and the Pierre shale, reported measured stratigraphic sections and accompanying fossils, and made many other valuable observations.

Serving as Nicollet's assistant, Lt. J. C. Fremont mapped the Des Moines River part of the area. The training under Nicollet formed the basis of Fremont's later brilliant career in western explorations and mapping.

Although not published until 1843, the book displayed was the result of work done from 1833 to 1838 and essentially finished by 1841. Another edition appeared in 1845 with a map printed on thinner and whiter paper. The great and beautiful 30 by 36 inch map is too large to be displayed.

35. Mather, William Williams, 1804-1859.

First Annual Report of the Geological Survey of Ohio. 1838.

In the decade of the 1830's geological surveys of most of the eastern states were organized, and the pattern of state participation in geological investigation was begun. In the Ohio part of the Mississippi Valley the states of Virginia, Pennsylvania, Ohio and Indiana established surveys in this decade, but the other states did not do so until later — Illinois in 1851, Missouri in 1853, Kentucky in 1854, and Iowa in 1855.

The Geological Survey of Ohio was authorized in 1837, and Dr. W. W. Mather and his assistants, most of them well known and highly respected scientists, published the First Annual Report in early 1838 and the Second Report in late 1838, surely a record of prompt action. These reports are of careful studies of modern character. They are no longer reconnaissance reports, but some of them are of such detail that they are still valuable. They included measured sections and diagrams of measured columnar and longitudinal sections drawn to scale.

A copy of the First Report of the Ohio Geological Survey is opened to show the first folding plate which illustrates the variety of geological information in the volume. Figure three is a measured section of Pleistocene deposits containing vertebrate fossils; figure two is probably the first American illustration of mass movement — slumping of the Devonian shale and overlying drift along the Lake Erie bluffs at Cleveland in which the curved planes of movement and backward rotation of the blocks are accurately shown. Figure four is an accurate section of the strata in southern Ohio, showing the regional dip and the erosion levels.

36. Owen, David Dale, 1807-1860.

Report of a Geological Reconnoisance of the State of Indiana... Indianapolis, J. W. Osborn and J. S. Willets, 1839.

The Geological Survey of Indiana was established in 1837, but unlike Ohio, where several experienced scientists immediately set to work, the Indiana Survey was the sole work of the great David D. Owen. The Survey's first report is explicitly a summary, but a valuable and important one. Owen began this report with an elegant summary of American stratigraphy. Then he compared the Indiana rocks with those of

Ohio and very briefly described the strata of each county or group of counties. He was more inclined than his neighbors in Ohio to make correlations of Mississippi Valley rocks with those of Great Britain.

Owen's report was published in two parts, the first in 1838, the second in 1839. Charts, sections, and maps accompanied the manuscript report, but were not published, a regretable circumstance, because Owen was an expert draftsman and competent artist, and his illustrations must have been of real importance. The two parts of the report were reprinted by the State in 1859.

The copy of the first report displayed has been loaned for this exhibit by Dr. John B. Patton, State Geologist of Indiana.

37. Owen, David Dale, 1807-1860.

Report of a Geological Exploration of Part of Iowa, Wisconsin and Illinois. 1840.

Although state geological surveys were not established in Illinois, Iowa, and Wisconsin until 1851, 1855, and 1853 respectively, a survey of parts of the three states was made in 1839 by the Federal government because of the interest in determining the extent of lead-bearing rocks. The report of this survey is a landmark in American geology.

David Dale Owen and his assistants surveyed several thousand square miles, made the maps, illustrated the fossils, and wrote the report in less than twelve months, an unbelieveable accomplishment! With the assistance of Dr. John Locke, Owen accurately mapped, described and correlated the rocks, worked out the relation of ore deposits to stratigraphy and structure, and determined which lands were mineral-bearing. Even Owen's introduction with six diagrams on origin of strata and later structures is noteworthy.

The copy exhibited here is the earliest of several printings which appeared over five years with minor changes and differing numbers of plates and maps. A complete set of plates is reported to be twenty-five, but this copy has twenty-seven. The book is opened to Plate XII with Owen's beautiful drawings of the characteristic fossils of the "White Limestone," now called Cedar Valley and Wapsipinicon limestones of Middle Devonian age (Dr. Charles Collinson, personal communication, August 20, 1962).

38. Hall, James, 1811-1898.

"Notes, Explanatory of a Section from Cleveland, Ohio, to the Mississippi River, in a Southwest Direction; with Remarks upon the Identity of the Western Formations with Those of New York," *Association of American Geologists and Naturalists Transactions* (1843), 267-296.

The Association of American Geologists and Naturalists, which in 1848 became the American Association for the Advancement of Science, held its first meetings in 1840. Papers of the first three meetings were published in 1843. The volume of the Transactions, which shows the full attainment of modern geology, includes the Rogers brothers' paper on the structure of the Appalachian Mountains, as well as papers on mineral deposits, fossils, correlation, continental glaciation, and reports of discussions.

Many of the papers and reports dealt with the Mississippi Valley, and the most important paper of all was "Notes, explanatory of a Section from Cleveland, Ohio, to the Mississippi River, . . ." by James Hall. The correlation was in part by means of lithologic similarities, but greater stress was placed on similarity of the fossils. The accumulated mass of organized and detailed observations from the publications of the new state geological surveys was enabling such synthesis and the development of a uniform stratigraphic terminology. The pioneer days were coming to an end, and professional geologists were taking over.

The sheet of hand-colored sections displayed is in three parts. The lower section shows the rocks from the Precambrian at the St. Lawrence River to the strata of the southwest corner of New York. The center and upper sections are parts of a continuous section of the Paleozoic rocks from Cleveland across the Cincinnati Arch and thence across Indiana and the Illinois Basin to the Mississippi River. This section was based on Hall's personal knowledge and on information supplied by D. D. Owens and others.

REFERENCE BOOKS
FOR
HISTORY OF GEOLOGY

STECHERT-HAFNER

VOLUME XVIII, NUMBER 3, NOVEMBER 1963

Book News

Reference Books for History of Geology

By GEORGE W. WHITE

INTEREST in the history of science is rapidly increasing. Studies of the history of geology, which have lagged behind those of some of the other physical sciences, are now sharing in this development. Those responsible for building collections will be well advised to give particular attention in the immediate future to their holdings in books on the history of geology. These may be divided into three categories: (1) books on the history of geology; (2) bibliographies; and (3) classical items and important early books. This brief essay is mainly concerned with the first two groups.

Very modern methods of reproduction are now beginning to furnish us with facsimile reprints of some of the scarce standard works on history of geology and are already furnishing us with exact and beautiful facsimile reprints of some of the outstanding classics in early geology. These classics are the books which only the major libraries can hope to obtain in their original condition, but it is now possible for all libraries to secure fine facsimile copies. The favorable response to those now available leads us to hope that additional facsimiles will be welcomed and will be made available by publishers.

Histories of Geology

It is inevitable that the author of a history of geology emphasize the contributions of his own culture. The German histories emphasize the contributions of the Germanic-speaking peoples, English those of the British, the American those of the American geologists; and the Russians have a whole institute on the history of geology, which, however, cannot be discussed in this brief article.

One of the earliest histories of geology in English, and still a very useful one, is that comprehended in the first chapters of *Principles of Geology* (1830) by Sir Charles Lyell. This three-volume work (1830, 1832, 1833) is one that all libraries should seek to acquire, although good copies are now in the "scarce" category. Fortunately, the historical chapters in Volume I were continued without much change in succeeding editions, which are still available at very reasonable prices. A number of other histories, by A. C. Ramsay (1849), J. B. Woodward (1911), T. Crook (1953), and others, are of some importance but of greater importance is the book by Sir Archibald Geikie, *The Founders of Geology* (second edition, 1905), which should be in every library and in the hands of every serious geologist. The English translation of Zittel's *History of Geology and Paleontology* (1901) is a very important reference. In late years a whole series of journal articles have appeared by the experts F. J. North, John Challinor, C. E. N. Bromehead, V. A. Eyles, and others, but they have not been collected in book-length disquisitions. *The Birth and Development of the Geological Sciences* (1938, reprinted 1954) by F. D. Adams is the single best source for very early history of geology. It contains generous quotations, many illustrations, and extensive references.

The Germans have paid much attention to histories of geology, beginning several hundred years ago with Georgius Agricola, whose *De Re Metallica* (1556) contains some historical material. The great Hoover translation (1912 and facsimile reprint 1950) has so much added historical material and such extensive references that it is an important historical reference book in its own right. Friedrich Hoffmann's *Geschichte der Geognosie* (1838) contains some references not easily found elsewhere, but its lack of an index makes it hard to use, even by a specialist. Much the same can be said of Franz von Kobell's *Geschichte der Mineralogie* (1864). *Philosophie der Geologie und mikroskopische Gesteinsstudien* (1867) by H. Volgelsang deserves to be better known; it contains the first color plates of rock thin sections. K. A. von Zittel's *Geschichte der Geologie und Paläontologie* (1899), translated into English two years later, is one of the great histories. The German original contains some references not included in the English edition. *Geschicte der Geologie und des geologischen Weltbildes* (1954) by C. C. Beringer is an important little book; it concentrates on fewer subjects than most but goes into detail on some otherwise little-discussed topics and men. The recent elaborate book by Helmut Hölder, *Geologie und Paläontogie* (1960), contains long quotations and elaborate bibliographies.

The earliest American history of geology is contained in *Brief Retrospect of the Eighteenth Century* (1803) by Samuel Miller, a book that should be reprinted, preferably with critical analyses and notes. American geologists in the nineteenth century were too busy studying the geology of their rapidly expanding country to write much on the history of their subject. However, short summaries of history of geology were included in some of the textbooks, such as those of Jeremiah van Rensselaer (1825), W. W. Mather (1833), and especially in that of Edward Hitchcock, which, under various titles, was published in at least thirty editions from 1840 to 1855.

The most important histories of American geology are the three books by G. P. Merrill. His *Contributions to the History of American Geology* (1906), published by the U.S. National Museum, was revised as *The First Hun-*

dred Years of American Geology (1925). Long out of print, it is now happily to be reprinted. Merrill's *Contributions to a History of the American State Geological and National History Surveys* (1920) is more specialized; it contains useful information, but for lists of the early state publications the Meisel bibliography *(see below)* must be consulted. Somewhat more popular, but still very useful, is *The Story of the Great Geologists* (1945) by C. L. Fenton and M. A. Fenton, and somewhat enlarged and revised as *Giants of Geology* (1952). *The Earth We Live on; the Story of Geological Discovery* (1957) by Ruth Moore deserves mention for its sprightly writing and excellent illustrations. Summaries of the history of the U.S. Geological Survey are available in official and other publications. Two recent books, *Army Exploration in the American West 1803–1863* (1959) by W. H. Goetzmann and *Great Surveys of the American West* (1962) by R. A. Bartlett, tell in some detail of the government surveys, both military and civilian, that preceded the present U.S. Geological Survey. The two biographies of one of the early directors of the Survey, the well-known Major James Wesley Powell, should be included in any list of histories of American geology. They are *Powell of the Colorado* (1951) by W. C. Darrah and *Beyond the Hundredth Meridian. John Wesley Powell and the Second Opening of the West* (1953) by Wallace Stegner. The biography of a great American geologist whose life and active work spanned almost the whole of the nineteenth century is almost a must for any historical collection: *James Hall of Albany, Geologist and Paleontologist, 1811–1898* (1923) by J. M. Clarke. Of the many biographies of American geologists, this is probably the best of many good ones.

Bibliographies of geology inevitably carry much material for the study of its history. An early, but still the greatest of geological bibliographies is Emmanuel de Margerie's *Catalogue des Bibliographies Géologiques* (1896). All of the histories of geology mentioned earlier have references and some have very important lists of references. The "Bibliographical Essay" in *Genesis and Geology* (1951 and recent paperback) by C. C. Gillispie is a charmingly written essay as well as being a useful list of early geological publications. *A Source Book in Geology* (1934) by K. F. Mather and S. L. Mason contains many useful excerpts and serves also as a bibliography of important publications. *The Checklist of the E. De Golyer Collection in the History of Science and Technology* (1954) compiled by Arthur McAnally and D. H. D. Roller is a very useful list of early books in geology. It stresses European publications more than early American ones. *The Lying Stones of Dr. Johann Barthomew Adam Beringer* (1963), a translation by M. E. Jahn and D. J. Woolf of *Lithographiae Wirceburgensis* (1726), contains a long list of early geological and paleontological works and discussions of some of them in the various notes.

Palaeontologi. Catalogus bio-bibliographicus (1938) by K. Lambrecht and W. and A. Quenstedt contains sentence-length biographical notes and references to biographies and memorials of over three thousand paleontologists and geologists of all countries; it is an important reference work and the point of beginning for biographical searching for non-Americans and even for some Americans.

A very useful and sometimes the only source for Spanish and Latin American geological references is *Apuntes para una Biblioteca Española . . .* (Madrid, 1871, 1872), two volumes, by Eugenio Maffei and Ramón Rúa Figueroa. It contains biographical sketches and bibliographical lists arranged in several categories.

A work that is indispensable for references to American geological publications up to 1865 is Max Meisel's three-volume *Bibliography of American Natural History* (1924, 1926, 1929). It has many classifications of scientific materials and includes in chronological order lists and tables of contents of all scientific journals and other publications, including the geological ones.

Of course the great series of bibliographies of the United States Geological Survey, which list American publications or publications about America from 1732 and continuing to the present, is a standard working tool for all geologists. These are published as U.S. Geological Survey Bulletins, 746–747, 823, 937, 1049, 985, 1025, 1035, 1065, 1095, 1115, and 1145. Many of the state geological surveys have bibliographies of geological publications referring to the particular state. These are of uneven quality and completeness. One of the best for the recognition of the early publications of historical interest is Bulletin 24 of the Indiana Geological Survey, *Annotated Bibliography of Indiana Geology through 1955* (1962).

The study of the various histories and bibliographies noted here will soon show that certain books are mentioned in almost all the publications, and those responsible for building libraries can determine which of these classical items they already possess and which ones should be looked for in the future. The lists will help determine which reprints should be considered and which additional ones should be recommended to publishers. The reprints (facsimiles are greatly preferable) will be useful not only to those collections which do not have the originals but also to those that do, in order to preserve the now rare originals for the really serious student who must use original copies.

Everyone knows of the importane of Steno, Agricola, Hutton, Playfair, William Smith, Lyell, Murchison, Lewis Evans, Maclure, and Eaton, but everyone would also be well advised to consider—naming only Americans or those reporting on American geology—Chastellux, Peter Kalm, John and William Bartram, Volney, Mease, Edwin James, William Keating, Parker Cleaveland, Thomas Nuttall, H. R. Schoolcraft, Featherstonhaugh, Samuel Mitchill, as well as a whole host of others, including Agassiz, Frémont, Nicollet, Owen, James Hall, Mather, and Hitchcock. These are all dealt with in the various histories and bibliographies in such a way that the librarian and his advisors can form some estimate of the relative importance and desirability of their books.

[GEORGE W. WHITE, *A.B., Otterbein College; M.S. and Ph.D., Ohio State University, has been head of the Department of Geology at the University of Illinois, Urbana, since 1947. His geological field studies and publications and those of his graduate students have been mainly in glacial geology and stratigraphy of glacial deposits. He works closely with the University of Illinois Library in expanding its outstanding collection of early works in geology. He personally collects geological bibliographies, biographies, dictionaries, and histories of geology. Several chapters of a book on "Early American Geology, 1555–1787" have already been published as separate articles.*]

ANNOTATED BIBLIOGRAPHY
OF BOOKS
FOR HISTORY OF GEOLOGY

ANNOTATED BIBLIOGRAPHY OF BOOKS FOR HISTORY OF GEOLOGY

George W. White

University of Illinois

Presented at the University of Nevada, Reno, Conference on History and Philosophy of Geology, August, 1964

INTRODUCTION

This list of books consists of bibliographies, dictionaries, and histories of geology useful for the study of history of geology. Only regularly published books have been included in this list, except for a very few special issues of journals and one or two journal articles. The vast number of journal articles concerned with history of geology can be found listed in the various bibliographies included here and especially in the very exhaustive list now being compiled by A. LaRocque.

Many of the books on history of geology contain very extensive bibliographical lists, and many of the bibliographies have comments and annotations which give them historical interest. Altho the list of bibliographies is limited to bibliographies for the history of geology, inevitably some well-known general bibliographies of geology are included because they contain important historical sections. Biographies, as such, are not included; an annotated list of biographies will be found in the Wells and White reference. Many general works in geology, paleontology, and allied subjects contain from one to several historical chapters, but (except for Lyell) have not been included.

This list is based principally on my own "working apparatus" and on reference books in the University of Illinois Library, with the addition of 19 titles selected from those suggested by those acknowledged below. One only has to look at the bibliographic part of a great library to see how many hundreds or thousands of feet of shelf space are necessary for a complete working collection in the whole field of bibliography. It is obvious, therefore, that a rigid and even arbitrary selection has been necessary among the general works and catalogs.

The list is divided into three parts. The first consists of histories of geology or of some field of geology, the second of bibliographies, and the third of dictionaries. It has been difficult to classify some of the books because some of the bibliographies are actually biobibliographies and are almost histories in themselves. Some of the histories have such very extensive references that they are quite bibliographies in themselves.

From the earliest stages of interest in history of geology, which means also interest in early books in geology, the geologist should make the acquaintance and form firm friendships with sympathetic and patient librarians who are skilled in such matters. Specialized knowledge in bibliography is required to run down sometimes obscure references. For very early references, particularly those in Latin and in abbreviated forms, detective work of the highest order may be necessary (see Jahn and Woolf, The lying stones of Doctor Barringer, for an example of this very specialized kind of literary work). Many scientific book dealers are experts in this field and some are recognized scholars themselves, producing catalogues of permanent value.

The geologist who becomes a collector or who advises on library acquisitions must immediately become aware of editions and states of early books. He should recognize that variations exist and know where to verify them. The variations may range from those (known to all) of an early government publication that exists in Senate and House editions--and possibly in more than one of each--to the 1580 Palissy with the dropped page number on page 219 of some copies (known to almost no one). Early books on geology are particularly apt to have unknown or unrealized incompleteness, because maps, especially folded ones, diagrams, and plates may have been removed or misplaced. Maps issued in separate folders or atlases are notoriously absent from some of the early publications. The bibliographies will usually help to determine what plates and maps should be present but sometimes several different ones must be consulted to get a complete description.

Various geologic and bibliographic friends have contributed valuable suggestions and additions. The annotations contributed by them are indicated by initials. If two or more mentioned a book I had decided to omit, I restored the title to the list. Most of the titles listed without annotation have been suggested to me and I have not seen most of these. I have not included all the books mentioned to me, but have (perhaps arbitrarily) decided that some are too general or too specialized. I deeply appreciate the helpful comment and suggestions so generously given me on this, as well as on many other occasions, by C. C. Albritton, L. E. Bamber, R. E. Banta, I. B. Cohen, R. B. Downs, J. A. Eyles, V. A. Eyles, F. C. Haber, A. LaRocque, A. O. Liebstaedter, R. Rappaport, H. Reichner, R. R. Schrock, H. W. Smith, C. K. Swann, K. L. Taylor, H. M. Welch, J. W. Wells, and A. O. Woodford.

Note added in January, 1977.----Since this list was compiled in 1964, additional histories and bibliographies have appeared. No attempt has been made to include these in this compilation. However, these valuable bibliographies should be mentioned: Pestana, H. R., Bibliography of Congressional Geology: New York: Hafner Press, 1972, 285 p.; Hazen, R. M. and Hazen, M. H., Bibliography of American-published Geology: 1669 to 1850: Geological Society of America Microform No. 4, 1976, 984 p. (also to be published in G.S.A. Memoir); and Sarjeant, W. A. S. Bibliography of Histories of Geology *and Biographies of Geologists*, soon to be in press in several volumes. Also must be mentioned the magnificent definitive history of petroleum geology, Edgar W. Owen, *Trek of the Oil Finders: A History of Exploration for Petroleum*: American Association of Petroleum Geologists, Memoir 6, 1975, 1647p.

PART I

Histories of Geology

ADAMS, F. D. The birth and development of the geological sciences. Williams and Wilkins, Baltimore, 1938; facsimile reprint, Dover Publications, New York, 1954.

The standard work on ancient, medieval and early modern geology. Generous illustrations, quotations, and translations. Fine pace and style. References not always exact.

ALBRITTON, CLAUDE C., JR., Editor. The fabric of geology. Addison-Wesley Publishing Company, Inc., Reading, Mass., Palo Alto, London, 1963.

This book contains the sixteen articles which review the history and philosophy of geology produced on the occasion of the seventy-fifth anniversary of the founding of the Geological Society of America. Includes an elaborate bibliography on philosophy of geology that contains many historical items of importance. Some articles are of more historical importance than others.

AMERICAN ASSOCIATION FOR ADVANCEMENT OF SCIENCE. Symposium on the history of American geology. Journal of Washington Academy of Sciences, Washington, Vol. 49, No. 7, pp. 197-260, July, 1959.

Eleven papers on development of American geology from 1756 to recent times. Many illustrations and extensive bibliographies.

AMERICAN PHILOSOPHICAL SOCIETY. The early history of science and learning in America with especial reference to the work of the American Philosophical Society during the eighteenth and nineteenth centuries. Papers read before the American Philosophical Society Midwinter Meeting, February 13-14, 1942. The American Philosophical Society, Philadelphia, 1942.

Several papers on history of geology and related sciences. By far the longest and most important is G. G. Simpson's on "Beginnings of vertebrate paleonology in North America", which is well illustrated and fully documented.

ALCOCK, F. J. A century in the history of the Geological Survey of Canada. National Museum of Canada Special Contribution No. 47-1, 1947.

Brief history and interesting illustrations of this important survey. More references would be welcome.

d'ARCHIAC, E. J. A. Introduction a l'etude de la paleontologie stratigraphique. 2 vols. Paris, 1864.

"The author who does best with the French 18th century." (R.R.).

BAILEY, EDWARD. Geological Survey of Great Britain. Thomas Murby and Co., London, 1952.

"Unofficial" and spritely history of the oldest geological survey. Useful early references and brief bibliographies. Sir Edward has always enjoyed doing and saying things his own positive way.

BALD, ROBERT. A general view of the coal trade of Scotland....to which is added, an inquiry into the condition of those women who carry coals under ground in Scotland.... A. Neill & Co., Edinburgh, 1808.

This is a brief history of coal mining in Scotland and a discussion of the current mining practices from an economic and social point of view. Contains description of the pathetic conditions of the time under which the men, women, and children miners worked.

BARTLETT, RICHARD A. Great surveys of the American West. University of Oklahoma Press, Norman, Okla., 1962.

Best connected summary, descriptions, and analyses of the Hayden, King, Powell, and Wheeler surveys which preceded the United States Geological Survey. Contains some interesting illustrations, useful maps, and a good bibliography.

BELL, W. J., JR. Early American science, needs and opportunities for study. Institute of Early American History and Culture, Williamsburg, Virginia, 1955.

Chiefly bibliographies and sketches of 50 early American scientists, 22 of whom made more or less important geological observations.

BERGBAU UND BAULEUTE. Neue Beitrage zür Geschichte des Bergbaus und der Geologie. (Freiberger Forschungshefte. Kultur und Technik. D11). Akademie-Verlag, Berlin, 1955.

BERINGER, CARL CHRISTOPH. Das werden des erdgeschichtlichen weltbildes im spiegel grosser naturforscher und denker aus zwei jahrhunderten. Ferdinand Enke, Stuttgart, 1939.

Mainly philosophy of geology from historical basis.

BERINGER, CARL CHRISTOPH. Geschichte der geologie und des geologischen weltbildes. Ferdinand Enke, Stuttgart, 1954.

History of geology from ancient times to the present. Gives lengthy emphasis to men, events, and concepts usually very briefly treated in English works. Deserves wide circulation.

BERTRAND, ELIE. Recueil de divers traites sur l'histoire naturelle de la terre et des fossiles. Avignon: Louis Chambeau, 1766.

This collection of essays is included because it contains a great deal on history of theories and observations and includes extensive references.

BROWNING, WILLIAM. The relations of physicians to early American geology. Annals of Medical History, New Series, Vol. 3, pp. 547-567, 1931.

Many early geological observers had medical training. A very useful (if uneven) list of 100, with references to biographies and some mention of publications unlisted in other bibliographies.

CAILLEUX, A. Histoire de la géologie (Collection Que sais-je?) Paris: Presses Universitaire de France, 1961.

CHALLINOR, J. The early progress of British geology. From Leland, to Playfair, 1802. Annals of Science, Vol. 9, pp. 124-153, 1953; Vol. 10, pp. 1-19, 1954; Vol. 10, pp. 107-148, 1954.

Important series of papers of geological observations from earliest times to 1802. Exact references with quotations from the authors themselves. Well done and deserves republication in book form.

CHORLEY, R. J., DUNN, A. J., and BECKINSALE, R. P. The history of the study of landforms, or the development of geomorphology. Volume one: Geomorphology before Davis. London, Methnen, and Company, 1964.

This book deals with history of surface forms, surficial geology, and their geologic controls from Leonardo to Davis. This first volume (of over 600 pages) of a promised three volume study takes immediate rank with Zittel, Geikie, Adams, and Merrill. Its 132 illustrations include portraits and reproductions of "classical" diagrams and maps. Lengthy quotations of original sources advance the argument. Early--as well as later--American geomorphology is treated most perceptively and in detail. It is also the first history of quantitative geomorphology, which goes back to 1697.

COLLIER, KATHERINE BROWNELL. Cosmogonies of our fathers, some theories of the seventeenth and eighteenth centuries. Columbia Univ. Press, New York, 1934.

This book describes the theories of the origin of the earth and of the solar system from early times to the end of the eighteenth century. Contains an exhaustive bibliography. No other book deals so completely with this phase of history of geology.

COPPER DEVELOPMENT ASSOCIATION. Copper through the ages. Copper Development Association, Radlett, Herts., 1934, and many later impressions.

Very brief discussion of history of use and fabrication of copper followed by discussion of modern uses.

CROOK, THOMAS. History of the theory of ore deposits with a chapter on the rise of petrology. Thomas Murby and Co., London, 1933.

History of geology (and partly of petrology) from the standpoint of ore deposits. Quite disrespectful of the modern great authorities who are "repeating Agricola in modern terminology."

CUVIER, G. Éloges historiques, précédes de l'éloge de l'auteur par M. Flourens. E. Ducrocq, Paris, 1835.

Several geological writers included in this collection--Cuvier, Pallas, Hauy.

DANA, EDWARD SALISBURY, and others. A century of science in America with special reference to the American Journal of Science 1818-1918. Yale Univ. Press, New Haven, 1918.

Thirteen men each read all 194 volumes of the first hundred years of the Journal and wrote 13 chapters--including 9 well-written chapters on history of American geology, especially as reflected in the pages of the Journal. Gregory's well-documented chapter on landforms is still the best history of American geomorphology. Can 13 more do as well in 1968 for the third half century of the Journal and American geology? (Has geology itself done as well as in the first century??)

DAVIS, JAMES W. History of the Yorkshire Geological and Polytechnic Society, 1837-1887. With biographical notices of some of its members. Whitley and Booth, Halifax, 1889.

Interesting history of a local society that had (and has) many prominent members and did good work historically of more than local importance.

DAVISON, CHARLES. The founders of seismology. University Press, Cambridge, 1927.

History of seismology as well as of seismologists. Good on early origins and men. Good references.

DRACHMAN, J. M. Studies in the literature of natural science. Macmillan, New York, 1930.

Mainly a history of ideas, science as literature and science influences in literature. Much on evolution and geology. Extensive bibliography of unimportant books. Gillispie does the job much better.

DUNCAN, P. M. Heroes of science; botanist, zoologists, and geologists. Society for Promoting Christian Knowledge, London, 1882.

About half the book deals with geology and geologists, especially Hutton, Smith, Murchison, and Lyell. Derivative, but well done for its popular purpose.

DUPREE, A. H. Science in the Federal Government, a history of policies and activities to 1940. Belknap Press (Harvard), Cambridge, 1957.

Federal government in science and science in federal government from 1787 to 1940 inclusive. Much geology. A big canvass is covered by a wide brush.

EAVENSON, HOWARD N. Coal through the ages. American Institute of Mining Engineers, New York, 1935, 1942.

Good summary of history of coal mining in Great Britain and America. Contains bibliography of the many works on history of coal mining.

EAVENSON, HOWARD N. The Pittsburgh coal bed: Its early history and development. The American Institute of Mining Engineers, New York, 1938.

Exhaustive history of the world's greatest coal bed; important for its 282 references.

EDINBURGH MEETING OF THE BRITISH ASSOCIATION (LOCAL EDITORIAL COMMITTEE). Edinburgh's place in scientific progress. W. & R. Chambers, Ltd., Edinburgh and London, 1921.

This little book consists of twenty-five chapters on the scientific achievements related to the city of Edinburgh. Several of the chapters are on geology or related earth science subjects.

EDWARDS, W. N. Guide to an exhibition illustrating the early history of palaeontology. British Museum (Natural History) Special Guide No. 8, London, 1931.

Excellent brief history of paleontology with illustrations and fine, partly annotated, lists of almost all the important early books.

ELLIS, ARTHUR J. The divining rod, a history of water witching, with a bibliography. U.S. Geological Survey Water-Supply Paper 416, 1938.

A brief, but very objective and scholarly study of the divining rod from earliest times to the present. Includes exhaustive bibliography. This is the most objective book about the divining rod and its history (but also see Vogt, E. Z., and Hyman, Ray, Water Witching U.S.A., Univ. of Chicago Press, 1959).

FAIRCHILD, HERMAN LEROY. The Geological Society of America, 1888-1930. A chapter in earth science history. Geological Society of America, New York, 1932.

Brief history of geology, of American geology, of the AAAS and of several other things, as well as a longer history of the first 50 years of the Geological Society of America by an early secretary and president.

FAY, H. Sur l'origine du monde. Theories cosmogoniques des anciens et des modernes. 4th edition. Gauthier-Villars, Paris, 1907.

Discussion of ideas of the origin of the solar system and of the earth from ancient times almost to the present. Primarily astronomical emphases.

FENTON, CARROLL LANE and FENTON, MILDRED ADAMS. The story of the great geologists. Doubleday, Coran and Co., Inc., Garden City, New York, 1945.

History of geology from earliest times, with emphasis on personalities. Brightly written for popular audience. Brief but useful lists of references.

FENTON, CARROLL LANE and FENTON, MILDRED ADAMS. Giants of geology. Doubleday and Col, Inc., Garden City, New York, 1952 (Paper back, 1960).

Revised and enlarged edition of Story of the great geologists.

FISCHER, WALTHER. Mineralogie in Sachsen von Agricola bis Werner. C. Heinrich, Dresden, 1939.

This book discusses the history of mineralogy in Saxony and in neighboring regions in Germany. Contains a long index and list of references, but the elaborate list of references is not alphabetized and therefore does not constitute a real bibliography. A work of considerable importance to the specialist.

FISCHER, WALTHER. Gesteins-und lagerstättenbildung im wandel der wissenschaftlichen anschauung. E. Schweizerbart, Stuttgart, 1961.

Elaborate history of petrology and of mineral deposits.

FLETT, JOHN SMITH. The first hundred years of the geological survey of Great Britain. His Majesty's Stationary Office, London, 1937.

The official history of the first national geological survey. More inhibited than Bailey's history. Contains roster of geologists and historical bibliography.

FORBES, R. J. Bitumen and petroleum in antiquity. E. J. Brill, Leiden, 1936.

Well illustrated work on ancient uses of liquid and solid petroleum and references to them.

FORBES, R. J. Studies in early petroleum history. E. J. Brill, Leiden, 1958.

Ancient and medieval uses of petroleum, mainly in Europe. Shorter references to Latin America and to Asia. Instructive to Americans who think Col. Drake invented oil at Titusville in 1859.

FORBES, R. J. Studies in ancient technology, Volume VII. E. J. Brill, Leiden, 1963.

This volume is a survey of history of ancient geology and mining, dealing with pre-classical and classical antiquity. Many references and a chronological bibliography of ancient authors. Keyed to current editions.

FREYBERG, B. V. _Die geologische Erforschung Thüringens in alterer Zeit_. Geb. Borntraeger, Berlin, 1932.

Non-critical reviews of work of F. C. Lesser, J. C. Lehmann, K. E. A. von Hoff, and others, Portraits. (A.O.W.). Valuable for history of geology in Germany and for the extensive bibliography (V.A.E.).

FORTIS, A. _Memoires pour servir a l'histoire naturelle et principalement a l'oryctographie de l'Italie et des pays adjacens_. 2 vols. Paris, 1802.

FURON, RAYMOND. _Histoire de la Géologie de la France d'Outre-Mer_. 1955.

FURON, RAYMOND. _La paleontologie. La science des fossiles, son histoire, ses enseignements, ses curiosites_. Nouvelle edition. Payot, Paris, 1951, (First edition, 1943).

History of paleontology from Moses to the present; including related evolution and genetics. Very brief summary of descriptive paleontology followed by sections on living fossils and on fossils in art, folklore, industry, etc. Many illustrations; wide ranging, easily read, recommended to students for historical review and for "easy" French reading.

GARBOE, AXEL. _Geologiens historie i Danmark_. Two volumes. C. A. Reitzels Forlag, Copenhagen, 1959, 1961.

Elaborate, complete, and well-documented history of geological investigations in Denmark from the earliest times to the present. The illustrations of men, plates, diagrams, manuscipts, specimens, and important exposures alone make this so interesting and valuable that it deserves a place in every important library.

GEIKIE, ARCHIBALD. _The George Huntington Williams lectures on the principles of geology. Volume One: The founders of geology_. The Johns Hopkins Press, Baltimore, 1901.

See next entry.

GEIKIE, ARCHIBALD. The founders of geology. Second edition. Macmillan, London, 1905. (Also Dover, 1964.)

The second edition is expanded and is to be preferred. One of the four outstanding general works on history of geology, told mainly through "the lives and work of some of the masters to whom we mainly owe the foundation and development of geological science." Must be supplemented by Merrill, Zittel, and Adams.

GEOLOGICAL SOCIETY OF AMERICA. Geology, 1888-1938, Fiftieth Anniversary Volume. Geological Society of America, 1941.

Twenty-one papers celebrating the fiftieth anniversary of the Geological Society of America deal with the development of 21 different fields of geology in America and other parts of the world, mainly since the Society was founded in 1888. Varying in importance and unequal in coverage, but generally valuable reviews. (Note also special March 1939 number of G.S.A. Bulletin for other important papers not included here.)

GILLISPIE, CHARLES COULSTON. Genesis and geology, a study in the relations of scientific thought, natural theology, and social opinion in Great Britain, 1790-1850. Harvard University Press, Cambridge, 1951; Dover, New York, 1959.

Pre-Darwinian intellectual history of Great Britain based on what was then the most generally known science--geology. Written with youthful exuberance and obvious scholarship. The "Bibliographic Essay" is a gem and proves book lists need not be dull.

GOETZMANN, WILLIAM H. Army exploration in the American West 1803-1863. Yale University Press, New Haven, 1959; Second Printing 1960.

"The aim of this study is to describe, analyze, and evaluate the role played by the U.S. Army in exploring the trans-Mississippi West, and in particular the role of the Topographical Engineers between the years 1838-1863." Many of the expeditions had geologists. Summarizes many important travel reports that include geological observations and descriptions. Fine bibliographical essay and excellent bibliography section.

GOODE, G. B. *A memorial of George Brown Goode, together with a selection of his papers on museums and on the history of science in America*. Report of the U.S. National Museum for 1897, Part II, 1901.

Memoir of Goode and a series of his papers. The two longest and most important are on early natural history in America and early American science. Many portraits. The starting point for study of early American geology and natural history.

GREENE, JOHN C. *The death of Adam; evolution and its impact on western thought*. Iowa State Univ. Press. Ames, 1959. (Paperback by Mentor.)

Excellent and lengthy history of early geological and biological explanations of development and evolution before Darwin, followed by new light on Darwin. Excellent references and illustrations.

GROTH, P. *Entwickungsgeschichte der mineralogischen wissenschaften*. Julius Springer, Berlin, 1926.

Altho this is mainly a history of crystallography with valuable references and excerpts, it includes a briefer history of mineralogy that stresses modern developments. A short, but important section gives excellent, brief biographies of important mineralogists.

HABER, FRANCIS C. *The age of the world, Moses to Darwin*. The Johns Hopkins Press, Baltimore, 1959.

"....history of the transformation in Western though from the Mosaic view of a finished world created about 6,000 B.C. to the Darwinian view of an evolved world of indefinite antiquity." Contains excellent sections on early geological ideas and theories.

HABER, FRANCIS C. *Fossils and early cosmology; fossils and the idea of a process of time in natural history*. Chapter 1 and chapter 9 in Bentley Glass and others, *Forerunners of Darwin: 1745-1859*. The Johns Hopkins Press, Baltimore, 1959.

Excellent discussion of early ideas of fossils in ancient times and in the eighteenth century.

HECK, N. H. Earthquake history of the United States. Part I. Continental United States (exclusive of California and Western Nevada) and Alaska. U.S. Coast and Geodetic Survey, Serial No. 609, 1938.

Regional lists and brief descriptions of all major earthquakes starting with that at Plymouth in 1638. (Of earthquake interest are also L. D. Leet, Causes of Catastrophe, New York, 1948; and M. L. Fuller, New Madrid Earthquake, U.S.G.S. Bull. 494, 1912.)

HERDMAN, WILLIAM A. Founders of oceanography and their work, an introduction to the science of the sea. Edward Arnold & Co., London, 1923.

Brief history of early oceanography; lives of several oceanographers; but over half is discussion of oceanographic topics with only slight historical emphasis. A modern book would usefully displace this one.

HINDLE, BROOKE. The pursuit of science in revolutionary America 1735-1789. University of North Carolina Press, Chapel Hill, 1956.

General history of early American science with just enough geology to warrant inclusion here. Contains some good references.

HOFFMAN, ARNOLD. Free gold, the story of Canadian mining. Rinehart, New York, 1947.

Interesting and popular stories of Canadian mining told in a series of episodes and biographies.

HOFFMAN, FRIEDRICH. Geschichte der geognosie und schilderrung der vulkanischen erscheinungen. Nicolaischen Buchhandlung, Berlin, 1838.

Excellent early summary of history of geology based on discussions of workers and their observations and theories. The second half of the book is a description of volcanic features, theories of volcanism and observations of various men who wrote on this subject.

HÖLDER, HELMUT. Geologie und paläontologie in texten und ihrer geschichte. Karl Alber, Freiburg, 1960.

Exhaustive work on history and philosophy of geology organized around concepts rather than individual workers. Includes a very important bibliography. A very useful book.

HOLLAND, JOHN. The history and description of fossil fuel, the collieries, and coal trade of Great Britain. Whittaker and Co., London, 1841.

History of fire, a little history of early geology, history of theories of coal origin, of coal mining, and of the coal trade. Thoroughly and well done.

HOOVER, HERBERT CLARK, and HOOVER, LOU HENRY. Georgius Agricola, de re metallica: translated from the first Latin edition of 1556: with biographical introduction, annotations and appendices upon....mining methods....geology, mineralogy...from the earliest times to the 16th century. Mining Magazine, London, 1912; Dover, New York, 1950.

This great book must be included in any working apparatus for history of geology. It includes important biographical and bibliographical material as well as extensive discussions of history of early geology and mineral industry. The recent reprint is a service to geology and history of science, as the original edition is now beyond common acquirement.

HOOYKAAS, R. La naissance de la cristallographie en France au XVIIIe siecle, etc. Paris, 1953. (Les Conferences du Palais de la Decouverte. Ser. D. no. 21.)

HOOYKAAS, R. Natural law and divine miracle. A historical-critical study of the principle of uniformity in geology, biology and theology. E. J. Brill, Leiden, 1959.

A book on the history of the doctrine of uniformitarianism and of its application in science and in theology.

HORNBERGER, THEODORE. Scientific thought in the American colleges 1638-1800. The University of Texas Press, Austin, 1945.

Interesting book on the teaching of science in the early colleges, but contains only a few mentions of geology.

IRELAND, H. A. History of the development of geologic maps. Geological Society of America Bulletin, Vol. 54, pp. 1227-1280, 1943.

This paper lists and discusses early geologic maps by countries and regions. Particularly good disucssion of coloring of early maps. (Also see Marcou.)

JAFFE, BERNARD. Men of science in America. Simon & Schuster, New York, 1944 (and later).

Popular, but well-written and documented biographies of 19 men who worked from the 16th to 20th centuries, of whom nine were geologists or made geological contributions.

JAHN, MELVIN E. and WOOLF, DANIEL J. The lying stones of Dr. Johann Bartholomew Adam Beringer being his "Lithographiae Wirceburgensis." University of California Press, Berkeley and Los Angeles, 1963.

Not only a translation of Beringer's famous and ill-starred book, but contains extensive indices, notes, and bibliographies which illuminate the history of paleontology and theology of Beringer's time. Some of the references are very important.

JILLSON, WILLARD ROUSE. Big Bone Lick, an outline of its history, geology, and paleontology, to which is added an annotated bibliography of 207 titles. Standard Printing Co., Louisville, 1936.

History of the discovery, exploitation, and study of this great vertebrate fossil deposit. Valuable annotated bibliography of over 200 titles. (See also reference to paper by Simpson in earlier Am. Phil. Soc. entry.)

KEFERSTEIN, CHRISTIAN. Geschichte und litteratur der geognosie, ein versuch. Johann Friedrich Lippert, Halle, 1840.

Brief early history of geology, becoming much briefer for 19th century to 1839. Lists of geological books and articles contain some useful references but with many omissions and inaccuracies, particularly to English and American geology.

KEYES, CHARLES. Annotated bibliography of Iowa geology and mining. Iowa Geological Survey, Vol. XXII, 1913.

The first 155 pages deal with the history of geological observations in Iowa and in surrounding states, with especially good treatment of very early observations. The long bibliography is useless for historical purposes because the early works tantalizingly referred to in the historical part are not even listed in the bibliography!

KOBELL, FRANZ VON. Geschichte der mineralogie von 1650-1860. Muenchen, 1864.

A classic. The first part is history of mineralogy, mineral chemistry, mineral physics, and crystallography to 1860. The second part is a history and origin of mineral names to 1860, an indispensable source for early names.

KOBELL, FRANZ VON. Mineral-namen und die mineralogische nomenclatur. J. O. Cotta'schen Buchhandlung, Munchen, 1853.

Earlier work on origin of mineral names.

KOKEN, E. Die Deutsche Geologische Gesellschaft in den jahren 1848-1898 mit einem lebensabriss von Ernst Beyrich. J. F. Starcke, Berlin, 1901.

History of the first half century of the German Geological Society.

KUPSCH, W. O. Pioneer geologists in Saskatchewan. Sask. Dept. of Min. Resources, 1955.

Early history of geological travels and studies in Saskatchewan, with portraits and early illustrations.

LACROIX, A. Figures de savants. Two Vols. Gauthier-Villars & Cie, Paris, 1932.

"Short studies of the lives and work of 31 French mineralogists and geologists, from Desmarest to Haug, with portraits and facsimiles of handwriting. Important source of information on French scientists."

LAPPARENT, A. DE. La philosophie minerale, les theories de la matiere, la cristallographie, les vicissitudes de le pre-histoire, l'anciennete de l'homme et les glaciers. Bloud & Cie, Paris, 1910.

This book deals mainly with the philosophy of mineralogy and geology but contains a few historical references.

DE LAUNAY, L. La science géologique, ses methodes, ses resultats--ses problemes, son histoire. Paris, 1905 and later.

Excellent summary of history of geology in first 92 pages. Special emphasis (naturally) on French contributions. Brief bibliography of important post-1779 references. (Hauntingly reminiscent of Lyell's treatment of some of the subjects.)

DE LAUNAY, L. L'histoire de la terre. Ernest Flammarion, Paris, 1906.

Brief history of geological theories, followed by summaries of philosophical implications of geological processes.

LEET, L. D. Causes of catastrophe, earthquakes, volcanoes, tidal waves, and hurricanes. Whittlesey House, New York, 1948.

Historical description of catastrophic earthquakes and volcanoes, written in a popular style with very few references.

LENOBLE, ROBERT. La Géologie au milieu du XVIIe siècle. Paris, 1954. (Conferences du Palais de la Decouverte. Ser. D. no. 27.)

LENZ, HARALD OTHMAR. Mineralogie der alten Griechen und Römer, deutsch in Auszugen aus deren Schriften, nebst Anmerkungen. Gotha, 1861.

LESLEY, J. P. Historical sketch of geological explorations in Pennsylvania and other states. Second geological Survey of Pennsylvania, Report A, 1876 (reprinted with added note, 1878).

Contains a brief review of early geological observations in Pennsylvania from 1780 to 1833. Most of the book deals with history of geological observations and reports in Pennsylvania after 1833. The bibliographic references are generally very sketchy.

LOEWINSON-LESSING, F. Y. A historical survey of petrology. Translated from the Russian by S. I. Tomkeieff. Oliver & Boyd, Edinburgh, 1954.

"Concise and critical survey of petrology by a distinguished petrologist," translated by a well-known historian of geology. Emphasizes geochemical petrology. Lengthy bibliography includes some items--old and recent--otherwise little known.

LYELL, CHARLES. Principles of geology. Three Vols. John Murray, London, 1830, 1832, 1833, and many later editions.

About the first 100 pages of Volume One contained the first extensive summary in English of history of geology. It still has value and it is recommended to present day students. A brief glossary is included in the Appendix to Volume Three. (The history is retained in all the later 12 editions.)

MACNAIR, PETER and MORT, FREDERICK, Editors. History of the Geological Society of Glasgow, 1858-1908, with biographical notices of prominent members. Geological Society of Glasgow, Glasgow, 1908.

The society is minor indeed which does not have faithful members to compile semi-centennial--and then a centennial--history. This, one of the better ones, deals with local geology and geologists, some of whom were also of national importance. The many vintage portraits are unusually precious.

MARCOSSON, I. F. Anaconda. Dodd, Mead, and Co., New York, 1957.

History of the Anaconda Company and of mining at Butte, "The Richest Hill on Earth."

MARGERIE, EMMANUEL DE. Critique et géologie, contribution a l'histoire des sciences de la terre (1882-1947). Four Vols. Armand Colin, Paris, 1943, 1946, 1946, 1948.

Unique "personal scientific history" in which are reproduced photographs, maps, facsimiles of letters and manuscripts concerning de Margerie's voluminous writings and his long scientific career. In part, almost a publication of his personal files. Contains elaborately extensive references and much original material. Illuminates the history of geology of a time still too recent to be in the standard histories and hence of much value.

MARGERIE, EMMANUEL DE. Études Americaines, géologie et géographie. Two Vols. Armand Colin, Paris, 1952, 1954.

Lengthy discussions of the history of the publications of the U.S. Geological Survey and of some preceding publications, followed by the correspondence of de Margerie with various authors concerning some of the publications. In part, a reproduction of what must have been his whole filing cabinet labelled "U.S. Geology", and hence an important contribution.

MARGERIE, EMMANUEL DE. La Société Géologique de France de 1880-1929. Centenaire de la Société. Livre Jubilaire, Paris, pp. 1-82, 1930.

Introduction to the two volumes in honor of the 100th anniversary of the Society. (The history of the first half century by A. de Lapparent is in 3rd Series, Bull. 8, 1880.) History of the meetings, the officers and the activities, with excellent portraits.

MATHER, KIRTLEY, F. and MASON, SHIRLEY L. A source book in geology. McGraw-Hill, New York, 1939.

One to more than ten page excerpts from the works of 126 men from Leonardo to Van Hise who wrote on geology. Chronologically arranged. Very widely selected by topic, by country, and in time. Convenient topical list and index. A good anthology.

MATHEWS, EDWARD BENNETT, Editor. Fifty year's progress in geology 1876-1926. The Johns Hopkins University Studies in Geology, No. 8, 1927.

Several papers on progress in various kinds of geology with special reference to Johns Hopkins, given at 50th anniversary celebration.

MERRILL, GEORGE P. Contributions to the history of American geology. Report of the U.S. National Museum for 1904, 1906.

MERRILL, GEORGE P. The first one hundred years of American geology. Yale University Press, New Haven, 1924. (Reprinted in 1964.)

Merrill is the great historian of American geology. The second book includes, with some rearrangement, most of the material in the first, in addition to several added chapters. Merrill stresses men--their lives and geological activities. He has strong preferences but tries to be objective. He knew little, and cared less, about pre-Maclurian geology in America, and is at his best from the time of the founding of the state surveys onward. His references must always be checked carefully. The valuable biographical appendix of the earlier volume is omitted from the later one.

MERRILL, GEORGE P., Editor. Contributions to a history of American state geological and natural history surveys. Smithsonian Institution United States National Museum Bulletin 109, 1920.

Histories of the founding and development of the state surveys to about 1890. Arranged alphabetically rather than chronologically. Many of these data were used in his earlier book above. For exact references to early publications of surveys you must consult Meisel. (Some additional material on state surveys can be found in U.S.G.S. Bulletin 465, 1911.)

METZGER, HELENE. La Genese de la science des cristaux. Paris, 1918.

MEUNIER, STANISLAS. L'evolution des theories géologiques. Paris, Felix Alcan, 1911.

MIELEITNER, K. Geschichte der Mineralogie in Altertum und Mittlalter. 2 vols. Jena, 1922.

MILLER, SAMUEL. A brief retrospect of the eighteenth century. T. and J. Swords, New York, 1803.

This first American history of science was a notable production of the new country. Contains a brief but reasonably complete review of mineralogy, emphasizing Werner and Kirwan; probably based on information derived from his friend S. L. Mitchill. The chapter on geology contains reviews of theories of Burnet, Woodward, Whiston, de Maillet, and many others including several now almost unknown. Mentions no American observations in geology, but an impressive early American performance by a young clergyman who had important scientific acquaintances.

MILLIN, AUBIN-LOUIS. Minéralogie homerique, ou essai sur les minéraux, dont il est mention dans les poemes d'Homere. Garnery, Paris, 1790.

Surprising number of earths, rocks, gems, and metals named in the Illiad and Odyssey are listed and identified. References are included to their mention in other ancient writings. Very brief, but pleasant introduction.

MOORE, N. F. Ancient mineralogy; or, an inquiry respecting mineral substances mentioned by the ancients with occasional remarks on the uses to which they are applied. G. and C. Carvill, New York, 1834.

Minerals, metals, and gems mentioned by the ancients are classified mainly according to Pliny. Primarily of interest to specialists.

MOORE, RUTH. The earth we live on: the story of geological discovery. Knopf, New York, 1956.

A popular, somewhat "wide-eyed," but accurate, book on development of geological ideas and on the geologists responsible for them, written by a "star reporter." Several modern geologists and geophysicists treated at same length as more ancient worthies.

NORTH, F. J. Centenary of the glacial theory. Proceedings of the Geologists' Association, Vol. 54, Part 1, p. 1-28, 1943.

Brief, but very useful survey of history of glacial theory. Includes references not included by other authors and some original notes of Wm. Buckland's early observations.

NORTH, F. J. From Giraldus Cambrensis to the geological map; geological map to geological survey; W. D. Conybeare; geology in South Wales. The Cardiff Naturalists' Society's Transactions, Vol. 64, p. 20-97 (6 Pl.), 1933; Vol. 65, pp. 41-115, 1934; Vol. 66, pp. 15-68, 1935; Vol. 67, pp. 31-103, 1936.

Early geological observations, particularly of South Wales. Interesting illustrations.

NORTH, F. J. Geological maps, their history and development with special reference to Wales. National Museum of Wales, Cardiff, 1928.

Descriptions, with some illustrations, of early maps. Complete list of geological maps of Wales. Useful bibliography.

PARSONS, A. B. The porphyry coppers. American Institute of Mining Engineers, New York, 1933. The porphyry coppers in 1956. Ibid., 1957.

"An amazing record of engineering and technologic achievement" in profitable exploitation of a dozen of the great disseminated copper deposits of the Western Hemisphere. The second volume brings several earlier deposits up to date and adds important new ones. Geological control is incidental, but basic.

PARSONS, A. B., Editor. Seventy-five years of progress in the mineral industry, 1871-1946. The American Institute of Mining Engineers, New York, 1947.

This seventy-fifth anniversary volume contains several long articles dealing more or less directly with history of economic geology in the United States, as well as elaborate history of the AIME.

PRICE, PAUL H., and others. The West Virginia Geological and Economic Survey--its accomplishments and outlook. West Virginia Geol. and Econ. Survey, Centennial Edition, 1963.

History of the West Virginia Geological Survey in general and by special fields. It may be noted here that most state surveys have one or more summary histories (in addition to Merrill's), all of which to 1928 are listed in Meisel. These state survey histories may contain local information not found elsewhere.

RAMSAY, ANDREW C. Passages in the history of geology. Two small vols. Taylor, Walton and Maberly, London, 1848, 1849.

Brief, readable summaries with some mention of observers usually slighted or omitted by other authors.

READ, THOMAS THORNTON. The development of mineral industry education in the United States. The American Institute of Mining Engineers, New York, 1941.

Brief history of early European mineral industry education; scholarly summary of early Americans trained at Freiberg and Paris; detailed histories of American schools of mineral, petroleum and ceramic engineering. The role of geology is given careful attention.

REVILLIOD, PIERRE. Physiciens et naturalistes Genevois. Kundig, Geneve, 1942.

Includes brief sketches and portraits of several geologists, including Argand, Deluc, Favre, and de Saussure.

RICKARD, T. A. A history of American mining. AIME Series. McGraw-Hill, New York, 1932.

Interesting summaries of early discoveries and later mining activities. Useful references.

RICKARD, T. A. Interviews with mining engineers. Mining and Scientific Press, San Francisco, 1922.

Curiously illuminating directly quoted interviews with mining engineers, some of whom had geological interests and knowledge.

RICKARD, T. A. The romance of mining. The Macmillan Company of Canada, Ltd., Toronto, 1945.

This fascinating book tells of mining exploits from Jason and the Argonauts to the diamond mines of Kimberly. Contains enough geological allusions to be included in this list.

RICKARD, THOMAS ARTHUR. Man and Metals. A history of mining in relation to the development of civilization, etc. 2 vol. McGraw-Hill Book Co., New York and London, 1932.

An important book for history of mining with many geological references.

RIOLLOT, JULES. Les abeilles de la mine, reines et ouvrieres. Révélatrices ou Butineuses du suc des pierres. Charles-Lavauzelle & Cie, Paris, 1934.

This very interesting, curiously illustrated book lists all the women of myth and history (except Medea!) who had anything to do with mining, mineral resources, or applied geology, either directly or through assistance to their husbands, relatives, or friends. It is difficult not to be enthusiastic about this unimportant book.

RIZER, H. C. The United States Geological Survey, its origin, development, organization, and operations. U.S. Geol. Survey, Bulletin 227, 1904.

An historically unimportant report of this great organization. A real history is rumored to be in preparation. More useful is J. & M. Rabbitt's history of the U.S.G.S. in Science, v. 119, pp. 741-748, 1954.

ROGERS, A. W. The pioneers in South African geology and their work. Transactions of the Geological Society of South Africa, Annexure to Volume 39, 1937.

History of geological observation from 1654 to recent times organized on biographical basis. Several "interesting" portraits.

ROUFF, MARCEL. Les Mines de charbon en France au XVIII siecle, 1744-1791. Étude d'histoire economique et sociale (with a bibliography), Paris, 1922.

ROY, ANDREW. History of the coal miners of the United States. From the development of the mines to the close of the anthracite strike of 1902, including a brief sketch of early British miners. Columbus, Ohio (1903?).

Brief history of early coal mining in Great Britain and in United States. Detailed discussion of conditions at the turn of the century. Much on labor conditions, strikes, and "troubles."

SAINTE-CLAIRE DEVILLE, CH. *Coup-d'oeil historique sur la géologie et sur les travaux d'Elie de Beaumont*. G. Masson, Paris, 1878.

Partly history and philosophy of geology and partly a review (not a biography) of the scientific work of Elie de Beaumont.

SCHINDEWOLF, OTTO HEINRICH. *Wesen und Geschichte der Paläontologie*. Berlin, Wissenschaftlische Editionsgesellschaft MBH, 1948.

A general treatise on the scope and history of paleontology; includes discussion of the definition of paleontology, its relation to other disciplines, and states in the development of paleontologic research (A.LaR.).

SCHNEIDER, KARL. *Zur Geschichte und Theorie des Vulkanismus*. Prag, 1908.

SEMPER, MAX. *Die geologischen studien Goethes. Beitrage zur biographie Goethes and sur geschichte und methodenlehre der geologie*. Veit & Comp., Leipzig 1914.

This book is a discussion of geological history and philosophy of the time of Goethe. Constructed around the core of Goethe's really important but not well-known geological observations, thoughts, and writings. This important book gives insights into many early theories which are quite at variance with those contained in most other histories. The biographical part could be called the geological biography of Goethe.

SIMPSON, GEORGE GAYLORD. *Beginnings of vertebrate paleontology in North America*. See American Philosophical Society.

SINGER, CHARLES, and others. *A history of technology*. Five Vols. Oxford Press, New York and London, 1954-1958.

Each of these great volumes contains both separate chapters and other references to geological topics. The development is traced from prehistoric times to 1900 of mining, quarrying, and water supply, building materials and many other technologies using geology.

STAMPA, MANUEL CARRERA. Gacetas historicas. Congreso Geologico Internacional, 20 Sesion, Mexico, 1956.

The history of Mexican geology told in 20 sketches of geological organizations, mineral districts and geologists. Useful references.

STUDER, B. Geschichte der physischen geographie der Schweiz bis 1815. Stampflische Verlagshandlung, Bern, 1863.

Topography, glaciers, geology, regions and men are discussed. A useful book; good index.

SWEETING, G. S. The Geologists' Association 1858-1958. A history of the first hundred years. Benham and Company, Ltd., Colchester, England, 1958.

A centennial history of this unique association which joins professionals and amateurs in their interest in geology. Shows well the great popular interest in geology in the nineteenth century.

THWAITES, F. T. The development of the theory of multiple glaciation in North America. Trans. Wisconsin Academy of Sciences, Arts, and Letters, Vol. 23, pp. 41-164, 1927.

History of discovery of buried soils and peat, multiple tills and other evidences of multiple ice advances. Classified by years and major districts. Elaborate references. A valuable summary by one who has seen many of the "classical" localities. For later references see his "Outline of Glacial Geology."

UNKLESBAY, A. G., Editor. Biographical sketches of men important in the history of geology to 1900. History of Geology Class, University of Missouri, 40 pp. (mimeo), 1949.

Little sketches of almost 200 men who were geologists or who wrote on geological subjects. Many ancient writers listed.

VOGELSANG, H. Philosophie der geologie und mikroskopische gesteinsstudien. Max Cohen & Sohn, Bonn, 1867.

Relation of geology to other sciences, historical development of geology, philosophy of geology, and history of microscopic petrography. The chemical approach of "der geistreiche und weitblickende Vogelsang" (Zittel), who wrote this book at 29 and died at 36, is very modern and metasomatic. The ten colored plates of thin sections are among the very earliest published and may be the first ones.

WELLS, JOHN WEST. Early investigations of the Devonian system in New York, 1656-1836. Geol. Soc. Am. Special Paper No. 74, New York, 1963.

A real display of geologic, stratigraphic, paleontologic, literary, historical, and linguistic erudition done in a spritely and charming manner. Covers far more territory than indicated by the title; skillfully compresses into 76 pages material which most writers would have required a much larger book to deal with. It includes a good bibliography.

WHEWELL, WILLIAM. History of the inductive sciences from the earliest to present time. Three Vols. John W. Parker and Son, London and New York editions, 1837, and later. (Reprinted 1963.)

The first great history of science. Volume three contains lengthy history of mineralogy and of geology.

WHITE, GEORGE W. Early American geology. Scientific Monthly, Volume 76, pp. 134-141, 1953.

A summary of geological observations in America from 1497 to 1787 with references to original sources.

WOODWARD, HORACE B. History of geology. Watts & Co., London, 1911.

This useful little book covers a broader field than many larger ones.

WOODWARD, HORACE B. The history of the Geological Society of London. Geological Society, London, 1907.

An important history of British geology as well as of the oldest geological society. By far the best and most readable of the society histories. Finely printed portrait plates.

YOUMANS, WILLIAM JAY. Pioneers of science in America; sketches of their lives and scientific work. D. Appleton and Co., New York, 1896.

A useful collection of biographies of 50 early American scientists of whom 30 did geological work and 23 are known primarily for their geological observations and writings. A portrait of each is included. A book of real importance.

ZITTEL, KARL ALFRED VON. *Geschichte der geologie und paleontogie bis ende des 19 yahrhunderts*. R. Oldenbourg, Munich, 1899.

The greatest history of geology so far published. Contains footnote biographical sketches and excellent references.

ZITTEL, KARL ALFRED VON. *History of geology and palaeontology to the end of the nineteenth century, translated by Maria M. Ogilvie-Gordon*. Walter Scott, London, 1901. (Now available in reprint.)

The English translation is important, but the original should be consulted for exact references and for certain material not included in the translation.

PART II

Bibliographies for History of Geology

ACADEMY OF NATURAL SCIENCES. Catalogue of the Library of the Academy of Natural Sciences of Philadelphia. Philadelphia, 1836.

In 1836 the library of the Academy possessed 6890 volumes and 435 maps, of which William Maclure had presented 5232 volumes and nearly all the maps. The list contains many hundreds of geological books, many of the greatest rarity. French and German works are well represented, and many Spanish works are included. These reflect Maclure's travels and interest in Latin America, Spain and the Continent, and his disinterest in geology of England.

AGASSIZ, LOUIS. Bibliographia zoologiae et geologiae. A general catalogue of all books, tracts, and memoirs on zoology and geology. Corrected, enlarged, and edited by H. E. Strickland. Four volumes. Ray Society, London, 1848-1854.

Lists about 40,000 references. Useful for checking early publications. Especially valuable for references to reviews.

BUCK, SOLON JUSTUS. Travel and description 1765-1865, together with a list of county histories, atlases, and biographical collections and a list of territorial and state laws. Ill. State Historical Library, Springfield, Ill., 1914.

Most geological travelers entered or crossed Illinois and thus are listed in this bibliography, which is of far greater importance than the title indicates.

BÖHMERSHEIM, AUGUST BÖHM VON. Geschichte der Moränenkunde. R. Lechner, Wien, 1901.

This book is far more than the title indicates. It is an excellent historical bibliography of glacial geology. The major part is a chronological description followed by alphabetical listing and ending with a complete index.

BRITISH MUSEUM (NATURAL HISTORY). Catalogue of the Books, Manuscripts, and Drawings in the British Museum (Natural History). 5 vols. 1903 to 1916. With a Supplement of 3 vols.

CLARK, THOMAS D., Editor. Travels in the old South--A bibliography. Three volumes. University of Oklahoma Press, Norman, Oklahoma, 1956, 1956, 1959.

A specialized bibliography of travels, which includes many geological ones. Fairly complete descriptions, brief to lengthy annotations, and summary statements by authors.

COTTA, BERNHARD V. Beitrage zur geschichte der geologie. Erster abteilung: Geologisches repertorium. Mit einem vollstandigen index (all published). J. J. Weber, Leipzig, 1877.

Bibliography in chronological order of geological publications from 1530 (Agricola) to 1876. Some annotation. References annoyingly incomplete and with many omissions; furnishes some good leads, but bibliographically inept.

DARTON, NELSON HORATIO. Catalogue and index of contributions to North American geology 1732-1891. U.S. Geological Survey Bull. 127, 1896.

Of some use for early references, but use Nickles and successors for references after 1785. The chronological lists by states are useful, but few of these references are earlier than 1809.

DRYANDER, J. Catalogue Bibliothecae Historico-Naturalis Josephi Banks. Vol. IV: Mineralogi, 1799.

The geological volume of a huge five volume work. The set is now being reprinted by Asher of Amsterdam and single volumes will be available (A.O.L.).

ENGELMANN, WILHELM. Bibliotheca Historico-Naturalis Verzeichniss der Bucher uber Naturgeschichte welche in Deutschland, Scandinavien, Holland, England, Frankreich, Italien, und Spanien in den Jahren 1700-1846 erscheinen sind. Leipzig, 1946.

GRONOVIUS, LAURENTIUS THEODORUS. Bibliotheca regni animalis atque lapidei, seu recensio auctorum et librorum, qui de regno animali & lapideo methodice, phycice, medice, chymice, philologice, vel theologice tractant, in usum naturalis historiae studiosorum. Lugduni Batavorum, Auctoris, 1760.

One of the earliest bibliographies of natural history including geology. The references are generally incomplete but give useful clues which may be followed up. Includes references to both books and early journal articles.

HOWELL, J. V. and LEVORSEN, A. I. Directory of geological material in North America. Second Edition. American Geological Institute, Washington, 1957.

Lists of institutions and libraries, publications, maps and materials by continent, country, and state with a brief general bibliography. A very useful reference work.

HOWES, WRIGHT, U. S. Iana (1650-1950). A selective bibliography in which are described 11,620 uncommon and significant books relating to the continental portion of the United States. R. R. Bowker, New York, 1962.

Although this great list deliberately omits most scientific books, it does include a few of the early ones. It is particularly useful for the early travel books and especially so for the early government publications in which so much geology was published. It is the most convenient guide to the often bewildering variations in early government publications which may vary greatly in completeness and value; indispensable for anyone working with early American materials, especially western.

KOCH, MANFRED. Geschichte und entwicklung des bergmannischen schrifttums. Hermann Hubener Verlag K. G., Goslar, 1963.

Discussion on an historical basis of mining literature from classical times to the present. Includes much on applied geology. Concludes with complete references to the various editions of all works mentioned.

KUPSCH, W. O. Annotated bibliography of Saskatchewan geology (1823-1958). Saskatchewan Department of Mineral Resources, Regina, 1959.

Contains references to historical studies and to early geological observations, but omits some travel accounts with geological notices (for these see Kupsch, 1955, under "Histories").

LAMBRECHT, K., QUENSTEDT, W. and A. Fossilium catalogus I: Animalia. Pars 72. Dr. W. Junk, 's-Gravenhage, 1938.

This book is technically a serial and not likely to be indexed by libraries under the authors' names. It contains sentence biographies of over 3,000 paleontologists and geologists from many countries and elaborate references to published biographies in books and journals. This is the most convenient--and sometimes the only--place to secure references for more detailed information on the lesser known men.

LOMSKY, JOSEF. Soupis periodik geologickych ved. Periodica geologica, palaeontologica et mineralogica. Nakladatelstvi Ceskoslovenske Akademie Ved. Jirina Bydzovska, Praha, 1959.

Alphabetical bibliography of 3,582 items of geological, paleontological and mineralogical periodicals, from the beginning of the 18th century to 1958. Useful for Slavic items, but incomplete for others.

McNALLY, ARTHUR, and ROLLER, DUANE, H. D., Compilers. A check list of the E. DeGolyer Collection in the history of science and technology as of August 1, 1954. University of Oklahoma Press, Norman, 1954.

A very useful list (but not catalogue descriptions) of one of the great collections, now at Oklahoma, of classical books on science and especially on geology. A detailed catalogue of the geological books would be of great service.

MAFFEI, EUGENIO, and RUA FIGUERO, RAMON. Biblioteca espanola de...la mineralogie y geologia...accompanados de biographicas.... Two volumes. J. M. Lapuente, Madrid, 1871, 1872.

The great Spanish bio-bibliography. Bibliographic sketches and bibliographic notes; includes not only Spain and Ibero-Americo, but also some material from other parts of the world. Contains some biographies and other notes not available elsewhere. Two elaborate indexes.

MARCOU, JULES, and MARCOU, JOHN BELKNAP. Mapoteca geologica Americana. A catalogue of geological maps of America (North and South) 1752-1881, in geographic and chronologic order. U.S. Geol. Survey Bull. 7, 1884.

Brief history of geologic maps, followed by useful chronological lists by geographic regions. (See also Ireland, under "Histories".)

MARGERIE, E. DE. Catalogue des bibliographies géologiques. Paris, 1896.

This great reference work contains almost 4,000 annotated references to bibliographies and other works containing literature lists. Includes section on biographies. Citations are complete and precise. Any desired reference is almost certainly included, but it may be under a classification different from the expected one. Contains three indexes.

MASON, BRIAN. The literature of geology. The American Museum of Natural History, New York, 1953.

Lists works by general topics and by continents and countries. Brief list for each U.S. state and Canadian province.

MEISEL, MAX. A bibliography of American natural history, the pioneer century, 1769-1865. An annotated bibliography of the publications relating to the history, biography and bibliography of American natural history and its institutions, during colonial times and the pioneer century, which have been published up to 1924; with a classified subject and geographic index; and a bibliography of biographies. Three volumes. Premier Publishing Co., Brooklyn, 1924, 1926, 1929.

Annotated lists of biographies, bibliographies, organizations, including Surveys. Gives complete tables of contents of journals for natural history, including papers. Chronological list of separate publications from Hariot, 1590 (should be 1588) to 1865. Most early American geology books are included, but there are some omissions. Includes foreign publications about U.S. Well indexed. Its importance is well recognized by historians of American biology, but geologists will also find this the indispensable starting point for references on early American geology.

MONAGHAN, FRANK. French travellers in the United States 1765-1932. Antiquarian Press Ltd., New York, 1961.

A useful list for those dealing with geology in early travel books. Includes some comments on books and authors difficult to find elsewhere.

NEVERS, G. M., and WALKER, R. D. Annotated bibliography of Indiana geology through 1955. Bloomington, Indiana, January 1962.

One of the best of the bibliographies prepared by state geological surveys. This one has excellently indexed references to early publications, including the travel books which contain geology.

NICKLES, J. M. Geologic literature on North America 1785-1918. U.S. Geol. Survey Bulls. 746, 747, 1923, 1924.

This and following bibliographic bulletins are indispensable for all geologists. Most early American works are listed, but always also check Darton for additional references and Meisel for other references and for further details.

NICKLES, J. M., and later authors. Bibliography of North American Geology 1919-1928; 1929-1939; 1940-1949; and annual volumes from 1950. U.S. Geol. Survey Bulls. 823, 937, 1049, 985, 1025, 1035, 1054, 1075, 1095, 1115, 1145.

These bulletins bring the bibliographic record almost down to date. Each one contains valuable lists of articles (and sections in other articles and books) written in the period covered on history of geology.

PANGBORN, M. W., JR. Earth for the layman. A list of nearly 1400 good books and pamphlets of popular interest on geology, mining, oil, maps, and related subjects. AGI Report 2, Second edition. American Geological Institute, Washington, 1957.

"The professional geologist and student of geology will find this annotated bibliography a convenient source of references, particularly to literature pertaining to the history of geology and early geologists." The section on geological novels is unique. Very high in quality and usefulness. Very low in price.

PEARL, R. M. Guide to geologic literature. McGraw-Hill, New York, 1951.

Deals mainly with information and sources of information about current publications. Has some good leads to non-North American current sources otherwise not easy to find (but see also Mason).

SCHMECKEIBIER, L. F. Catalogue and index of the publications of the Hayden, King, Powell, and Wheeler surveys. U.S. Geol. Survey Bull. 222, 1904.

SWANSON, E. B. A century of oil and gas in books. A descriptive bibliography. Appleton-Century-Crofts, Inc., New York, 1960.

Contains a long chapter on "History and Biography". Excellent annotations.

WELLS, J. W., and WHITE, G. W. Biographies of geologists. Ohio Journal of Science, Volume 58, pp. 285-298, 1958.

Annotated list of 210 biographies of geologists, all (with one or two exceptions) of book length.

WHITE, G. W., and SLANKER, B. O. Early geology in the Mississippi Valley; an exhibition of selected works held in the University of Illinois Library at Urbana, November, 1962. Urbana, 1962.

Comments on 38 books in which the development of observation and explanation of the geology of the Middle West can be traced from 1588 to 1843.

WROTH, L. C. An American bookshelf. University of Pennsylvania Press. Philadelphia, 1934.

The references to early American science, including a few to geological observations, provide the excuse for including this delightful book.

PART III

Dictionaries and Glossaries

ARKELL, W. J. and TOMKEIEFF, S. I. English rock terms chiefly as used by miners and quarrymen. Geoffrey Cumberlege, London, 1953.

Origins of ancient geological words taken into later technical usage. Very useful to Americans reading early British geology; contains some terms not in New English Dictionary; lists references to many terms.

BERTRAND, ELIE. Dictionaire universel des fossiles propres, et des fossiles accidentels.... Pierre Gosse, Jr. et Daniel Pinet, La Haye, 1763.

Geological, mineralogical, and paleontological dictionary with paragraph to several page length discussions of many terms and with many useful references.

BRISTOW, H. W. A glossary of mineralogy. Longman, Green, Longman, and Roberts, London, 1861.

An alphabetical list of mineral names, including obsolete ones. For each is given a description, composition, and locality and, for most, one or more references.

CHALLINOR, JOHN. A dictionary of geology. University of Wales Press, Cardiff, 1962. Oxford University Press, New York, 1962.

Contains not only definitions, but brief discussions of geological and stratigraphic terms. Gives examples of the first or early uses of many of the terms, but without page references for those cited. A very useful book, especially for determining British usages of standard terms.

CHANEL, EMILE. *Dictionnaire de géologie etymologie, paléontologie, minéralogie a l'usage des eleves des lycees et colleges; des écoles normales primaires; des écoles primaires superieures; des écoles professionnelles et de tous les etablissements d'instruction*. Imprimerie du "Courrier de L'Aim," Bourg, 1897.

A useful little dictionary for definitions of French geological terms and stratigraphic units. It includes some historical references.

CHESTER, A. H. *A dictionary of the names of minerals, including their history and etomology*. Wiley, New York, 1896.

Lists obsolete and current (1895) mineral names, with history of names and references. Useful historical introduction and extensive literature list. Elaborate author index.

FAY, A. H. *A glossary of the mining and mineral industry*. Government Printing Office, Washington, 1920.

This glossary of about twenty thousand terms is primarily of mining terminology, but many geological terms, rock and mineral names, and rock units are included. The source of each definition is given. Includes an elaborate list of references not only to dictionaries and glossaries, but also to other works which include chapters on these subjects.

GREENWELL, G. C. *A glossary of terms used in the coal trade of Northumberland and Durham*. Bemrose & Sons, London, 1849, 1855, 1888.

Deals almost entirely with mining terms, but does contain an occasional related geological reference which may be difficult to find elsewhere.

HIMUS, G. W. *A dictionary of geology*. Penguin Books, Harmondsworth, 1954 (American, Baltimore, 11).

An outstanding value for fifty cents, although it contains very few historical references.

HOWELL, J. V., Coordinating Chairman. Glossary of geology and related sciences, a cooperative project of the American Geological Institute. The American Geological Institute, Washington, 1957.

Gives history of many of the terms listed, with references. Includes useful lists of dictionaries and glossaries. The most important single book in this list of dictionaries and glossaries.

HUMBLE, WILLIAM. Dictionary of geology and mineralogy. H. Washbourne, London, 1840. Third edition, 1860.

Useful source of information and definitions in semi-modern terminology of obsolete names found in early geological works. Includes biological and chemical terms related to geology and mineralogy. Third edition much expanded.

MURRAY, J. A. H., Editor. A new English dictionary on historical principles; founded mainly on materials collected by the Philological Society. Ten volumes in 17. Oxford, 1888-1928.

This greatest of all dictionaries (commonly known as "OED") gives the origin and history of each word and cites dates and examples of early uses at length and in detail. If you are not accustomed to using OED as a geological (and historical) tool, try "geology," "erratic," "fossil," and "rock" and become a believer.

PAGE, DAVID. Handbook of geological terms and geology. William Blackwood and Sons, Edinburgh and London, 1859, 1865.

Dictionary and glossary of geological, mineralogical and paleontological terms. Includes names of British rock formations. Gives derivation of each term, but includes little on the history of the origin of the term or of history of its first use.

RICE, C. M. Dictionary of geological terms (Exclusive of stratigraphic formations and paleontologic genera and species). 1956.

A very useful dictionary which gives origin of each definition, but contains almost no historical references.

STAMP, L. D., Editor. A glossary of geographical terms. John Wiley and Sons, Inc., New York, 1961.

This book is actually almost a dictionary and glossary of physical geology and geomorphology. It gives the origin of terms and the usages by different authorities and for the terms it includes it is often the most complete source. For terms which have been used in various ways the author concludes by giving his comment on the preferable current use. This book deserves to be much better known by geologists.

STOKES, W. L., and VARNES, D. J. Glossary of selected geologic terms with special reference to their use in engineering. Colorado Scientific Society Proceedings, Volume 16, Denver, 1955.

Useful as a dictionary, but contains no historical references.

WEBSTER, NOAH. Webster's new international dictionary of the English language, Second edition, Unabridged. G. & C. Merriam Co., Springfield, Massachusetts, 1957.

The special editors for geology, mineralogy, petrography, and physical geography have provided a wide and accurate coverage including some obsolete terms. It gives derivation of words, but not their history, for which see Murray, OED.

WILMARTH, M. GRACE. Lexicon of geologic names of the United States (Including Alaska). U.S. Geol. Survey Bull. 896, 1938.

Indispensable for references to first and subsequent descriptions of rock units. Includes obsolete as well as current stratigraphic names. (A later Bulletin 1056, 1957, 1959, continues the record from 1936, but is not of immediate historical interest.)

EARLY STATE GEOLOGICAL SURVEY PUBLICATIONS

STECHERT-HAFNER

Book News

VOLUME XXIII, NUMBER 2, OCTOBER 1968

Early State Geological Survey Publications

By GEORGE W. WHITE

A LARGE body of early American science is contained in the publications of the state geological surveys, the first of which was established almost 150 years ago. Although generally regarded as technical reports of only local interest, these early publications contain historical matter of wider interest and deserve much more attention than they have received from libraries, collectors, historians of science, and geologists.

Information on geology and mineral resources were only incidental parts of the reports of the government exploring expeditions of Lewis and Clark, Stephen H. Long, and others. The first federal government reports dealing primarily with geology were those of George W. Featherstonhaugh in 1835 and 1836, followed by those of D. D. Owen in 1840 and 1848. Even earlier several of the states made appropriations for the investigation of mineral resources and of the geological framework which is necessary to understand these resources. The first state survey was authorized in 1823 in North Carolina, but it was in the decade of the 1830's when almost all the states east of the Mississippi established geological surveys and published reports varying in length and scope. Their state geologists and associates were generally outstanding scientists, some of whose most important work appeared in the state survey reports.

The history of the various state geological surveys will be found in G. P. Merrill, *Contributions to a history of American state geological and natural history surveys*, Smithsonian Institution, U. S. National Museum, Bulletin 109, 1920. This contains information about the establishment of each of the state surveys, their staffs, and portraits. However, the information on publications of these surveys is meager and scattered through the text without bibliographic citations. The material on the early surveys that Merrill included in his book, *The first one hundred years of American geology*, 1924 (reprinted, 1964), is more lively and forms a good starting point for study of the early history of the surveys. (Additional information will be found in C. W. Hayes, *The state geological surveys of the United States*, U. S. Geol. Survey Bull. 465, 1911, and in W. B. Hendrickson, *Nineteenth-century state geological surveys . . .*, ISIS, Vol. 62, p. 357-371, 1961. See also G. W. White, *Reference books for history of geology*, Stechert-Hafner Book News, Vol. 18, Nov. 1963, p. 29-30).

The indispensable reference work is by Max Meisel, *A bibliography of American natural history, the pioneer century, 1769-1865 . . .*, 3 vols., 1924, 1926, 1929. Happily, this great work has just been reprinted and is again available. It is the best source for references of history and early publications of state geological surveys, of scientific societies, of journals and of many other important matters. Volume I includes a bibliography of publications and histories of institutions with an elaborate index and a bibliography of biographies (additions will be found in Volume III, p. 703-730). Volumes II and III give a brief history of each institute, survey, expedition and organization, and a critical bibliography of histories of each. For each organization there is an annotated list of publications with table of contents of each. Volume III includes an index of authors and naturalists with reference to all of their books and journal articles, and an index of institutions.

Meisel gives bibliographical details for most of the survey publications—pages, maps, and illustrations. The earliest state (as well as federal) reports present bibliographic complexities. The number of illustrations varies—some maps and sections are colored, some are not; some have extra plates, both colored and uncolored; some have extra additions; some have parts published together or separately. An example is the *Massachusetts Report*, 1833, of which some copies have 692 pages and some (as listed in Meisel) have 700 pages, because a final "Animal List" is included. Therefore, a library or a collector located in a given state could perform a real bibliographic service by collecting not just one copy of the early reports, but several of each in order to compare them. I think there must be at least six variants of *Ohio Report 2*, 1838; perhaps there are more. The problem with some of the reports with many plates is not to find variants, but to find two copies exactly alike!

Illustrations in the reports are not only interesting but valuable. Many are wood engravings, some are steel engravings. The maps and sections were colored by hand, as were all maps before 1850 and therefore copies of the same map differ in greater or lesser degree—some are brighter than others, and some are only partly colored and some have no color at all. Some illustrations, such as those of the Arkansas Survey by D. D. Owen, who was an accomplished artist, are charming lithographs which have considerable merit and are desired by print collectors.

The writing in many of the early reports was personal and

narrative in form. In modern geological writing all personal references recounting the circumstances of field work, transportation, living facilities and so forth are rigorously excluded. The modern geologist is taught from his undergraduate days to write organized description and analysis—any personal allusions are sharply excised. If an unwary young professional includes anything about scenery or field difficulties (or pleasures) in a report, he receives a sharp reprimand. (In my younger days I once wrote that a certain delightful upland of geological significance in the Allegheny Plateau "presented a pleasant picture of pastoral prosperity." It not only never got printed, but brought a personal note of criticism, "Write geology, not *belles lettres.*") But in the early days there was more freedom. John Locke, noted Ohio physician, chemist, teacher, and geologist, wrote the section on the geology of the southwestern quarter of the state of Ohio in the *Second Report of the Geological Survey,* 1838. Interspersed between his excellent descriptions of the geology of various parts of this area are comments on methods of travel, of his field work, and of his experiences with the people he met, and of the faithfulness of his horse, all as related to his geological work. His opinion of precious metal mines is shown by these statements (pp. 208, 250):

> Many persons suppose that it is the business of the geologist to find gold and silver mines, and that he renders no service to the State unless he does that. While on my excursions, I got out of all patience with being asked if I was hunting for "goold." The State of Ohio has the richest gold mines in the world in the great fertility of her soil, and any geologist who should suggest the means of perpetuating or improving so invaluable a blessing, would confer a greater benefit on the State than by discovering a gold mine to turn the brains and corrupt the morals of the community. . . .

Locke's region included that sparsely settled part of southern Ohio east of the fertile limestone region, some of which is still almost a wilderness. One of his methods of estimating soil quality of the valley of Scioto Brush Creek in this wilderness area is unique in geological literature (p. 256, 270):

> As we came upon these waters, sheltered thus from the scorching sun in a hot day, several boys were enjoying the luxury of a cool bath in one of the lagoons of the stream. This fact, not precisely geological, in itself, showed that the soil in the neighborhood was habitable, and productive in young shoots of the human species. . . .

Locke's high opinion of his horse, which he fondly thought of as his "geological assistant" is shown by the following (p. 258):

> I met with several personal incidents in which even life was endangered, but I have studiously avoided a narrative of them in this report. As the following disabled me somewhat for a time from the usual very active mode of prosecuting my pursuits, I venture to present it.
>
> In preparing to ascend we had driven the wagon quite out of the road and tied the horse securely to a tree. On descending we found the horse had been untied and led into the road, while a long-nosed, lank-sided species of hog common to this region, was a little ahead with a sheaf of oats in his mouth. It appeared as if the plan had been laid by some "native" to have the hog with his oats entice the horse, and the horse pursue and impel the hog. . . . The horse being too well disciplined in his duty of eighth geological assistant to be guilty of any such insubordination, kept his ground.

Although some were not, most of the residents of the territory Locke surveyed were hospitable to the geologist and his horse (p. 273):

> In avoiding personal narrative I have omitted to make due acknowledgments for the various favors which I received in my several journeys. Traveling with a wagon in by-roads, often into the woods destitute of roads, laden with specimens of the rocks examined, and always encumbered with instruments for observation, leaving my horse frequently, to make examinations on foot, and occasionally obliged to drive in the night, I was liable to numerous accidents which did not fail to occur. In one instance my horse being left was attacked by the united force of twenty-eight swarms of bees—broke his harness, drew the wagon a quarter of a mile bottom upwards, and strewed the road with specimens; it was a sad labor to get things righted again. On another journey I was overtaken by darkness in the woods, and a tree having fallen across the road my horse leaped over it and drew the wagon into the knarled branches from which it was impossible for me to extricate it; a journey on foot through mud knee deep, in the blackness of midnight, to the nearest inhabitant to procure assistance, was the only chance of escape. Mr. Williams, near College Corner, Butler county, although his family were all sick with the measles, enabled me to get forward to the next inn.
>
> In several instances I was entertained for several days free of expense, and as my salary was so small that I did not receive, clear of expense, the wages of a day laborer, this was a substantial kindness. The weather-beaten, hedge, and crag-worn appearance of the travelling geologist is not calculated to give common people the impression of a commissioned "officer of state," sent by the governor to serve them. Some brief printed proclamation signed by authority, to be distributed among those people who have never heard of the survey, might be useful, to dispel those suspicions which honestly arise.

One early 20th century report has some of the flavor of the pioneer reports of 60 years earlier. In *Fourth Series Bull. 1, Petroleum and natural gas in Ohio,* 1903, Dr. J. A. Bownocker described various oil fields in Ohio. One of the recently important fields in eastern Ohio was that located at Scio, a village which was the seat of Scio College. As a prelude to the geological description Dr. Bownocker wrote (p. 235):

Stechert-Hafner Book News is published by Stechert-Hafner, Inc., international booksellers, 31 East 10th Street, New York, N.Y. 10003, with branch offices in Bogotá, London, Paris, Rio de Janeiro and Stuttgart; and by Hafner Publishing Company, Inc. and Ltd., New York and London; SH Service Agency, Inc., New York. Issued monthly September through May and distributed free to libraries. Annual index is available for $2.00.

Before the discovery of oil Scio had a population of about 900, composed largely of retired farmers and others in well to do circumstances. It had long been a favored place for retired Methodist ministers who move there because of the advantages which the college offered their children, as well as for the healthfulness of the location, the moral tone of the community, and the pleasant social conditions.... Once only in 18 years was there a saloon in the town, and that one was soon driven out by the angry citizens. Billiard halls and bowling alleys were conspicuously absent, while card parties and dances were in bad repute. The discovery of oil worked the most sudden revolution ever seen in an Ohio town. The population of 900 in 1898 increased with amazing rapidity, and it is estimated that in the following January and February it had risen to 12,000....

The people came principally from the oil fields of Pennsylvania, West Virginia and Northwestern Ohio, and consisted of operators, drillers and teamsters; with them came those who supplied the necessities of life, boarding-house keepers, hotel men, butchers, etc. But with those necessary to develop the territory came a large number who maintained themselves by preying on the weaknesses of men, —the fakir, gambler, saloon keeper and dissolute woman were conspicuously present. There were 11 saloons in the town and an unknown number of "speak-easies." Three or four variety theaters were started, and the nights were made hideous by the barkers of all kinds. These features so horrified the staid citizens that they sold or rented their property and moved away. The village was unable to care properly for one-tenth of the people who had so suddenly gathered there. They lived in tents, shanties, reconstructed stables, boiler houses, or wherever they could find shelter. Many slept in chairs....

It is hardly necessary to state that the general conditions then existing were very disastrous to the college. In 8 weeks time the attendance dropped from 300 to about 25, and all departments of the school except two were moved to New Philadelphia, 24 miles distant. After about 4 months, however, these were taken back to Scio.

As has been seen, the early survey reports contain history, art, and human geography, as well as geological description and analysis. In these publications can be traced the discovery and development in America of many important geological concepts. One for example is the tracing of early observations of glacial drift and views of its origin. The earliest recognition of the glacial origin of the drift is in Hitchcock's *Final report on the geology of Massachusetts,* 1841. His great report, already in press in early 1841, contained extensive descriptions of topographic forms of uncertain origin. When he read Agassiz' *Études sur la glaciers,* which had appeared in late 1840, and saw the magnificent illustrations in the accompanying atlas (for a very recent translation of Agassiz' text and reproduction of the plates see *Studies of Glaciers* by Louis Agassiz edited by A. V. Carozzi, 1967), he added eight pages of "Postscript" and reproduced some of Agassiz' plates. Hitchcock's *Report* thus becomes an *incunabulum* of American glacial geology.

Any library can collect the publications of the geological survey of its own state and should consider including those of neighboring states. The early publications turn up in "old collections" that come to all libraries and in the past have tended to be overlooked. This is a field in which local dealers can be particularly helpful. The geological and historical importance of the pamphlet-sized preliminary legislative documents dealing with survey establishment, and the usually pamphlet-sized earliest survey reports which preceded the larger volumes, should be carefully noted and retained. All of the reports should be collected in multiple copies and no apparent duplicate copy should be disposed of until the fact of its correspondence with another copy is checked page by page, figure by figure, and map by map.

The size of a collection of early publications of a single state survey can range from the single volume Rhode Island *Report* of 1840 with no recorded preliminary documents; through the little, several variant Indiana 1838-1839 and 1854 reports and the larger one of 1862; to the New York reports, those great treasures of American geology, the very listing of which requires 8 pages in Meisel and a complete collection of which requires a whole large bookcase of its own.

Although the Meisel *Bibliography* is indispensable, it is but a starting point for a specialized bibliography of the publications of the early state surveys, because the different issues of the earlier reports are not all listed—and indeed no one at present knows of all these issues because no definitive study has been made of them for any state. Librarians, bibliographers and collectors located near the headquarters of a state survey would find a cordial welcome and high cooperation in developing a complete and definitive bibliography of the early publications. Some of the state surveys have recently published bibliographies of the geology of their state and other surveys have such bibliographies in process, but no one of them has a special section on early survey publications as such.

American historians, historians of science, and bibliographers will bless that library or collector who can provide multiple copies of these early reports. The history of the early state surveys is a promising field for "in depth" study. These surveys form an important part in the development of American science and their importance is becoming ever more evident. But the history will be incomplete until special and adequate collections are formed and made available to scholars.

[GEORGE W. WHITE, Research Professor of Geology, University of Illinois, is Vice-President for North America of the International Committee for the History of Geological Sciences. Dr. White was State Geologist of Ohio. He is the author of more than 100 published papers and bulletins on glacial geology and on the history of early American geology. He is a leading member of numerous professional societies, including the American and British History of Science Societies.

We are pleased to announce that Dr. White has joined the staff of Hafner Publishing Company as Consulting Editor. He is the editor of a series, Contributions to the History of Geology, *two volumes of which have appeared and two more to be published soon by Hafner.*]

THE HISTORY OF GEOLOGY
AND MINERALOGY
AS SEEN BY AMERICAN WRITERS
1803-1835

A Bibliographic Essay

The History of Geology and Mineralogy as Seen by American Writers, 1803-1835: A Bibliographic Essay

*By George W. White**

INTRODUCTION

AT LEAST SEVEN EARLY AMERICAN attempts at a general history of geology were made between 1803 and 1835. Some of these cannot properly be called histories, as they are recitations of facts without much interpretation or evaluation, but others are well-reasoned contributions to intellectual history, even if to a narrow segment of it. These early histories of geology must be assessed in the background and position of the authors: Samuel Miller, a clergyman, theologian, and later church historian; DeWitt Clinton, a lawyer, statesman, and supporter of science; Jeremiah van Rensselaer, a physician with broad scientific training; G. W. Featherstonhaugh, Fellow of the Royal Society, land-holder, linguist, editor, and field geologist; William Gibbons, young printer-naturalist who later became a physician and medical botanist; and N. F. Moore, a classical scholar.

Geologists can see in these works the development of their science in America and how it was regarded by the several writers of these histories. Historians can see the American attitude toward geology and can assess the degree of provincialism or cosmopolitanism in American geology and those who wrote about it.

SAMUEL MILLER—1803

The greatest production in intellectual history in the early United States was that of Samuel Miller, whose book published in 1803, *A Brief Retrospect of the Eighteenth*

Received Mar. 1971: revised/accepted Oct. 1972.
* Department of Geology, University of Illinois, Urbana, Illinois 61801. This paper was presented in an abbreviated form at the Constituent Assembly of the International Committee on the History of Geology, Yerevan, Armenian S.S.R., in June 1967. The Russian text of the shorter paper (translation by T. N. Sofiano) is "Rannie Amerikanskie Naukchnie Trudy po Istorii Geologii 1803–1835gg," *Izdatel'stov Akademii Nauk Armianskoi S.S.R.*, Yerevan, 1970: 114–129 (with abstract in English). I am indebted to three anonymous reviewers for their critical comments and particularly to Dr. Robert P. Multhauf for his generous help in the revision of the original manuscript.

Century,[1] contained chapters on the history and current state of mechanics, chemistry, natural history (which included mineralogy and geology), medicine, geography, hydrology, mathematics, navigation, agriculture, mechanic arts, fine arts, literature (divided by countries, both Oriental and European), linguistics, history, biography, poetry, journals, literary and scientific associations, education, and separate discussions of several countries.

Samuel Miller, who was born in 1769, was a Presbyterian minister in New York from 1793 to 1813, after which he was professor of church history at Princeton Theological Seminary until his death in 1850. He participated in a great variety of scholarly and academic affairs for over fifty years. As a young clergyman in New York he was closely associated with a circle of friends that included his older brother, Dr. Edward Miller (1760–1812), a noted physician, editor, and mineralogist; Dr. Elihu Hubbard Smith (1771–1798), a young physician of extensive scientific interests; Charles B. Brown (1771–1810), the American novelist and editor;[2] and especially with Samuel Latham Mitchill (1764–1831), that remarkable man who was physician, lawyer, politician, jurist, senator, mineralogist, geologist, and botanist.[3] It was through these men, and especially Mitchill, that Miller had access to the large collection of late-seventeenth- and eighteenth-century works to which he refers. Although all of these associates evidently encouraged and guided Miller in the production of his book, it is likely that Mitchill was the most influential in helping his friend.

The first volume of Miller's book has separate chapters (sections) on mineralogy (pp. 145–156), on geology (pp. 156–189), and on hydrology (pp. 198–200). In the chapter on mineralogy he distinguished between those who "did little more than make large collections of mineral substances" and those who constructed systems which were useful in furthering the science. He had a high regard for the Swedes Linnaeus and Wallerius. He thought the system of mineralogy of Vaugel,[4] "a respectable philosopher of Germany," to be of considerable value, as likewise that of Dr. Hill,[5] who in 1748

[1] Samuel Miller, *A Brief Retrospect of the Eighteenth Century. Part First; in Two Volumes; Containing a Sketch of the Revolutions in Science, Art, and Literature, during that Period* (New York: T. & J. Swords, 1803; also 3 vols., London, 1805; facsimile reprint of 1803 ed., New York: Lennox Hill Publishing Co. – Burt Franklin, 1970). A definitive edition, edited by Rodman D. Rhodes, with extensive notes and new biographical material on Miller, is in the final stage of preparation for publication by the University of Illinois Press, Urbana.

[2] For Brown's scientific interests see G. W. White, "Introduction," *A View of the Soil and Climate of the United States of America by C. F. Volney, translated with occasional remarks by C. B. Brown, Facsimile of the Philadelphia 1804 Edition* (New York: Hafner, 1968), p. xi–xii.

[3] C. R. Hall, *A Scientist in the Early Republic, Samuel Latham Mitchill, 1764–1831* (New York: Columbia University Press, 1934).

[4] I cannot identify any mineralogist by this name. Rodman D. Rhodes suggests (personal communication) that "Vaugel" may be a misprint for Vogel. Rudolph Augustin Vogel (1724–1774 was a well-known German physician and scien tist, whose book *Practisches Mineralsystem* (Leipzig, 1762, 1776) is mentioned favorably in the standard reference works. Miller refered approvingly in several other places to the work of "R. A. Vogel" in chemistry, medicine, and botany.

[5] "Dr. Hill" was very probably the John Hill who wrote *A General Natural . . . History of the Animals, Vegetables, and Minerals . . . of the World . . .* 3 vols., folio (London: Thomas Osborn, 1748). Vol. I, *The History of Fossils*, describes and illustrates minerals, rocks, and organic fossils.

Except in this and the preceding footnote and in occasional comments in the text, no attempt will be made to identify the writers discussed by Miller and by the others who are the subjects of this essay. To do so would result in a lengthy bio-bibliographic report which would be essentially a history of geology to 1835. Most of the writers are well known and are listed in the widely available histories of geology by K. A. von Zittel, gave an account of the minerals of England. He commented on Pott, Margraaf, and Neumann, among others, who had advanced the investigation of minerals "through the medium of chemical analysis" rather than entirely through the description of external character. He had a high regard for the investigations of Frederick Cronstedt and particularly for his "natural classification," as well as for Scheele and Bergman.

Considerable space in the mineralogy chapter is devoted to Abraham G. Werner of Freiburg, of whose work he gave a reasoned and approving description. He noted that the classification of Werner was followed by Dr. Walker of Edinburgh, by Daubenton, Patrin, and Monge of France, and by Kirwan of Ireland.[6] His opinion of the work of Kirwan was particularly favorable. He noted the work on crystallography by Romé de l'Isle, on specific gravity by Brisson, and on crystallography by Haüy. He named over a score of the various Europeans who had mentioned minerals in any way in their reports. He stated that by 1800 "the number of species now known is between seven and eight-thousand" and that within the century "seven new kinds of earth have been discovered . . . and 10 new metallic substances." He concluded his discussion with a paragraph on mineralogy in America, which he regretfully said has been "but little explored," although he recognized that coal, gypsum, marble, metals, and other mineral riches were abundant in America and that Mitchill had described some of the minerals of New York.

Miller devoted a whole chapter to hydrology, which he believed "holds so important a place among the objects of human knowledge." He listed those who had contributed to knowledge of the "qualities," including taste and specific gravity, of various kinds of fresh water and sea water. He thought highly of the "numerous and very useful investigations of mineral waters" and gave credit to Boyle as "the first person who pointed out the method of examining mineral waters." He noted that Black was the first to discover carbonic acid, which was so important in the content of mineral waters. He was, perhaps, led into this subject by information derived from Kirwan's *Essay on the Analysis of Mineral Waters*, 1799, and by Mitchill's interest in chemistry of mineral waters. Most of the very brief chapter is devoted to mineral waters rather than to hydrology in the modern sense of origin, movement, and emergence of underground and surface waters. He was not familiar with the work of Pierre Perrault or of

---

Geschichte der Geologie und Paleontologie bis Ende des 19 Yahrhunderts (Munich: R. Oldenbourg, 1899); English translation by M. M. Ogilvie-Gordon, *History of Geology and Paleontology to the End of the Nineteenth Century* (London: Walter Scott, 1901); A. Geikie, *The Founders of Geology* (London: Macmillan, 1905; reprinted, New York: Dover, 1962); or F. D. Adams, *The Birth and Development of the Geological Sciences* (Baltimore: Williams and Wilkins, 1938; reprinted, New York: Dover, 1954). Biographies may also be found in J. C. Poggendorff, *Biographisch-Literarisches Handworterbuch* . . . (Leipzig: 1863—); or among the 3,000 capsule biographies of geologists in K. Lambrecht and W. and E. Quenstedt, "Paleontologi. Catalogus Bio-bibliographicus," *Fossilium Catalogus, I, Animalia, Pars 72* ('s-Gravenhage: Junk, 1938). Where Miller and the others used first names I have included these but not otherwise. The forthcoming edition of *Brief Retrospect* by Dr. Rodman Rhodes contains a biographical list of all the writers discussed by Miller. Biographical material on practically all the American writers mentioned will be found in G. P. Merrill, "Contributions to the History of Geology," *U.S. National Museum Report for 1904*, 1906, No. 135:189–733. A revised text is *The First Hundred Years of American Geology* (New Haven: Yale University Press, 1924; reprinted, New York: Hafner, 1964).

[6] Miller treated Werner only as a mineralogist and said almost nothing about his geological theories on the origin of basalt and classification of rocks.

Edmé Mariotte of the late seventeenth century, of which editions were also published later in the eighteenth century. Miller's very brief treatment of water as a resource is understandable in a new country in which water supply was not yet a problem and in which necessary water could be found from surface or groundwater sources close to most of the centers of population.

In the chapter on the development of geology in the eighteenth century, Miller asserted that most of the progress in the science had been made in that century because little had been done "prior to the commencement of the 18th century." He did go back to the latter part of the seventeenth century and noted the works of Burnet, Whiston, and Woodward. He described the theory of Dr. Thomas Burnet in *Telluris theoria sacra*, who believed that "the earth was first a fluid heterogeneous mass. The heaviest parts descended and formed a solid body." He agreed that Burnet's work, "if it did not command the assent of the philosophic mind, will be found to display much learning, and a very vigorous imagination." He felt that Woodward's theory as displayed in the book published in 1695, *Natural History of the Earth*, displayed much more knowledge of minerals than Burnet's, but was far less ingenious and interesting. The *New Theory of the Earth*, 1696, by William Whiston, supposed that the earth had consolidated into a number of "coats or shells arranged according to density with the atmosphere on the outside, with the water covering the surface of the solid earth." Although Miller used from one to three pages to describe each of the three theories, he decided that "the fanciful and untenable theories which have been briefly stated serve little other purpose than to amuse the curious, and excite to new, and for the most part unsuccessful modes of speculation on this interesting branch of natural history." He felt that authors "have been too often covered by imagination more than judgement; and have laboured rather to support a favorite hypothesis, than to consult the voice of authentic history, or patiently to examine the materials and structure of the fabric which they undertook to describe."

Miller described the theory of John Hutchinson, which today is little known, as one which proposed that the earth was formed by precipitation from "a thick, muddy, chaotic mass." Some consideration was given to the theories of Moro of 1740 and of le Cat of 1744. He felt that the *Telliamed* of de Maillet, which he apparently had read either in the 1750 English translation or the Baltimore 1797 reprinting of it (the original French had appeared in 1748), in which "the earth was once wholly covered with water, which by means of strong currents raised on its bosom all those mountains which different countries bear on their surface; and that this water has been ever after gradually diminishing." He was particularly opposed to the long time assumed by de Maillet and felt that "this atheistical and absurd theory, if it deserve the name, not more hostile to revelation than to all sound philosophy, seems to have gained but few adherents, and but little celebrity."

The theory of Buffon was described at some length—"a comet falling into the body of the sun with great force, struck from its surface a large mass of liquid fire . . . this fragment forms the globe we inhabit. It assumed its present figure when in a fluid state. As the heated mass gradually cooled, the vapours which surrounded it condensed, fell down in the form of water upon the surface, depositing at the same time a slimy substance."

Miller briefly reviewed the theories of Raspé which were published in 1733: he "maintained, that the strata, of which the shell or surface of the earth is composed were originally formed at the bottom of the sea, by the constant agitation of the waters, and the continual production of plants and shells."

As described by Miller, Dr. William Worthington's theory, published in 1773, was so little based on observation that others have justifiably paid no attention to it. According to Miller, Worthington "maintained that the earth, in its primitive state was plain and uniform; and that all mountains and everything irregular and rugged in the surface of it are the result of the curse pronounced on the ground after the fall; that the melancholy lapse of our first parents was immediately followed by earthquakes and every species of convulsion, which produced these dreadful effects in the surface of the earth."

Miller described in several pages John Whitehurst's theory, first published in 1778, that the earth was "once in a state of fluidity, and that the earth acquired its oblate spheroidical form by revolving round its axis in that state. In this fluid state, the component parts of the earth were suspended in one general undivided mass." He gave Whitehurst credit for his assertion "that the *marine shells* found in various places, on and below the surface of the earth were for the most part generated, lived and died in the places in which they were found; that they were not brought from distant regions as some have supposed; and consequently, that these beds of shells, &c., were originally the bottom of the ocean." In the case of Whitehurst, as with all of the others with whom Miller dealt, he was more impressed by the theory than by the description of rocks and structures which could be seen on the earth's crust or on its surface. He did not mention Whitehurst's beautiful plates of rock structures and mineral deposits.

Miller gained his knowledge of the theory of James Hutton through the memoir published in the first volume of the *Transactions of the Royal Society of Edinburgh* in 1788. He gave a brief but fair description of Hutton's theory, concluding:

> . . . according to this theory, the continents which we now inhabit must, in process of time, be worn away and destroyed, and others be forced up to supply their place. The length of time to be allowed for this excessive destruction and reproduction, Dr. Hutton supposes to be far greater than is generally imagined. His system, therefore, is to be arranged, of course, among those which are hostile to the sacred history; and the best judges have pronounced it equally hostile to the principles of probability, to the results of the ablest observations on the mineral kingdom, and to the dictates of rational philosophy.

The objections were probably derived from those of Kirwan, the violent antagonist of Hutton.

As with most of the other writers that Miller had mentioned, his discussion of the work of John Williams, contained in the book published in 1790, is on his theory of the earth—Williams' long discussion of rocks, stratification, and mineral deposits being scarcely mentioned. Williams' theory was, like so many of his predecessors, one in which it was assumed the earth had been formed from material "originally mixed with water into a fluid or a chaotic mass" which had later been precipitated.

Similarly the theories of Delametherie, 1795, and of Philip Howard, 1797, which supposed the earth had been formed by precipitation from water, were described in several pages; as was the theory of Bertrand. Miller mentioned about ten other theories including that of Benjamin Franklin, but chose not to discuss them. As Miller was a believer in a general deluge, he summarized all the evidence he could in favor of such an event, including the presence of shells and rocks at various elevations and the fossil elephants and rhinoceri in polar regions; he attempted to equate the Neptunian doctrines with orthodoxy and the correct explanation of geological features. He was also certain that the present earth was of very slight age, presumably only about six thousand years.

It must be remembered that Samuel Miller was a clergyman and a firm supporter of the orthodox beliefs of the day. He was not a geologist and apparently had had little experience in looking at rocks or minerals in the field. It is evident that he was far more impressed by theories than by actual observations. A really important aspect of Miller's history is the complete list of writers of the eighteenth century which he included and the little sketches he gave of so many of their theories. He omitted no important mineralogist or geologist and included many who have been almost lost to sight in the present day and indeed a few who are not listed at all in any modern history of science. If he was biased in his interpretation and approvals, at least he was fair is listing all the workers and their chief publications.

DeWITT CLINTON—1814

DeWitt Clinton (1769–1828) was one of those "universal" men of the early republic along with Thomas Jefferson and Samuel Latham Mitchill. Clinton was successively lawyer, politician, statesman, U.S. Senator, Mayor of New York City, and Governor of New York. He was a supporter of the arts and sciences and writer on scientific subjects. He is now probably best known for his part in bringing the Erie Canal to successful completion.

When the Literary and Philosophical Society of New York was established in 1814 Clinton was elected president. He delivered "An Introductory Discourse" at the opening meeting of the Society on May 4, 1814, which was published in Volume I of the *Transactions* of the Society and in two separate book publications in 1815.[7] The discourse dealt with the intellectual history of the American colonies and the United States from their founding, and the history of the various sciences. The part concerning geology dealt exclusively with the history of geology of the new country. Clinton referred to the classification of William Maclure which divided the country into "the *primitive*, *transition*, *floetz*, and *alluvial* rocks," and to that of Mitchill which divided the country into "the granite . . . the schistic, the limestone, the sandstone, and the alluvial." Clinton asserted that Volney had "borrowed the ideas of Mitchill," but it may well be that the interchange of ideas between the two had been mutual.

Clinton thought the "three kinds of alluvial formations: one arising from the retreat of the ocean, another from the subsidence or extinction of lakes, and another from the overflowing, retreat, and change of rivers . . . support the Neptunian theory." In a long "Note G" he discussed the earlier views of the former greater extent of Lake Ontario and Lake Erie, and then went on to a perceptive account of his own observations.[8]

[7] DeWitt Clinton, "An Introductory Discourse, delivered on the 4th of May, 1814," *Transactions of the Literary and Philosophical Society of New York*, 1815, *1*:21–284 (also New York: Van Winkle and Wiley, 1815; New York: David Longworth, 1815).

[8] The route of the Erie Canal followed in some places the ancient drainage lines of the higher Great Lakes when they discharged by way of the Mohawk Valley to the Hudson River, rather than to the St. Lawrence River, as at present. Already Clinton was thinking of possible routes for the

In "Note H" Clinton dealt at length with the discovery of "petrifactions" by Kalm and by others. He recognized that these had been described primarily in "Calcarious Stones," but he had "seen very curious ones in sandstone" and other places. The character of the fossils in both the limestone and the sandstone inclined him to say that he should "therefore pronounce them to be oceanic." He recounted the history of observations of medicinal and mineral springs and of bituminous springs "whence the famous Seneka oil is obtained" and discussed at length the history and character of "springs of petroleum" in "Note K."

Under the head of "Zoology" Clinton discussed the discovery of the fossil elephants both in the United States and in Russia. He extended his review of the earlier comments of Charlevoix, Kalm, Fabri, Buffon, Barton, Pownall, Peale, and, of course, Thomas Jefferson, in a fifteen-page "Note N" and recounted at length the discovery of a complete mammoth in 1799 on the River Lena. An important part of this long note is the printing of a letter from Silvanus Miller who had participated in the excavation of the Orange County, New York, mastodon. This account described the geographic and physiographic setting, the appearance of the bones, and the method of excavation. As far as I know, this is the most detailed account of the finding and excavation of this famous material.

Clinton asserted that "the principal attempts that have been made to illustrate our geology, . . . although entitled to merit, are imperfect, and probably, to a considerable extent fanciful" and stated that a great deal of more detailed work would be necessary in order to discriminate among the different kinds of rocks which are "not only piled upon each other, but are frequently buried in the bowels of the earth." He noted the repeated reports of the abundance of iron, lead, building stones, and other industrial minerals and thought that there would be plenty of coal, although "no discovery of any consequence" had yet been made.

From a list of the members of the Literary and Philosophical Society in the first volume of the *Transactions*, it may be seen that Clinton was in close communication with many men who could have discussed the geology of the United States—Andrew Ellicott, Samuel Harris, Washington Irving, Samuel Mitchill, Benjamin Barton, Thomas Cooper, Joseph Correade de Serra, Timothy Dwight, Thomas Jefferson, Benjamin Silliman, and Archibald Bruce. It was from his own knowledge and from his conversation with these men that Clinton was acquainted with all of the geological activity of the day and was able to give this summary of history of geology of the early republic.

JEREMIAH VAN RENSSELAER—1825

In 1825 Dr. Jeremiah van Rensselaer, a young physician of New York, presented a series of six lectures on geology to a large audience at the New York Athenaeum. These were published as a book the same year.[9] Van Rensselaer (1793–1871) was born

---

canal he envisioned, and he soon realized that the evidence of a former greater discharge of the Mohawk River required much higher lake levels to furnish the amount of water required. The former greater extent of the Great Lakes had been recognized as early as 1751 by John Bartram and Lewis Evans; G. W. White, "Early American Geology," *Scientific Monthly*, 1953, *76*:134–141; G. W. White, "Lewis Evans; Contributions to Early American Geology—1743–1755," *Illinois Academy of Sciences Transactions*, 1951, *44*:152–158.

[9] Jeremiah van Rensselaer, *Lectures on Geology; Being Outlines of the Science, Delivered in New York Athenaeum. In the year 1825* (New York: E. Bliss & E. White, 1825).

in New York, graduated from Yale, and then studied medicine with his uncle, Archibald Bruce, noted physician and leading mineralogist of his time.[10] Receiving the degree of M.D. in 1817 he spent three years in Edinburgh, London, and Paris and travelled widely on the continent. He is said to have been the first American to have climbed Mt. Blanc. Returning to the United States about 1820 he practiced medicine in New York until his death. His interest in geology was soon submerged in his growing medical practice, and he published nothing on geology after 1825.[11]

As is often done in today's courses in geology, van Rensselaer's first lecture took up the history of geological observations. He asserted that "to the admirers of Nature's works, Geology offers new treasures of enjoyment." He also included a section on the "utility of Geology," pointing out its "value to miners, to agriculturists in fertilizing and draining lands, to those seeking water, to the architect and to the general citizen in the enjoyment of scenery."

The aim of his first chapter was to "give a rapid sketch of the rise and progress of this Science, and the various Theories that have been published in illustration of it." The earliest work he mentioned was that of Burnet of 1680, whose theory he described in some detail. His interesting comment about its possible origin I have never seen elsewhere: "It has been asserted, I know not upon what authority, that this theory had been previously published by Francisco Patrizio, a professor at Rome, in a dialogue called 'Il Lamberto.' It is generally viewed as an elegant Romance; the product of mere imagination unsupported by any observed phenomena."

His discussion of Leibniz' *Protogaea* indicates that he was familiar with the work at first hand. In his review of Woodward's theory he realized that "the mere existence of beds of light substances placed beneath many of greater specific gravity is at once an objection that cannot be removed." Woodward had insisted that all matters had been precipitated in a universal ocean in order of their specific gravity. He noted Hooke's *Discourse on Earthquakes* of 1688, but did not discuss this important work. The theories of Halley, Whiston, Moro, de la Pryme, King, and de Luc were only mentioned briefly. His recognition of the importance of Jean-Etienne Guettard deserves complete quotation:

> In 1746 Guettard first complied with the wishes of Lister, who in 1684, recommended the adoption of geological maps. He was one of the first writers who seems to have forsaken cosmogony, and applied himself to geological enquiries. He possessed, for that period, much important information, which led him to pretty correct conclusions: for instance, he formed 3 divisions of the earth's surface. The first he called *Schistose*, which embraced the primary rocks; secondly the *Marls*, which included the secondary, pretty generally; and thirdly, *Sand*, coinciding with what is now termed tertiary. His views were too extensive for the state of science at the time in which he lived.

Van Rensselaer did not think so highly of Lehmann, although he recognized that Lehmann had distinguished between primary and secondary rocks. He was more impressed by Whitehurst's theory than by his observations of strata and structure, although he said "there is some romance and much valuable information in his book."

[10] For Bruce see J. C. Greene, "Introduction," *The American Mineralogical Journal Conducted by Archibald Bruce, M.D. Facsimile of the 1814 Edition* (New York: Hafner, 1968), p. vii–xii.

[11] H. A. Kelly, "Jeremiah van Rensselaer," *Cyclopedia of American Medical Biography* (Philadelphia: W. B. Saunders Co., 1912), Vol. II, p. 469.

He wrote favorably of Kirwan's explanation of the origin of earth features in terms of the Noachian Deluge. He devoted several pages to the theory of Buffon, but did not regard favorably that part which dealt with the long period of time necessary to wear away mountains, to excavate valleys, and to make new deposits.

Van Rensselaer was familiar with Hutton's theory of the earth through its 1788 publication. He gave an objective summary and stated that "in this theory there is a happy union of the agency of both fire and water; the one collecting and depositing, the other consolidating and elevating." He spoke of the "elegant illustrations of Professor Playfair" and noted that "this book continues to be the textbook of the best English geologists."

Although van Rensselaer devoted several pages to Werner, it appears that his information about him was derived from some summary, rather than any original source. He repeated the often-told story of Werner's antipathy to writing and of his final unwillingness even to open letters. He concluded that "no one but his most attached pupils upholds his theories, which he carried beyond the bound of philosophy."

In his summary of the history of geology of America, van Rensselaer mentioned the importance of the great mineralogical cabinet of Colonel Gibbs and spoke highly of the work of Dr. S. L. Mitchill and Dr. Akerly. It is strange that he did not mention the important work of his uncle, Dr. Bruce, nor of Parker Cleaveland. Although he was favorably impressed by the essays of Maclure, he did not have the inordinate admiration for Maclure's work that some others of the time had, as he says "no great standard work has yet been given on the geology of America."

He touched lightly, generally with approval, the work of Pierce, Hitchcock, Major Long, Major Delafield, Dr. Bigsby, and Dr. Schoolcraft, but he did not mention Amos Eaton or any of those who wrote on geology in Bruce's *Mineralogical Journal*, the various publications of the American Philosophical Society, or the American Academy of Arts and Sciences. Although his chapter gives the impression that he did not have access to as complete a library as that enjoyed by Samuel Miller twenty-five years earlier, the many references throughout the rest of the book make it evident that van Rensselaer actually did have access to a large collection of publications on geology from Europe and America.

We must not be too harsh in judging the insufficiency of van Rensselaer's history; it had to be fitted into a restricted space and time allotted to one lecture. It appears to be almost a transcript of an oral presentation and gives no evidence of expansion for later printing; indeed it may have already been printed for distribution to those attending the lecture. Slight as it is, we may be grateful that he made the attempt and left a record of his knowledge of the history of geology and his evaluation, however brief, of some of the theories.

G. W. FEATHERSTONHAUGH—1831-1832

In July 1831 appeared the first number of Featherstonhaugh's *Monthly American Journal of Geology and Natural Science*.[12] It was destined to last for only one year,

[12] *Monthly American Journal of Geology and Natural Science . . . Conducted by G. W. Featherstonhaugh.* Philadelphia: Vol. 1, No. 1, July 1831 to No. 12, June 1832; 576 pp. (all published). Facsimile reprint (New York: Hafner, 1968). For a biography of Featherstonhaugh, see G. W. White, "Introduction" to 1968 ed., pp. vii–xix.

during which a volume of twelve numbers appeared. George W. Featherstonhaugh (1780–1866) was an Oxford graduate who had travelled widely in Europe and who came to America in 1807. Successful in business affairs, he turned to an interest in geology and, as he was well acquainted with British geology and geologists (he was both Fellow of the Royal Society and of the Geological Society of London), he particularly emphasized the comparison of American and British geology. In the 1830s he examined the geology of the "western country," but soon afterward returned to England where he had a long, fruitful, and interesting career in the diplomatic consular service.

In the first number of the *Journal* the editor said that "the state of the Atlantic natural sciences will be reviewed in each number.... Each number will contain a continuous *Essay on Geology* as a science . . . so that the great principles, from which philosophical views of the arrangements and operations of nature are drawn, may be lucidly brought forward." The "Introduction," which Featherstonhaugh regarded as the first of his essays on geology, immediately assumed an historical character, beginning with a discussion on the relation of geological time to the Biblical six days of creation. In this number was also an extensive and glowing report on the first award of the Wollaston medal of the Geological Society of London to William Smith, whom Featherstonhaugh had known in England.

Some remarks on the history of geology were included. The author took up the history of the Deluge as recorded in the various ancient literatures and gave some account of ancient cosmographies. He dealt at some length with the opinions of Herodotus and gave Ovid's long summary of the opinions of Pythagoras.[13] A second part of the essay again dealt with the ancient Greeks, but it was concerned more with biology than geology. The third part dealt with the scientific ideas of the Romans and set the historical background for this time and for the Italian part of the Middle Ages.

Featherstonhaugh then went on to describe the effects of floods and volcanic eruptions "on the industry and the social existence of man" and indicated that the recognition of these from the time of earliest history lends some credence to the principle of uniformitarianism. A part of the essay in No. 6 was devoted to the general development of science throughout the Middle Ages and the Renaissance rather than specifically to geology. A later part dealt with the early recognition of fossils in Italy and with the opinions of Fracastoro, Fallopio, Mercati, Palissy, Colonna, Steno, Scilla, Quirini, Plot, and Lister. He reviewed in great detail the opinions of Hooke, Ray, Woodward, Burnet, and Whiston, but said "for further details we refer our reader to Mr. Lyell's 'Principles of Geology', an eloquent modern work in the possession of every naturalist." Featherstonhaugh recognized the importance of the stratigraphic organization of Arduino, Lehmann, Fuchsel, and others. He acknowledged the importance of Werner in mineralogy, but after an extended summary of Werner's stratigraphic system and of his views of aqueous origin of basalt, gave his opinion that "the classification made by Werner, and published in 1787, appears to be far short of the progress already made by Fuchsel."

Featherstonhaugh presented an extensive summary of Hutton's theory, which he

[13] A translation of Ovid's summary is also given by Charles Lyell, *Principles of Geology*, Vol. I (London: John Murray, 1830, 1832, 1833), pp. 13–14, but the two accounts are in different order and Featherstonhaugh, a competent classical scholar, presented his own translation of Ovid.

apparently knew from the 1788 paper. While generally approving of many of Hutton's ideas, he could not completely approve of the immensity of the geological time proposed by Hutton and of the complete uniformity of geological activity throughout that time. The tone of objection, however, is quite moderate and he goes on to say that we may "infer the future establishment of a geological theory of a progressive character, rather than one of uniform mutations."[14]

The work of William Smith was reviewed with high approval, and the summary of Smith's discovery and use of index fossils is an accurate one. Featherstonhaugh closed this section with a mention of the establishment in 1807 of the Geological Society of London, of which he was not only a Fellow, but also a member of the Dining Club, a prestigious, small "in group," which still flourishes.

In an article "On the Encouragement Given to the Study of Natural History in Europe and North America" (p. 391) the editor listed and rated the major scientific journals of Europe and the United States. He listed the relative standings of countries in support of scientific journals, indicating his wide knowledge of these journals.

In the first half of the nineteenth century various organizations sought geological information by questionnaire, and the extent and character of this method of collecting geologic data would be an interesting topic for investigation. Featherstonhaugh reprinted (p. 425) a lengthy series of "Queries on Geology of Pennsylvania from the Pennsylvania Geological Society," which had just been circulated by that organization. These show just what kind of knowledge was regarded important for geological studies at that time.

In No. 12, for June 1832, the last number of the volume, the editor states he is closing the volume and "When the heats of summer shall have past over . . . the editor proposes to resume the publication of his Journal, upon a more enlarged plan." His intent, however, did not come to fruition and no further issues appeared.

In addition to its articles on history, the single volume of this *Journal* is of added importance for present-day historical study because each number contains reviews which give the opinions of the time about the geological ideas expressed. Some of the reviews were intensely personal and some of them were from a quite different standpoint from reviews of the same works appearing in the much better known and still continuing *American Journal of Science*. The ordered material on history of geology in eight of the twelve numbers provides a valuable summary of the knowledge of the subject as known in America in 1831. In addition, it provides us with an insight into the background from which Featherstonhaugh viewed his fieldwork and his later publications on geology of the Western frontier.

WILLIAM W. MATHER—1833

William W. Mather (1804–1859) was a brilliant young army officer who became interested in geology while teaching at West Point and went on to become a distinguished American geologist. His wide knowledge of geological literature and history is well shown in the chapter on history of geology in his little textbook of geology for the

[14] Here Featherstonhaugh seems to embrace "actualism" as opposed to strict "uniformitarianism"; see M. J. S. Rudwick, "The Strategy of Lyell's *Principles of Geology*," *Isis*, 1970, *61*: 32–33.

use of schools published in 1833.[15] He listed thirty "principal facts of geology," a succinct statement of objective geological knowledge of the time.

He stated that "geology is a science of facts and has no necessary connection with any of the various theories...." He contrasted the Huttonian or volcanic theory with the Wernerian or Neptunian theory. His description of the latter was quoted directly from Parker Cleaveland,[16] and as Cleaveland was a Wernerian, the summary is therefore as favorable as possible. However, Mather observed that Werner's theory is "actually so interwoven with hypothetical points, which are not confirmed by succeeding observations in other parts of the world, that it cannot be received. It is not all true, and it does not account for all the facts." Mather's summary (p. 109) of the Huttonian theory was the neatest, most compact and objective statement to appear in America to that time:

> "The Huttonian theory" consists of inferences drawn from facts, but the deductions are sometimes carried farther than the facts warrant. It supposes the present continents derived from the wrecks of older ones, because we find a large portion of the rocks upon the surface of the earth formed of the fragments of others cemented together, and often filled with the remains of animals that once inhabited the sea. But whether volcanic agency has caused those lands to be elevated, which have evidently been for a long period under the surface of the ocean, is a point which is not conceded by all. It is however rendered highly probable by some facts which have been mentioned under the head of volcanos and earthquakes. The Huttonian theory supposes the secondary rocks, after they were formed in the bed of the ocean, to have been indurated by subterranean heat, before they were raised to become islands and continents. Some experiments by Sir James Hall, which have been already mentioned, render this supposition probable.

The final chapter, "Sketch of the History of Geology," began with "the priests of Egypt" and continued through Thales, Zeno, Pliny, Aristotle, and Strabo, the mineralogical observations of Avicenna of the tenth century, and briefly the ideas of Alessandro, Fracastoro, and Agricola.

Mather was acquainted with George Owen's work in England in the late sixteenth century on the successive order of rocks. He thought little of the work of Burnet, but more highly of Llwyd and of Lister. He also mentioned the works of Woodward, Hooke, Ray, Whiston, Packe, and Strachey. Mather continued with very short paragraphs about de Maillet, Guettard, Lehmann, Mitchill, Fuchsel, Whitehurst, Douglass, Hutton, Saussure, and Werner, whom he had discussed in detail earlier. He thought highly of the work of William Smith and agreed that "his labors have done more to advance this science in England, than those of any other individual." He mentioned fifteen more English, seven French, and five German geologists who were active in the earlier part of the nineteenth century. He listed eleven Americans who "with some others have been active in developing the geology of the United States. Mr. Maclure and General S. van Rensselaer deserve notice particularly.... These two men have given impulse to the science, that must result in the development of the geology and resources of this country." Mather concluded his little history by a statement (pp. 121–122) that earlier geology was

> ... directed to the wild and ample region of pure theory, connected with the origin of the

[15] W. W. Mather, *Elements of Geology of the Use of Schools* (Norwich: William Lester Jory, 1833).

[16] Parker Cleaveland, *An Elementary Treatise on Mineralogy* (Boston: Cummings and Hilliard, 1816), p. 591–598.

globe; the connection of scriptural history with physical events, and the traces and effects of the Noachian deluge.

The descriptive part of geology was then a blank.—For their soaring views, it would have appeared beneath the dignity of science to classify rocks and minerals, or descend into a minute comparison of the organic remains entombed in the rocky strata. . . . Werner and Hutton, both of whom gave much weight to the fact of observation, raised a new and very superior class of geologists. . . . Hutton's doctrine of the consolidation of earthy materials by heat and pressure, and Werner's theory of universal formations, were brought to tests of experiment and observation. The one was confirmed more satisfactorily than could have been expected under factitious circumstances; the other was shewn to have originated in its utmost generality in the narrow views of its ingenious but untravelled author. . . .

The acrimony, which the controversy between the disciples of these two men produced, caused a reaction, and geologists resolved for a time to banish hypotheses, and unbiassed by theory, endeavor to view nature as she is; and thus, men have become habituated to using well their hands and eyes, and amassing facts before theorizing.

Mather strongly favored objective treatment of geology and had a distinct preference for the work of those who made actual observations to those who theorized from insufficient data. He soon became one of the outstanding field geologists in the United States, as well as a superior geological synthesizer. He was among the first of the professional geologists whose work from about 1835 was solidly based on observation and synthesis of American rocks.[17]

THE ADVOCATE OF SCIENCE—1833-1835

A short-lived journal that consisted of only two volumes, *The Advocate of Science* began publication in 1833 at Philadelphia.[18] It was published and edited by William Peters Gibbons (1812–1897), whose father was William Gibbons (1781–1845), the distinguished physician of Wilmington and Philadelphia. William P. Gibbons, who first became a printer, was always interested in science and later studied medicine and took a medical degree in 1847. He moved to California in 1862 where he was not only a successful physician but also a prominent medical botanist. The young editor was aided by his older brother, Dr. Henry Gibbons (1808–1884), who was already on the way to becoming a distinguished physician and naturalist and a voluminous writer.[19]

A series of twenty articles on "Geology," signed "S," are almost certainly by Dr. Henry Gibbons, as they are so similar in content and wording to later articles signed by him. These articles dealt in a way generally equal in standard to many other publications of the time with descriptions of rocks and positions and relations of the strata. Of particular importance is the precise definition of the term "formation" in No. 3:[20]

[17] *Cf.* K. R. Aalto, "Specialization and Professionalism of Pre-Civil War North American Geology," *Journal of Geological Education*, 1969, *17*:91–94.

[18] *The Advocate of Science*, Philadelphia, Vol. 1, Nos. 1–48, 1833–1834, 192 pp. (all published). It was continued as *The Advocate of Science and Annals of Natural History*, Philadelphia, Vol. 1, Nos. 1–9, 1834–1835, 440 pp.

[19] For William P. Gibbons see H. A. Kelly and W. L. Burrage, *Dictionary of American Medical Biography* (New York: Appleton, 1928), p. 464. For Henry Gibbons see *National Cyclopedia of American Biography*, Vol. VII (New York: James T. White, 1897), p. 287.

[20] The history of the use of "formation" in geology would extend to an essay in itself. It was used as a restricted geological term as early as 1773 by Fuchsel. This definition by Gibbons is the earliest I have seen in an American publication. For a current definition of "formation" see J. V. Howell, ed., *Glossary of Geology and Related Sciences* (Washington: American Geological Institute, 1957), p. 114.

> The word *formation* being used in Geology as a technical term, it is proper to define it. Any stratum or bed of rock, sand, clay, or other substance, is thus designated, on account of its having been formed at one time, and by the same causes.

"S" divided rocks into six classes, "Primary, Transition, Secondary, Tertiary, Diluvial, Alluvial, Volcanic and Basaltic" and described the classification in detail. He then dealt with the contrast between the Neptunian or Wernerian and the Plutonic or Huttonian theories. He described and gave the history of observations of avalanches and landslides, with examples from Europe and America, including the famous Willey slide of 1826 in the White Mountains of New Hampshire. The history of deltas, lakes, and the retreat of Niagara Falls led him to believe that Lake Erie would eventually be drained.

A series of articles, presumably by the editor, William P. Gibbons, on "History and Study of the Natural Sciences" ran concurrently with the "Geology" series by "S." After considering science in general in classical and medieval times, the second part of this series took up the history of geology very briefly, mentioning Linnaeus, Haüy, Werner, Burnet, and Buffon. In the third part the editor-author discussed Werner, mainly by quotations (without credit) of anecdotes taken directly from van Rensselaer, without including the geological summary of that writer. In this part the author dealt mostly with the subject matter of mineralogy and geology, highly approving of the current recognition of the usefulness of fossils for deciphering history of the earth. He was insistent on the usefulness of the fossil record to demonstrate the progress through geologic time from early primitive plants and animals to later more complex forms.

After discussing the history of volcanic eruptions and accompanying earth movements the author took up the history of coral growth, and coral islands as well as peat. On the history of the concepts of valley cutting he thought ordinary streams incapable of cutting large valleys, but "original subsidence" was also unsatisfactory. He favored occasional huge floods and quotes Jefferson's explanation in *Notes on Virginia* for the cutting of the Potomac River across the Blue Ridge at Harpers Ferry by sudden breaking out of a great lake, the drainage of which cut the gorge. The author reviewed theories of veins and asserted that the theory of Hutton of vein filling from below was preferred over that of Werner of filling from above. He discussed theories of the origin of coal and concluded that some coal is formed from plants and some from "primary minerals."

Volume 1 of *The Advocate of Science* closed with Nos. 47–48 on February 20, 1834. It was continued under the same editor as *The Advocate of Science and Annals of Natural History*, which again started as Volume 1 in August 1834 in a new and more formal format and with illustrations. This form of the *Advocate* was a volume of 440 pages, the last issue, No. 9, appearing in April 1835. It contains many papers of geological importance by Conrad, Morton, and others, and an historical paper, "A Sketch of the History of Geology" (pp. 112–120), signed, in this case, by Dr. Henry Gibbons. In this "sketch," Gibbons asserted that until the time of Werner only "baseless systems" had been proposed. He gave in some detail the examples of Burnet, Leibniz, Woodward, Halley, Whiston, Whitehurst, Buffon, and Erasmus Darwin but again said these are all "absurd romances." Werner's anecdotes are repeated from the preceding volume and again the Neptunian system is described. Hutton's theory is reviewed and favored over Werner's. It is the similarity of this article to those in the preceding year by "S" that leads me to believe that "S" was actually Dr. Henry Gibbons.

The first bibliography of American geology, entitled "Chronological Analysis of Various Papers hitherto published on the Geology of the Upper Secondary and Tertiary Formations of the United States," appeared in the last issue of the volume (pp. 395–400). In it works on stratigraphy and paleontology are exactly cited and briefly but expertly annotated.

The historical material in the two forms of Gibbons' *Advocate* is generally of slight importance, but it does have its usefulness, and the bibliography of American geology is valuable. The articles on history show the final eclipse of Neptunian doctrines, even among those who were not primarily working geologists. The working geologists, such as Mather, Featherstonhaugh, Conrad, and the "western men," Keating, Schoolcraft, Long, and Bradbury, had long since abandoned Wernerianism. Many of these were in the "Philadelphia circle," and it was from them that William and Henry Gibbons learned that the strata were most logically explained as slow deposits of materials eroded from other pre-existing rocks. Here is another example of the fact that the decade of the 1830s in America was a time of reassessment and reorganization of geological thinking—the final abandonment of theorizing and the willingness to welcome new observations of careful fieldwork and the conclusions drawn from it.

NATHANIEL F. MOORE—1834

An unusual work on the history of earth science to appear in the first third of the nineteenth century was the very specialized study by Nathaniel Fish Moore on the mineralogy of the ancients.[21] Moore (1782–1872) was born on Long Island, New York, and graduated in 1802 from Columbia, where his uncle, Bishop Benjamin Moore, was president. He trained for the law but preferred classical studies. In 1817 he was appointed to a position in Greek and Latin at Columbia College and in 1838, after two years' travel in Europe, he became librarian, having sold to the college his own great library of classics. In 1842 he became president of the college but resigned in 1849. Throughout his career he continued to write books and articles on classics, grammar, and academic affairs. He combined a thorough knowledge of the classics with an interest in science, especially mineralogy. He disclaims in the brief preface of his study any pretension to original discovery[22] and asserts:

> The author is aware that in venturing upon a scientific theme he transcends the limits of his province, but relies for indulgence on the fact, that his subject is considered chiefly in the light in which it presents itself in the writings of those ancients whom it is his proper business to interpret.

---

[21] N. F. Moore, *Ancient Mineralogy; or an Inquiry Respecting Mineral Substances Mentioned by the Ancients: with Occasional Remarks on the Use to Which They Were Applied* (New York: G. & C. Carvill, 1834; 2nd ed., 1859).

[22] Moore had at least a working knowledge of geology and mineralogy. He was an enthusiastic mineral collector who understood the necessity of careful packing of specimens exactly labelled for locality and date of collecting. In a trip to St. Anthony Falls in 1845, he collected minerals at the lead mines near Galena, Illinois, geodes from the famous locality near Keokuk, Iowa, and minerals from Iron Mountain, Missouri. The shipment of his minerals required a trunk in which the specimens were packed in "cotton wadding." He noted in his travel diary the limestone along the Mississippi in the Prairie du Chien region and some of the geological features elsewhere in his travel on the Mississippi and Ohio rivers. The diary of this trip has been published: Nathaniel Fish Moore, *A Trip From New York to the Falls of St. Anthony in 1845*, Stanley Pargelis and Ruth Lapham Butler, eds. (Chicago: University of Chicago Press for the Newberry Library, 1946).

In his book Moore discussed the descriptions of mineral substances by a great many of the ancient authors, especially Dioscorides, Pliny, and Theophrastus. He considered Biblical references to minerals and discussed at length the foreign and domestic sources of minerals known to the Israelites and the foreign trade and travel that these indicate. The section on "Metallic Substances, Earths, &c., Employed as Pigments by the Ancients" is a valuable source of reference to terms for colors and pigments in classical literature, for localities from which the materials were mined, and for the methods of preparation.

Other features useful to present-day students are lists of Latin names with their modern interpretation and lists of books in several languages that include material on ancient mineral terms and references. Moore not only catalogued the references to minerals and rocks in the ancient writings, he also wrote on their origin and uses and discussed the references to earth materials as indications of degree of civilization, of economic development, and of artistic advancement. He was well aware that the mineralogy of the ancients should not be judged on the basis of current knowledge and systems, but he states that

> In the writings of the ancients we should look in vain for any such well grounded comprehensive system. They were, for the most part, content with describing individually the minerals they knew, by some more striking external characters, or some obvious property, or the uses to which they were applied. And in this branch of Natural Science, as well as in others, there is often great difficulty in ascertaining to what species objects mentioned by the ancients are to be referred; because they were apt to dwell upon features, striking perhaps in the individual before them; but in no wise characteristic of its kind. And they were besides, so poetical, often, and imaginative, that they could hardly bring themselves to give a plain unvarnished description of any thing remarkable, whether in nature or in art.

Moore recognized that the primitive state of mineralogical knowledge of the ancients actually continued into the seventeenth century, or even beyond, as shown by his statement, "If we should light in Pompeii on a cabinet of minerals; which would be one of Pliny's time; it might be found, perhaps, full as well arranged as was that of Sir John Hill, the translator of Theophrastus, sixty years ago" (p. 8).

Moore's book is notable for the citations of almost all of the ancient writers from Homer to Vitruvius and for the exact references in the text and in footnotes. The index, in which he lists all of the minerals discussed in the text or in the footnotes, adds much to the importance of the work. The book excited little attention in America, which is understandable because Americans were still in the early phases of mineralogical investigations, but it received wider recognition in Great Britain, where mineralogical studies were farther advanced. The book was well enough regarded to require a second edition in 1859.

CONCLUSION

Some of those Americans who made early attempts at histories of geology were young men at the beginning of their careers, one being only twenty-one. Clinton, Featherstonhaugh, and Moore were mature men with wide scholarly knowledge, and Samuel Miller, although only thirty-four, really belongs with the group of mature men, for his knowledge of history and literature was already extensive. A detailed study of each writer will eventually disclose the sources each used (one such a study is now underway[23]). In this concluding section only some opinions and speculations can be offered on sources for the history of geology each wrote.

The first extensive history of geology in English, that of Charles Lyell in 1830, was known in America by 1831, when Featherstonhaugh mentioned it, and it would have been available to Mather, the Gibbons, and Moore. Some earlier historical material is in Parkinson,[24] but there is no indication that it was used by any of the American writers.

Samuel Miller's work on history of geology, part of a much larger work on intellectual history, foreshadowed his later career as professor of church history at the Princeton Seminary. His work appears to be from original sources which were in the extensive library of Samuel L. Mitchill and other associates of the time. He had a broad knowledge of world history. His description of the works and theories of the eighteenth-century writers was always fair, but he did emphasize the way in which their theories conflicted with or confirmed revealed religion as he understood it.

DeWitt Clinton, the mature man of affairs, was a competent classical scholar who knew the ancient writers and was no mean naturalist himself. His "Discourse" was on American intellectual history and hence he confined his summary to American geology. He had seen many of the localities in which the authors based their reports. He was personally acquainted with most of the men and hence with their work. Although he was well acquainted with reports of scientific work in Europe, he said very little about the relation of American geology to European, except to compare American and European fossil elephants. It is a puzzle why he ignored the work of some men, especially that of Amos Eaton.

Jeremiah van Rensselaer, a young physician, had access to the library of his uncle Archibald Bruce, the noted mineralogist, and we may assume that he had access to materials of Benjamin Silliman and others at Yale. How much of his work is derived from that of Miller and Clinton can only be determined by further study. Van Rensselaer had at least some acquaintance with actual rocks, as he had travelled and climbed in the Alps. Whether from original or secondary sources, he had an adequate grasp of the history of geology as understood at the time, shown not only in his chapter on the subject, but also through references at various places in his book.

G. W. Featherstonhaugh, the man of many practical, literary, and scientific qualities, was fully acquainted with the original geological literature in classical and modern languages. His rating of the relative positions of scientific journals in Europe and America indicated his wide knowledge of the recent and current writings. He knew the English geologists and had probably been in the field with some of them and could assess the quality and importance of their work. At the time he wrote, he was the most geologically knowledgeable and critical of all the men discussed in this essay.

W. W. Mather, the young army officer, was just beginning his geological career when he wrote his little book with its historical summaries. He knew at first hand the writings of the Americans he discussed but had some, probably much, of his knowledge of European writings from secondary sources. He exhibited an overall grasp of the history of geology, as shown in the references in his later geological reports.

[23] See n. 1.

[24] James Parkinson, *Organic Remains of a Former World*. . . . Vol. I (London: J. Robson, 1804).

The historical summaries of William Gibbons illustrate the precosity of the twenty-one-year-old printer who later became a physician and naturalist. He must also have been greatly aided by his twenty-five-year-old brother, Dr. Henry Gibbons. It is probable that much of their summaries of early European work was derived from Lyell and from the earlier writings of Clinton and others, but their annotated bibliography of American geology indicates a detailed familiarity with the American sources. It is tempting to assume that they discussed these with Morton, Conrad, and others in the Philadelphia region.

N. F. Moore's book on ancient mineralogy is a very specialized work of a mature classical scholar and bibliographer. It is entirely based on original sources, most of which he had in his own personal library. Moore's work was factual reporting of the use of ancient terms and mineral description, devoid of any "presentism"; he clearly understood and emphasized the physical and mental framework in which his authors worked. His knowledge of the history of the classical time of which he wrote was probably the most complete of anyone in America.

It must be remembered that those who dealt with history of American geology were writing of the work of men they knew more or less personally, and the estimation of this work by contempories is a valuable part of the historical summaries. On the other hand, except for Featherstonhaugh (and possibly to a much slighter degree van Rensselaer), they were writing of recent and current work in England and the continent at a distance and could not assess its importance from personal knowledge of men and location. Even those reports derived from the original works were often influenced by the comments of others about them.

The historical summaries of these writers show that some American scholars and geologists were concerned with the history and also the philosophy of the science, and that there must have been a lively general interest in history of geology to warrant its repeated discussion in books and journals. These histories, covering a space of thirty years, also show the change in America from the general attitude in 1800 of approval of catastrophism and a short duration of geological time, through the uncertainty of these suppositions by 1825, to the beginning of general acceptance (not yet complete, however) by 1835 of actualism and a longer span of geological time.

BIOGRAPHIES
OF
GEOLOGISTS

BIOGRAPHIES OF GEOLOGISTS

JOHN W. WELLS AND GEORGE W. WHITE

Department of Geology, Cornell University, Ithaca, N. Y., and *Department of Geology, University of Illinois, Urbana, Illinois*

Ten years ago, one of us published *A list of books on the personalities of geology* in this journal (vol. 47, 192–200, 1947). At the time it was noted that such a list was certainly incomplete, and the intervening years have shown that this was decidedly an understatement. Since then we have found many more such books, new ones have been published, and interested friends (especially J. V. Howell and F. S. Colliver) have suggested important additions. Only 78 biographies were included originally, and since some 132 more can now be added, it seems worthwhile to present a more complete listing, even though it is probably still incomplete. We have added brief comments on all the biographies except those few we have not been able to examine. In the original list, books of geologists' travels, histories of geology, and "miscellaneous" books were included. These are omitted here, only biographical and autobiographical books being admitted. Even with this restriction, choice in some instances has not been easy. As in the first list, we have included only items that have appeared as books, deliberately excluding all but a very, very few biographical notices, memoirs, and obituaries that were published in various journals.

All those seriously interested in geology and its branches as a profession or avocation cannot fail to win something from the reading of their predecessors' struggles and triumphs. Who can work with the great quarto volumes of the *Palaeontology of New York* without wondering about the character of a man with enough fortitude to produce them? Who can notice the constant allusions to Sir Charles Lyell in hundreds of geological works without wondering how he lived and worked? Who can read of the first geological exploration of the Grand Canyon of the Colorado without a thought to the motives that pushed John Wesley Powell into the unknown? Who can read *The Old Red Sandstone* without wondering what sort of geologist could write such elegant and lucid prose? Who can discover that a mineral, such as dolomite, was named for one Deodat Dolomieu without wondering who this man was to merit such an honor? And who can contrast the great volumes on the zoology of the United States Exploring Expedition with the many editions of *A System of Mineralogy* without wondering at the versatility of James Dwight Dana? Or at the even greater versatility of Louis Agassiz—from ice to jungle exploration, from fossil fish to embryology?

We hope that geologists and other scientists will find this compilation a useful guide to books about the men who have labored on the main fabric or in the dark corners of the vast edifice of minerals, volcanoes, glaciers, fossils, mountains and rivers that is geology

BIOGRAPHIES

AGASSIZ, Alexander. 1835–1910. Agassiz, G. R. *Letters and Recollections of Alexander Agassiz.* Houghton Mifflin, Boston, New York, and London, 1913.—Son of Louis Agassiz, an accomplished zoologist, oceanographer and geologist, who made a fortune in Keweenaw copper and expended much of it on the Museum of Comparative Zoology and long geological and biological expeditions in the Pacific.

AGASSIZ, Louis. 1807–1873. Agassiz, E. C. *Louis Agassiz, his Life and Correspondence.* 2 vols., Macmillan, New York and London, 1885.—Student of glaciers and glaciation, author of the greatest work on fossil fishes, founder of the Museum of Comparative Zoology at Cambridge. The "official" biography.

Agassiz, E. C. *Louis Agassiz, sa Vie et sa Correspondence.* (transl. by A. Mayor.) Librairie Fischbacher, Paris, 1887.—French edition, with slight additions, of the preceding.

Cooper, L. *Louis Agassiz as a Teacher.* Comstock Publishing Co., Inc., Ithaca, 1945. (First edition 1917).—Illustrative extracts from writings of his students and a summary of his life.

Gould, Alice Bache. *Louis Agassiz.* Beacon Biographies: Small, Maynard & Co., Boston, 1901.

Holder, C. F. *Louis Agassiz, his Life and Work.* Putnam's, New York, 1893.—Parson Weemsish, fulsome, derivative, unimportant.

Marcou, J. *Life, Letters and Works of Louis Agassiz.* 2 vols. Macmillan and Co., New York and London, 1896.—Louis Agassiz seen through the eyes of his erratic but brilliant friend; valuable geological emphasis.

Robinson, M. L. *Runner of the Mountain Tops; the Life of Louis Agassiz.* Random House, New York, 1939.—A recent study, much less definitive than the prececeding.

Teller, J. D. *Louis Agassiz. Scientist and Teacher.* Ohio State Univ. Press, Columbus, 1947.—A history and analysis of Agassiz' influence on scientific education in America.

AGRICOLA, Georg. 1494–1555. Beecher, F. L. *Die Mineralogen Georg Agricola und A. G. Werner.* Freiberg, 1819.—An early study of the man who was physician, mining engineer, mining geologist and scholar. He was the first of the "moderns."

Darmstaedter, E. *Georg Agricola, Leben und Werk.* Muenchen, 1926.

Hartman, H. *Georg Agricola,* 1494–1555, *Begrunder dreier Wissenschaften, Mineralogie—Geologie—Bergbaukunde.* Wissenschaftliche Verlagsgesellschaft M.B.H., Stuttgart, 1953.—Not the definitive life, but an analysis of Agricola's place in science.

Hofman, R. *Dr. Georg Agricola, Ein Gelehrtenleben aus dem Zeitalter der Reformation* F. A. Perthes, Gotha, 1905.— Short, scholarly study of Agricola's life and work.

AMEGHINO, Florentino. 1854–1911. Cabarera, A. *El pensamiento vivo de Ameghino.* Editorial Losada, Buenos Aires, 1944.—A little on the life and much on the work of the self-taught Argentinian vertebrate paleontologist.

ANDREWS, Roy Chapman. 1884–. Andrews, R. C. *Under a Lucky Star, a Lifetime of Adventure.* Viking Press, New York, 1943.—Lively autiography of the leader of the American Museum's expeditions to the Gobi Desert.

ARGYLL, George Douglas, Duke of. 1823–1900. Duchess of Argyll, ed. *George Douglas, Eighth Duke of Argyll, autobiography and memoirs.* Dutton, New York, 2 vols., 1906.—"Few men of the duke's era displayed more versatility of intellect." In the field of science, especially geology, he endeavoured to harmonize the new discoveries and theories with religion.

BAGG, Rufus Mather. 1869–1946. Taber, Gladys. *Especially Father.* Victor Gollancz, London, 1950.—A professional writer produces a bitingly witty biography of her geologist-professor father without once mentioning his name.

BAILEY, Loring Woart. 1839–1925. Bailey, J. W. *Loring Woart Bailey.* J. & A. McMillan, St. John, N.B., 1925.—Son of early West Point professor of chemistry, mineralogy and geology and microscopist; student at Harvard and Brown who spent his entire professional life at the University of New Brunswick. Interesting comments on Canadian geologists of the time.

BALL, Valentine. 1843–1895. Ball, V. *Jungle life in India; or the journeys and journals of an Indian Geologist.* De La Rue, London, 1880.—Entertaining recollections of fifteen years' field work for the Geological Survey of India.

BECK, Theodoric Romeyn. 1791–1855. Hamilton, F. H. *Eulogy on the life and character of Theodoric Romeyn Beck,* M.D., LL.D. Albany, 1856.—Beck was the scientific leader of Albany in the early 19th Century, amateur geologist, and friend of Amos Eaton.

BELL, Robert. 1864–1934. Stendall, J. A. S. *Robert Bell, geologist: a biographical sketch.* Quota Press, Dublin, 1938.

BONNEY, Thomas George. 1833–1923. Bonney, T. G. *Memories of a Long Life.* Metcalfe and Co. Ltd., Cambridge, 1921.—Mainly reminiscences of university experiences with almost no mention of his important work in geology.

BRONGNIART, Alexandre. 1770–1847; and Adolphe. 1801–1876. Launay, L. de. *Les Brongniart, une grande Famille de Savants.* Rapilly et Fils, Paris, 1940.—A fine study of a brilliant family. Alexandre was a geologist and mineralogist, director of the Sevres procelain factory and son of an eminent architect. Adolphe, his son, was famous as a botanist and paleobotanist.

BROOM, Robert. 1866–. Broom. R. *Finding the Missing Link.* C. A. Watts & Co. London, 1950.—Broom, physician and vertebrate paleontologist, was the great discoverer of the African ape-men.

BROWN, Alfred. 1834–1920. Drennan, M. R. *Gogga Brown, the Life Story of Alfred Brown, South Africa's Hermit-Naturalist.* Maskew Miller, Capetown, n.d.—Brown was the Robert Dick of South Africa, but the author is no Samuel Smiles.

BUCKLAND, William. 1784–1856. Gordon, Mrs. *The Life and Correspondence of William Buckland, D.D., F.R.S.* John Murray, London, 1894 (Also a New York edition).—One of the best geological biographies, about one of the great figures in geology in the early years of the 19th Century, professor of geology at Oxford, later Dean of Westminster.

BUFFON, George-Louis Leclerc, comte de. 1707–1788. Roule, L. *Buffon et la description de la nature.* Flammarion, Paris, 1924.—A good sketch of the career of the celebrated naturalist, author of *Les Epoques de la Nature,* a remarkable attempt at a synthesis of geologic history based on the fossil record.

Condorcet, J.-A.-N. de *Eloge de M. le comte de Buffon.* Paris, 1790.

L. Bertin et al. *Buffon.* Paris (Mus. Nat. Hist. Nat.) 1952.

Anon. *Vie privee de M. le comte de Buffon.* Lausanne, 1788.

BUNBURY, Charles James Fox. 1809–1886. Lyell, Katherine M. *Life, letters, and journals of Sir Charles James Fox Bunbury, Bart.* Murray, London, 2 vols., 1906.—Although himself merely an amateur geologist and botanist, Bunbury's letters provide many interesting sidelights on his wife's sister's husband, Sir Charles Lyell.

CALVIN, Samuel. 1840–1911. Thornton, H. J. *Samuel Calvin.* U. of Iowa Press, Iowa City, 1947.—Professor at University of Iowa and state geologist.

CLARKE, Edward Daniel. 1769–1822. Otter, W. *The Life and Remains of Edward Daniel Clarke.* J. F. Dore, London, 1824. (other editions London 1825 and New York 1827).—Professor of mineralogy at Cambridge, famous traveller in the Levant.

CLOOS, Hans. 1885–1951. Cloos, H. *Gesprach mit der Erde.* R. Piper, Munchen, 1947; *Conversation with the Earth.* (translation of the above by E. B. Garside, edited by Ernst Cloos and Curt Dietz). Alfred Knopf, New York, 1953.—At once autobiography, geologic description and illustration. Only Germans can write personalized description like this; may well become a classic.

CONDON, Thomas. 1822–1907. McCornack, E. C. *Thomas Condon, Pioneer Geologist of Oregon.* Univ. Press, Eugene, Ore., 1928.—Life of a lesser figure who worked valiantly in a formerly remote region.

CONRAD, Timothy Abbot. 1803–1877. Wheeler, H. E. *Timothy Abbot Conrad, with Particular Reference to his Work in Alabama One Hundred Years Ago.* Paleont. Res. Inst., Ithaca, 1935.—The pioneer paleontologist of the coastal plain Tertiaries.

COPE, Edward Drinker. 1840–1897. Osborn, H. F. *Cope: Master Naturalist; the Life and Letters of Edward Drinker Cope.* Princeton Univ. Press, 1931.—Definitive life of the driving figure of one of the greatest vertebrate paleontologists.

CROLL, James. 1821–1896. Irons, J. C. *Autobiographical Sketch of James Croll, with Memoir.* Edward Stanford, London, 1896.—Croll is noted for his contributions to the problem of the causes of continental glaciation.

CUVIER, George L. C.F.D. 1769–1832. Lee, Sarah (Mrs. R.) *Memoirs of Baron Cuvier.* J. & J. Harper, New York, 1833. (also London).—Biography by one who knew him of the great student of natural history and founder of vertebrate paleontology.

Vienot, J. *Le Napolean de l'Intelligence: George Cuvier, 1769–1832.* Fischbacher, Paris, 1932.

DANA, James Dwight. 1813–1895. Gilman, D. C. *The Life of James Dwight Dana.* Harper's, New York, 1899.—Accurate, but dull and lifeless biography of one of the most important nineteenth century geologists; founder of modern system of mineral classification.

DARWIN, Charles. 1809–1882. Darwin, F., ed. *Life and Letters of Charles Darwin.* 3 vols., Murray, London, 1887, 1892; 2 vols; Appleton, New York, 1896.—Darwin needs no mention here. He was by all odds the greatest figure in biological science of the 19th century, for he finally focussed and gave meaning to the choas of paleontological and neontological facts.

Bettany, G. T. *Life of Charles Darwin.* Scott, London, 1887.—Short but well-written biography.

Geikie, A. *Charles Darwin as Geologist.* Univ. Press, Cambridge 1909.—Excellent short summary set against background of the times; valuable notes. Deserves reprinting.

West, G. *Charles Darwin, a Portrait.* Yale Univ. Press, New Haven, 1938.—One of the best modern studies of Darwin.

Kingston, R. W. G. *Darwin.* London, 1934.

Barlow, Nora, ed. *Charles Darwin and the Voyage of the Beagle.* Philos. Library, New York, 1946.—Contains much previously unpublished material on the celebrated voyage.

DARWIN, Erasmus. 1731–1802. Pearson, H. *Doctor Darwin.* J. M. Dent & Sons, London, 1930.—Grandfather of Charles Darwin who was an evolutionist in his own right. Lively biography of a positive person.

DAVID, T. W. Edgeworth. 1858–1934. David, M. W. *Professor David, the Life of Sir Edgeworth David.* Arnold, London, 1937.—Very readable life of Australia's great field geologist and teacher.

DAWSON, John William. 1820–1899. Dawson, J. W. *Fifty Years of Work in Canada, Scientific and Education.* (ed. by Rankine Dawson). Ballantyne, Hanson and Co., London & Edinburgh, 1901.—Discoverer of air-breathers in the Carboniferous of Nova Scotia, of Eozoon, professor of geology and principal of McGill University.

DERBY, Orville Andelbert. 1851–1915. Lamego, A. R. et al. *Orville A. Derby, 1851–1915. Algunos aspetos da sua obra.* Ministerio da Agricultura, Rio de Janeiro, 1951.—Derby accompanied Charles F. Hartt on his famous Brazilian expeditions in the early 1870's, and remained in Brazil the rest of his life, helping to found the Comissao Geologica, the Museu Nacional, and the Instituto Historico e Geografico.

DICK, Robert. 1811–1866. Smiles, S. *Robert Dick, Baker, of Thurso, Geologist and Botanist.* Murray, London, 1878.—Appealing and sympathetic account of the keen naturalist, discoverer of many new plants and fossil fishes, baker by trade, friend of Hugh Miller, Peach, and Geikie, but unknown to most of his contemporaries. Includes interesting engravings of Caithness and Thurso.

DOLOMIEU, Deodat (Dieudonne-Sylvain-Guy-Tancrede de Gratet de). 1750–1801. Lacroix, A. *Deodat Dolomieu: Sa vie aventureuse—sa captivaté—ses oeuvres—sa correspondence.* Perrier, Paris, 2 vols. 1921.—Definitive work on Dolomieu, who travelled widely in the Mediterranean region—"a careful and indefatigable observer, and provided copious material towards the establishemnt of the principles of geology" (Geikie).

EATON, Amos. 1776–1842. McAllister, E. M. *Amos Eaton, Scientist and Educator, 1776–1842.* Univ. Pennsylvania Press, Philadelphia, 1941.—Remarkably thorough study of one of the pioneer teachers of science in America, the first great American geologist, writer of geological textbooks and founder of Renssalaer Polytechnic Institute.

EHRENBERG, Christian Gottfried. 1795–1876. Laue, M. *Christian Gottfried Ehrenberg. Ein Vertreter deutscher Naturforschung in neunzehnten Jahrhundert.* Springer, Berlin,

1895.—Factual account of the life and work of Ehrenberg, famous for his pioneer studies of the coral reefs of the Red Sea and on Foraminifera and other protists.

ESCHER VON DER LINTH, Arnold. 1807–1872. Heer, O. *Arnold Escher von der Linth, Lebensbild eines Naturforschers.* Friederich Schulthess, Zurich, 1893.—Swiss geologist who did pioneer work in Alpine stratigraphy and structure.

EVANS, Lewis. 1700–1755. Gipson, L. H. *Lewis Evans, to which is added Evans' A Brief Account of Pennsylvania Together with Fascimiles of His Geographical, . . . Essays, Numbers I and II . . . an Analysis of a General Map of the Middle British Colonies. . . .* Historical Society of Pennsylvania, Philadelphia, 1939.—Evans was a surveyor, engineer, engraver, traveller and map maker. He made many important geological observations and gave the first description of physiographic divisions of the United States. This finely printed book includes facsimiles of Evans' Map and "Analysis."

FORBES, Edward. 1815–1854. Wilson, G. and Geikie, A. *Memoir of Edward Forbes,* F.R.S. Macmillan and Co., Cambridge and London, 1861.—Edward Forbes was a pioneer in the study of animal ecology and geography.

FORBES, James David. 1809–1868. Shairp, J. C.; Tait, P. G.; and Adams-Reilly, A. *Life and Letters of James David Forbes,* F.R.S. Macmillan, London, 1873.—Forbes, professor of natural philosophy at Edinburgh, later at St. Andrews, is famous for his studies of glaciers.

FORSTER, Westgarth. 1772–1835. Nall, W. (*Memoir of*) *Westgarth Forster,* in Forster, Westgarth, *A Treatise of the Section of the Strata from Newcastle-upon Tyne to Cross Fell with remarks on Mineral Veins . . . Third edition, revised and corrected to the present time by the Rev. W. Nall, M.A. . . .* Andrew Reid, Newcastle: Edward Stanford, London, 1883.—Forster was a mining engineer and geologist (financially unsuccessful) who first recognized cyclic sedimentation in Carboniferous rocks, produced elaborate stratigraphic columns and described relation to lead ore to stratigraphy and structure. His work is mainly known from the second edition (1821) as the first (1809) is very rare. The three editions vary considerably. Forster deserves a modern study and evaluation.

FRENCH SAVANTS. Lacroix, A. *Figures de Savants.* 2 vols., Gauthiers-Villars, Paris, 1932.—Short studies of the lives and work of 31 French mineralogists and geologists, from Desmarest to Haug, with portraits and facsimiles of handwriting. Important source of information on French scientists.

GEIKE, Archibald. 1832–1924. Geikie, A. *A Long Life's Work, an Autobiography.* Macmillian, London, 1924.—Geike was an outstanding figure in British geology, long director of the Geological Survey of Great Britain; prolific biographer and writer on history of geology.

GEIKE, James. 1839–1915. Newbiggin, M. I., and J. S. Flett. *James Geikie, the Man and the Geologist.* Oliver and Boyd, Edinburgh, 1917.—Brother of Archibald, professor at Edinburgh, glacial and structural geologist.

GILBERT, Grove Karl. 1843–1918. Davis, W. M. *Biographical Memoir Grove Karl Gilbert 1843–1918.* Memoirs of the National Academy of Sciences, vol. 21. Washington, 1927.—A geological biography of possibly America's greatest geologist by another great geologist who himself deserves a book length biography. One of the three outstanding American biographies.

GOETHE, Johann Wolfgang. 1749–1832. Magnus, R., with a foreword by Gunther Schmid, translated by Heinz Norden. *Goethe as a Scientist.* Henry Schuman, New York, 1949.—(Translation of *Goethe als Naturforscher.* Leipzig, 1906). Contains a chapter on Goethe's contribution to geology and mineralogy. Some asserted his priority in the idea of evolution and of the Ice Age.

GREENLY, Edward. 1861–1951. Greenly E. *A Hand Through Time: Memories—Romantic and Geological; Studies in the Arts & Religion, and the Grounds of Confidence in Immortality.* Thomas Murby & Co., 2 vols. London, 1938.—At once autobiography, biography of Mrs. Greenly and sketches of geologists and geologic activities of the time. Keen geological observations and maudlin biography. Must be seen to be believed.

GRESSLY, Amand. 1814–1865. Rollier, L. *Amand Gressly's Briefe: Lettres d'Amand Gressly, le Geologue Jurassien.* Soc. Jur. D'Emul., Actes, 1911.—The only book-size work on Gressly, whose study of the Swiss Jurassic is the foundation of paleoecology and the stratigraphic concept of facies. Gressly is almost unknown in America!

GUNN, John. 1801–1890. Woodward, H. B., ed. *Memorials of John Gunn.* Norwich, 1891.—Rector of Irstead, keen amateur geologist, leader in early geological studies in Norfolk, best known for his work on the Cromer Forest Bed.

HAAST, Sir Julius von. Haast, H. F. *The life and times of Julius von Haast.* Gov't Printer, New Zealand. n.d.

HALL, James. 1811–1898. Clarke, J. M. *James Hall of Albany, Geologist and Palenotologist, 1811–1898.* Albany, 1921. (reprinted 1924).—One of the best biographies of a scientist. Excellent study of the great Paleozoic paleontologist and lively account of the early days of geologic study of the classic ground in New York and adjoining areas. Much material on contemporary associates and geological activities.

HAMMOND, John Hays. 1855–1936. Hammond, J. H. *The autobiography of John Hays Hammond.* 2 vols. Farrar & Rinehart, New York, 1935.—Interesting and well-written accounts of mining ventures over the world by the Freiberg-trained successful engineer and consultant of the select group of Pumpelly, Penrose, and Hoover.

HARRISON, Benjamin. 1837–1921. Harrison, E. R. *Harrison of Ightam.* Oxford Univ. Press, London, 1928.—The Robert Dick of Kent, discoverer of the "eoliths." A must for all students of the Pleistocene.

HARTT, Charles Frederick. 1840–1878. Menezes, C. A. de *Biographia do Professor Carlos Frederico Hartt.* Rio de Janeiro, 1878.—Brief study of the first professor of geology at Cornell University and pioneer in the geology of Brazil.

Wright, A. H. *Pre-Cornell and Early Cornell II. Letters to C. F. Hartt, first Professor of Geology at Cornell. A cross-section of the Agassiz Period.*—Letters of 1863–1879 to Hartt, by many important scientists which pertain to science of the time.

HEILPRIN, Angelo. 1853–1907. Pollock, G. *Michael Heilprin and his sons.* Dodd, Mead, New York, 1912.—Michael was editor of the *Nation.* Louis, the encyclopedist, was his first son; Angelo, the geologist, his second.

HEIM, Albert. 1849–1937. Brockmann-Jerosch, Marie; Heim, A.; and Heim, Helene. *Albert Heim, Leben und Forschung.* Wepf & Co., Basel, 1952.—Definitive biography and summaries of scientific work of the great authority on Alpine tectonics and glacial geology.

HITCHCOCK, Edward. 1793–1864. *Reminiscences of Amherst College, historical, scientific, biographical and autobiographical; also of other and wider life experiences.* Bridgman and Childs, Northampton, Mass., 1863.—Professor of geology at Amherst, sometime state geologist of Massachusetts and Vermont. Did excellent work in many kinds of geology, including early work on glacial drift.

HOBBS, William Herbert. 1864–1953. Hobbs, W. H. *An Explorer-Scientist's Pilgrimage, the Autobiography of William Herbert Hobbs.* J. W. Edwards, Inc., Ann Arbor, 1952.—Hobb's activities in several fields of geology, in geography and in international affairs make us wish his book had been longer and more detailed.

HORNER, Leonard. 1785–1864. Lyell, K. M. *Leonard Horner.* 2 vols., Women's Printing Society, Ltd., London, 1890.—Horner was a Scottish geologist and merchant, the father-in-law of Charles Lyell.

HOUGHTON, Douglass. 1809–1845. Bradish, A. *Memoir of Douglass Houghton.* Raynor & Taylor. Detroit, 1889.—Includes reprints of geological writings of Houghton, pioneer Michigan geologist, who was also physician and explorer.

Rintala, E. K. *Douglass Houghton, Michigan's Pioneer Geologist.* Wayne Univ. Press, Detroit, 1954.—Supplements Brandish somewhat in personal details of Houghton.

HUBBARD, Bela. 1814–1896. Hubbard, B. *Memorials of a Half-Century in Michigan and the Lake Region.* Putnam's, New York, 1888.—Early Michigan geologist, naturalist and meterologist. Pleasant, rambling reminiscences.

HUMBOLDT, Alexander von. 1769–1859. Anonymous. *The Life, Travels and Books of Alexander von Humboldt.* With an Introduction by Bayard Taylor. Rudd & Carelton, New York, 1859.—There are many biographies of this German geologist, traveller, explorer, botanist and philosopher who wrote 74 books on these subjects. This is not the best one.

Assing, L., ed. *Letters of Alexander von Humboldt to Varnhagen von Ense, from 1827 to 1858* (Trasl. F. Kapp). Rudd and Carleton, New York, 1860 (from 2d German edition).—Interesting but of little geological import.

Banse, E. *Alexander von Humboldt. Erschliesser einer neuen Welt.* Wissenschaftliche Verlagsgeschellschaft M.B.H., Stuttgart, 1953.—Straightforward biography; emphasizes Humboldt's geographical importance.

Bruhns, K., ed. *Life of Alexander von Humboldt* (Transl. by J. and C. Lassell). Longman, Green, London, 2 vols., 1873.—Compiled in commemoration of the centenary of Humboldt's birth by J. Lowenberg, R. Arc-Lallemant, and A. Dove.

Buchner, W. *A. von Humboldt, ein Lebensbild.* Lahr, 1882.

De Terra, H. *Humboldt; the life and times of Alexander von Humboldt,* 1769–1859. Alfred A. Knopf, New York, 1955.—Well-written modern work on Humboldt's "long and fabulous life" with some summary of his scientific work; stresses American connections.

Klenche, H. and G. Schlesien. *Lives of the brothers Humboldt, Alexander and William,* transl. by Juliette Bauer. Ingram, Cooke & Co., London; Harper, New York (1854), 1852. *Eine biographisches Denkmal.* A. von Humboldt. Leipzig, 1859.

Wittwer, W. C. *A. von Humboldt, sein wissenschaftliche Leben und Wirken.* Leipzig, 1860.

HUTTON, James. 1729–1797. Playfair, J. *Biographical Account of the Late Dr. James Hutton, F.R.S., Edin.* Royal Soc. Edinburgh Trans., vol. 5, pt. 3, pp. 39–99, 1805; also collected works of John Playfair, Esq., vol. 4, pp. 33–118, 1822.—Sympathetic biography stressing Hutton's methods of geologic observation and induction.

Various Authors. *James Hutton* 1726–1797. *Commemoration of the* 150*th Anniversary of his Death.* Proc. Sect. B. Royal Soc. Edinburgh, vol. 63, pt. 4, 1950.—Hutton, "The Founder of Modern Geology," deserves a modern biography: the two listed are excellent for the areas they cover.

HUXLEY, Thomas Henry. 1825–1895. Huxley, L. *Life and Letters of Thomas Henry Huxley.* 2 vols., Macmillan and Co. London. 1900. (Also American edition 1901).—Huxley, celebrated biologist and forensic scientist, was the friend of Darwin and champion of the theory of evolution.

JEFFERSON, Thomas. 1743–1826. Martin, E. T. *Thomas Jefferson: Scientist.* Henry Schuman, New York, 1952.—Jefferson's scientific activities were remarkable. His interests in vertebrate paleontology are well known.

JOLY, John. 1857–1933. *Reminiscences and anticipations.* Fisher Unwin, London, 1920.—Professor of geology and mineralogy at the University of Dublin, Joly is perhaps best known to geologists as the author of the salt-content method of estimating the age of the oceans. This volume contains only scattered reminiscences.

JORDAN, David Starr. 1851–1931. *The days of a man: being memories of a naturalist, teacher and minor prophet of democracy.* World Book Co., Yonkers-on-Hudson, 1922.—Vivid autobiography of the founding president of Stanford University and great student of fossil and recent fish.

JUKES, Joseph Beete. 1811–1869. Jukes, Miss., ed. *Letters and extracts from the addresses and occasional writings of J. B. Jukes.* Chapman & Hall, London, 1871.—Professor in the Royal College of Science in Dublin, and local director of the Geological Survey of Ireland. An early student of coral reeefs.

JUNGHUHN, Franz. 1809–1864. *Gedenkboek Franz Junghuhn.* Nijhoff, 'S-Gravenhage, 1910.—Pioneer geologist and botanist in the Dutch East Indies.

KANE, Elisha Kent. 1820–1857. Elder, W. *Biography of Elisha Kent Kane.* Childs & Peterson, Philadelphia: Sheldon, Blakeman & Co., New York, 1868.—The Arctic explorer who made some geological and glaciological observations. His own books on explorations are more interesting than the biography.

KARPINSKY, Alexander P. 1847–1936. Lichkov, S. L. *Karpinskiy i Sovremennost.* Akademii Nauk SSSR, Moscow, 1946.—Brief outline of Karpinksy's life and work (in Russian).

KEITH, Arthur. 1866–1955. *An autobiography.* Wells and Co., London; Philos. Library, New York, 1950. Revealing account of a great comparative anatomist and student of fossil man.

KING, Clarence, 1842–1901. Various authors. *Clarence King Memoirs (The Helmet of Mambrino.)* Putnam, New York, 1904. (Publ. for King Memorial Comm. of Century Ass'n.)—A series of fulsome articles about this complex person who inspired many friendships. King was first director of U. S. Geological Survey. He flowered early and enjoyed life.

KOWALEVSKY, Vladimir Onufrieyevich. 1842–1883. Davidaschvili, L. S. *Biography,* in Russian. Acad. Sci., Moscow and Leningrad, 1946.—Life and work of the Russian vertebrate paleontologist.

LAMARCK, J. B. P. A. de M. de. 1744–1829. Packard, A. A. *Lamarck, the Founder of Evolution, his Life and Work, with Translation of his writings on organic Evolution.* Longmans, Green, New York, 1901.—Apparently the first book devoted to the life and work of the celebrated zoologist, paleontologist, and evolutionary pioneer.

Roule, L. *Lamarck et l'Interpretation de la Nature.* E. Flammarion, 1927.—An important evolution of Lamarck's work.

LECONTE, Joseph. 1823–1901. Armes, W. D., ed. *Autobiography of Joseph LeConte.* Appleton, New York, 1903.—Very readable reminiscences of the student of Louis Agassiz and early professor of geology at the University of California.

LEHMANN, Johann Gottlob. 1660–1709. Freyburg, B. v. *Johann Gottlob Lehmann (1719–1767) Ein Arzt, Chemiker, Metallurg, Bergmann, Mineraloge und grundlegender Geologe.* Universitatsbund Erlangen, 1955.—Exhaustive biography, genealogy, and bibliography of a man of whom little detail has been known. Lehmann made important observations in several sciences and especially on kinds of mountains, classification of strata and origin of ores.

LESLEY, J. Peter. 1819–1903. Ames, M. L. *Peter and Susan Lesley.* 2 vols., Knickerbocker Press, New York, 1909.—Geologist turned minister turned geologist. Consultant, state geologist of Pennsylvania, expert cartographer. Fine example of old-fashioned, slow-paced, but gradually revealing life and letters.

LHWYD, Edward. 1660–1709. Gunther, R. T. *Life and Letters of Edward Lhwyd,* Early Science in Oxford, vol. 14. University Press, Oxford, 1945.—Second keeper of the Ashmolean Museum; acquaintance of all the scientific men of his time; author of first English book of fossils.

LINNAEUS, Carl. 1707–1778. Fee, A. L. A. *Vie de Linne.* Paris, 1832.

Stoever, D. H. *The life of Sir Charles Linnaeus.* Transl. by J. Trapp from the Swedish. London, 435 p., 1794.—An early study of the great systematist whose work is the starting point for modern systematics of fossil and living organisms.

Fries, T. M. *Linne: lefnadsteckning af Th.M. Fries.* Fahlcrantz & Co., Stockholm, 2 vols. 1903.—According to Hagberg, a complete picture of the external life of Linne. *Linnaeus: the story of his life* (transl. by B. D. Jackson). Whitherly, London, 1923.

Hagberg, K. H. *Carl Linne, "Le Roi des Fleurs."* (transl. from Swedish by Hammer and Metzger.) "Je Sers," Paris, 1944.—An attempt to reveal the real man behind the celebrated systematist.

———. *Carl Linnaeus* (English translation by Alain Blair), Johathan Cape, London, 1952.

Smith, J. E. *A selection of the correspondence of Linnaeus and other naturalists.* Longmans, Hurst, Rees, etc., London, 2 vols., 1821.—Important source material on Linnaeus.

Carl v. Linne als Natorforscher u. Arzt. ed. by Svensk. Akad. Wissen., Jena, 1909.

LITTLE, George. 1838–1924. *Memoirs of George Little.* Tuscaloosa, 1924.—Professor of geology at the University of Mississippi.

LOGAN, William Edmond. 1798–1875. Harrington, B. J. *Life of Sir William E. Logan.* John Wiley & Sons, London, 1883.—Good biography of a pioneer in Canadian geology; first director of Geological Survey of Canada.

LONGYEAR, Edmund J. 1864–1954. Longyear, E. J. edited by Grace N. Nute. *Mesabi Pioneer, Reminiscences of Edmund J. Longyear.* Minnesota Historical Society, St. Paul, 1951.—Pioneer in Iron Range exploration and development.

LYELL, Charles. 1797–1875. Lyell, Mrs., ed. *Life, Letters, and Journals of Sir Charles Lyell, Bart.* 2 vols., Murray, London, 1881.—Thoroughly readable selections from the letters and journals of the celebrated British geologist, scientific descendent of Hutton and expounder of uniformitarianism.

Bonney, T. G. *Charles Lyell and Modern Geology.* Macmillan, New York, 1895.—Based upon the "Life and Letters," but a more connected biography.

LYMAN, Benjamin Smith. 1835–1920. Gonpei, K. *Biography of Benjamin Smith Lyman.* Sanseido, Tokyo, 1937.—A short appreciation of Lyman's life and geological work by one of his many Japanese friends.

MCGEE, William John. 1853–1912. McGee, E. R. *Life of W. J. McGee.* Farley, Iowa. 1915.—McGee was a self-educated geologist, anthropologist, hydrologist and scientific organizer. He was not noted for modesty.

MACLURE, William. 1763–1840. Morton, S. G. *Memoir of William Maclure, Esq.* Academy of Natural Sciences of Philadelphia, printed by T. K. and P. G. Collins, 1841.—Brief (37 pp.) biography of the so-called "Father of American Geology." It is high time for a comprehensive biography of Maclure, altho his thoroughly Wernerian work was no better than that of Volney, Mitchill, and others.

MACOUN, John. 1831–1920. *Autobiography of John Macoun, M.A., Canadian explorer and naturalist.*—Field Nat. Club, Ottawa, 1922.

MANTELL, Gideon Algernon. 1790–1899. Curwen, E. C. *The Journal of Gideon Mantell, Surgeon and Geologist,* covering the years 1818–1852. Oxford Univ. Press. London, 1940.—The revealing journal of the discoverer of Iguanodon, especially interesting for its candid remarks on famous geologists of the early 1800's.

Spokes, S. *Gideon Algernon Mantell.* John Bale, Sons & Danielsson, Ltd., London, 1927.—Life and letters of the surgeon-geologist; includes long American correspondence.

MARCY, Oliver. 1820–1899. *In memoriam . . . Oliver Marcy, LL.D.* n.p., n.d.—Professor of geology at Northwestern University.

MARSH, Othniel Charles. 1831–1899. Schuchert, C. and C. M. LeVene. *O. C. Marsh, Pioneer in Paleontology.* Yale Univ. Press, New Haven, 1940.—An excellent companion volume to Osborn's "Cope." Marsh and Cope, the outstanding American vertebrate paleontologists, were bitter rivals.

MICHELL, John. 1724–1793. Geikie, A. *Memoir of John Michell.* Cambridge Univ. Press, 1918.—Michell was "Fellow of Queen's College in Cambridge, 1749, Woodwardian Professor of Geology in the University, 1762" who understood continuity of strata and folded structures.

MILLER, Hugh. 1802–1856. Bayne, P. *The Life and Letters of Hugh Miller.* 2 vols. Gould and Lincoln, Boston, 1871. (Also London edition 1871).—A long, prosy, dull, poorly-documented, and generally unsatisfactory biography of the author of the classic, *The Old Red Sandstone*—Scottish stonemason, poet, editor, geologist, paleontologist and celebrant of the Old Red fishes.

Bingham, W. *The Life and Writings of Hugh Miller, an oration . . . G. W. Wood,* New York, 1859.—Unimportant.

Brown, T. N. *The Life and Times of Hugh Miller.* Rudd and Carleton. New York, 1859.—A poor biography.

Leask, W. K. *Hugh Miller.* Oliphant, Anderson and Ferrier, Edinburgh, 1896.—Excellent brief biography.

Miller, H. *My Schools and Schoolmasters; or, the Story of My Education.* Thomas Constable and Co. Edinburgh. 1858. (other editions, including Boston, 1863).—The best account of the early life of Hugh Miller.

Watson, J. L. *The life of Hugh Miller.* 1880.
The centenary of Hugh Miller, being an account of the celebration held at Cromarty on **22nd** *August,* 1902. Univ. Press, Glasgow, 1902.

MITCHELL, Elisha. 1793–1857. Phillips, C. *A Memoir of the Rev. Elisha Mitchell, D.D., Late Professor of Chemistry, Mineralogy & Geology in the University of North Carolina.* Chapel Hill, 1858.—Mitchell was all-round naturalist; wrote textbooks in geology and died exploring present Mt. Mitchell.

MITCHILL, Samuel Latham. 1764–1831. Hall, C. R. *A Scientist in the Early Republic, Samuel Latham Mitchill,* Columbia Univ. Press. New York, 1934.—A fair study of one of the very important early "all-round" scientists of America, friend and helper of many others; the first real contributor to the geology of New York State.

MURCHISON, Roderick Impey. 1792–1871. Geikie, A. *Life of Sir Roderick I. Murchison.* 2 vols., John Murray, London, 1875.—Well-written, like all of Geikie's works. Murchison was one of the great figures of the formative years of geology and elaborator of the Silurian System.

NOPCSA, Franz Baron. 1877–1933. Tasnadi, A. K. *Granz Baron Nopcsa.* Ungarn. Naturwiss. Mus., Budapest, 1945.—Life, letters, and scientific work of the famous Hungarian student of fossil reptiles.

ORTON, Edward. 1828–1899. (various). *In Memoriam, Edward Orton, Ph.d., LL.D., Addresses delivered at the Ohio State University, Sunday, November 26, 1899.* Printed by the University, Columbus, 1899 (?).—Includes among other essays a biographical sketch (anon.) and essay on Orton as geologist by G. K. Gilbert. Orton was teacher, administrator, geologist and gentleman.

OWEN, David Dale. 1807–1860. Hendrickson, W. B. *David Dale Owen, Pioneer Geologist of the Middle West.* Indiana Historical Collections, vol. 27, Indiana Historical Bureau, Indianapolis, 1943.—Owen was one of the most tireless and able of the early American geologists.

OWEN, Richard. 1804–1892. Owen, R. *The Life of Richard Owen.* 2 vols., Murray, London, 1894.—The letters and journals of the great English 19th Century anatomist and vertebrate paleontologist, celebrated for his deductions from fragmentary fossil bones.

PALISSY, Bernard. 1510–1590. Audiat, L. *Bernard Palissy.* Fontanier, Saintes, 1864.—A local memorial to their famous fellow townsman.
La Rocque, A. *The Admirable Discourses of Bernard Palissy.* Univ. of Illinois Press, Urbana, 1957.—A critical translation of Palissy's great work of 1580, with extensive notes and introductory essay dealing with Palissy's life and scientific contributions.
Morley, H. *Palissy the Potter, the life of Bernard Palissy of Saintes.* Chapman & Hall. London. 2 vols. 1852: Ticknor, Boston, 1853. (Other eds., 1855, 1865, 1878.—Imaginative biography of the famous maker of "rustic figulines" and important forerunner of inductive science who had modern notions of origin of salt, of ground water and artesian wells and partly modern ideas of origin of fossils. Morley includes long passages from Palissy's works, varying from free translations to paraphrases.

PENGELLY, William. 1812–1894. Pengelly, H. *A Memoir of William Pengelly of Torquay, F. R. S., Geologist.* Murray, London, 1897.—Schoolmaster and keen "amateur" geologist.

PENROSE, Richard Alexander Fullerton, Jr. 1863–1931. Fairbanks, H. R. and Berkey, C. P. *Life and Letters of R. A. F. Penrose Jr.* Geological Society of America, New York, 1952.—Comprehensive biography of a man who not only made princely bequests to the Geological Society of America and American Philosophical Society, but was also a great scholar, teacher, scientific and business organizer, traveler, and positive person.

PERCIVAL, James Gates. 1795–1856. Ward, J. H. *Life and Letters of James Gates Percival.* Ticknor and Fields, Boston, 1866.—The complex history of a complex character; physician, poet, lexicographer, linguist, first co-state geologist of Connecticut and second state geologist of Wisconsin.
Legler, H. E. *James Gates Percival, an anecdotal sketch and bibliography.* Milwaukee, 2 vols., 1901.

PLAYFAIR, John. 1748–1819. Playfair, J. G. Biographical account of the Late Professor Playfair in vol. 1, pp. xi-xxvi, of *Collected works of John Playfair, Esq. with a Memoir of the Author.* Edinburgh. Arnold Constable and Co. 1822. 4 vols.—Playfair was professor of mathematics and later of natural philosophy at Edinburgh; friend of James Hutton and author of the classic "Illustrations of the Huttonian Theory of the Earth," 1802.

PLAYFAIR, Lyon. 1818–1898. Reid, W. *Memoirs and Correspondence of Lyon Playfair, first Lord Playfair of St. Andrews.* Harper and Brothers, New York and London, 1899.—Early organic chemist (studied under Liebig) who was chemist to Geological Survey, knew royalty and all the great and was politically rewarded. Distant connection of John Playfair.

POWELL, John Wesley, 1834–1902. Darrah, W. C. *Powell of the Colorado.* Princeton Univ. Press. Princeton, N. J. 1951.—Excellent and completely documented biography with geological emphasis.

Gilbert, G. K., ed. *John Wesley Powell; a Memorial to an American Explorer and Scholar.* Chicago. 1903. (Reprinted, with slight changes, from *The Open Court*, vols. 16, 17.)—Powell was one of the founders of the United States Geological Survey, famous for his geological explorations in the West.

Stegner, W. E. *Beyond the Hundredth Meridian; John Wesley Powell and the Second Opening of the West.* Houghton Mifflin. Boston, 1954.—Well documented biography stressing Powell's arid land studies, and his relation to history of the times.

PRESTWICH, Joseph. 1812–1896. Prestwich, Mrs. Joseph. *Life and Letters of Sir Joseph Prestwich.* William Blackwood and Sons, London and Edinburgh, 1899.—Curiously reticent biography of geologist who was first a business man and late in life was professor at Oxford.

PUMPELLY, Raphael. 1837–1923. Pumpelly, R. *My Reminiscences.* 2 vols. Henry Holt, New York, 1918.—One of the most readable of scientific autobiographies, well-known for its real adventure and anecdotes of this pioneer economic geologist.

RAFINESQUE, Constantine Samuel. 1783–1840. Rafinesque, C. S. *A Life of Travels.* Chronica Botanica, vol. 8, 1944. (1st ed., Phila., 1836).—Fascinating self-portrait of one of the eccentrics of natural history, prodigious laborer among the hosts of new animals, plants, and fossils of America.

Call, R. E. *The Life and Writings of Rafinesque.* John P. Morton and Company Printers to the Filson Club, Louisville, 1895.—The standard biography of Rafinesque.

Fitzpatrick, T. J. *Rafinesque, a sketch of his Life, with Bibliography.* Historical Department of Iowa, Des Moines, 1911.—Brief but excellent biography and bibliographic description of 941 Rafinesque publications and 134 items referring to him.

RAMSAY, Andrew Crombie. 1814–1891. Geike, A. *Memoir of Sir Andrew Crombie Ramsay.* Macmillan, London, 1895.—Excellent for its summaries of work of the period and portraits of geologists of the time.

RAO, B. Rama. 1893–. *Recollections of an Indian geologist.* Mysore Geologists' Assoc., Bangalore, 1953.—One of India's distinguished geologists, Director of the Mysore Geological Department and later Director of the Indian Bureau of Mines.

RASPE, Rudolph Erich. 1737–1794. Carswell, J. *The Prospector, being the Life and Times of Rudolph Erich Raspe.* Cresset Press, London, 1950.—Best known as author of "Baron Munchausen's Travels," Raspe was scholar, antiquarian, scientist—one of the first to recognize igneous origin of basalt in 1769—author, mining manager, expatriate, expelled Fellow of the Royal Society and petty scoundrel.

RAY, John. 1628–1705. Derham, W. *Select Remains of the Learned John Ray. M. A.* and *F. R. S., with his Life.* George Scott, London, 1740.—Summary of Ray's life with some additional material. Ray is known primarily as a botanist, but his geological observations and writings—not sufficiently recognized by historians of geology—were extensive, penetrating and essentially modern.

Lankester, E., ed. *Memorials of John Ray.* Ray Society, London, 1846.—Reprinting of Derham's biography with bibliography, and notices by J. E. Smith, Cuvier, and other materials.—*Correspondence of John Ray.* Ray Society, London, 1848.—Interesting letters of Ray, some originally published by Derham, in Latin and English. Explanatory notes would have increased their value.

Raven, C. E. *John Ray Naturalist, his Life and Works.* University Press, Cambridge, 1942. (second ed. 1950).—The definitive biography and summary of Ray's scientific contributions. A great scientific biography.

RAYMOND, Rossiter Worthington. 1840–1918. Rickard, T. A., ed. *Rossiter Worthington Raymond, a Memorial.* Am. Inst. Min. Engr., New York, 1920.—Trained at Frieberg as were Pumpelly and other Americans, Raymond was mining engineer, metallurgist, lawyer, churchman, author, editor, and organizer and long-time guiding officer of the American Institute of Mining Engineers. His work touched economic geology and geologists at many points.

Various. *Dinner Given to Rossiter Worthington Raymond by his Friends in Commemoration of his seventieth Birthday, April 27, 1840. Saturday April the thirtieth, one thousand nine hundred and ten, The Plaza, New York.* 1910.—Sumptuous report of laudatory speeches. (Copies may have guest's name imprinted on leather cover.)

RICKARD, Thomas Arthur. 1864–. *Retrospect, an autobiography.* Whittlesey House, New York, 1937.—Very readable recollections of an eminent mining geologist.

ROGERS, William Barton. 1804–1882. Rogers, Mrs. W. B., assisted by W. T. Sedgwick. *Life and Letters of W. B. Rogers.* 2 vols., Houghton, Mifflin and Co. Boston and New York, 1896.—Teacher of chemistry, natural philosophy and geology; state geologist of Virginia, noted for early work (with his brother H. D.) on Appalachian structures. Founder and first president of Massachusetts Institute of Technology.

SAUSSURE, Horace Benedict de. 1740–1799. Freshfield, D. W. *The Life of Horace Benedict de Saussure.* Edward Arnold, London, 1920.—Explorer, Alpinist, and great Alpine geologist and writer on the Alps. Good biography, well illustrated.

Sennebier, Jean. *Memoire historique sur le Vie et les Escrits de Horace Benedict De Saussure.* J. J. Paschoud, Geneve, 1800.—Brief biography by a contemporary.

SAY, Thomas. 1787–1834. Weiss, H. B., and G. M. Ziegler. *Thomas Say, early American Naturalist.* Thomas, Springfield, Ill., 1931.—Say, who thought little of anything but his invertebrates, was a wistful and gentle figure in the roaring times when the midwest was the frontier.

SCOTT, William Berryman. 1858–1947. Scott, W. B. Some Memories of a Paleontologist. Princeton Univ. Press, 1939.—Highly interesting, and in places equally diverting, recollections of one of our great vertebrate paleontologists.

SEDGWICK, Adam. 1785–1873. Clarke, J. W. and T. McK. Hughes. *The Life and Letters of the Reverend Adam Sedgwick.* Cambridge Univ. Press. 2 vols., 1890.—Exhaustive record of one of the great figures ("the first of Men") in British Paleozoic geology.

SHALER, Nathaniel Southgate. 1841–1906. Shaler, N. S. *The Autobiography of Nathaniel Southgate Shaler, with a supplementary Memoir by his Wife.* Houghton Mifflin, Boston and New York, 1909.—Student of Agassiz; long-time popular professor at Harvard; interesting commentary on people and times.

SILLIMAN, Benjamin. 1779–1864. Fisher, G. P. *Life Of Benjamin Silliman.* 2 vols., Scribner's, New York, 1866.—Life of the chemist, mineralogist, and founder of the American Journal of Science ("Silliman's Journal"), valuable for letters from other scientists of the early decades of the 19th Century. "Lacks perspective."

Fulton, J. F. and Thomson, E. H. *Benjamin Silliman 1779–1864 Pathfinder in American Science.* Henry Schuman, New York, 1947.—Biography in modern manner. Good references.

SMITH, William. 1769–1839. Phillips J. *Memoirs of William Smith. LL.D.* John Murray, London, 1844.—Biography of the "Father of English Stratigraphy," author of the classic map, by "his nephew and pupil."

Sheppard, T. *William Smith: his Maps and Memoirs.* A. Brown and Sons, Ltd., Hull, 1920. (Reprinted from Proc. Yorks. Geol. Soc. vol. 19, pt. 3).—Contains summary of pre-Smith geological maps and elaborate descriptions of Smith's maps and publications, and later references to them.

Cox, L. R. *New Light on William Smith and his Work.* Proc. Yorks. Geol. Soc., vol. 25.—Many interesting new details of Smith's life and work derived from study of a long neglected packing case in the Oxford University museum.

STENSEN, Niels (Nicholas Steno). 1631–1686. Plovgaard, K. *Niels Stensen. Anatom, Geolog og Biskop.* Pauluskreden, Copenhagen, 1953.—The first biography of Steno appeared in 1686. Plovgaard's is but the latest of many subsequent studies of one of the great pioneers of geology, but tells little of his geological work (for which see Sherz, 1956).

Scherz, G. *Vom Wege Niels Stensen.* Munkesgaard, Copenhagen, 1956. (Acta Hist. Sci. Nat. Mee., vol. 14).—The latest appraisal of Steno's scientific work, including a German translation of Steno's MS catalogue of the geological collection in the Pitti Palace, Florence.

STERNBERG, Charles Hazelius. 1850–1943. Sternberg, C. H. *The Life of a Fossil Hunter.* Henry Holt. N. Y., 1909.—Sternberg was the great professional fossil collector in the western United States.

Sternberg, C. H. *Hunting dinosaurs in the bad lands of the Red Deer River, Alberta, Canada.* San Diego, 261 p., 1932.—Further reminiscences.

STOPPANI, Antonio. 1824–1891. Tomaselli, R. *Stoppani.* "La Scuola," Editrice, Brescia (Italy), 1948.—Short study of a famous Italian geologist and paleontologist.

STOW, George B. William. 1822–1882. Young, R. B. *Life of G. B. Stow, South African Geologist and Entomologist,* Longmans, Green, London, 1908.—Stow was another Robert Dick of South Africa.

STRICKLAND, Hugh Edwin. 1811–1853. Jardine, W. *Memoirs of Hugh Edwin Strickland, M.A.; selection from the scientific writings.* Van Voorst, London, 1858.—Geologist, paleontologist, and ornithologist, oriental traveller, and leader in early attempts to establish rules of zoological nomenclature, he was tragically killed while studying a section in a railroad cut.

SUESS, Eduard. 1831–1914. Suess, E. *Eduard Suess: Errinerungen.* Hirzel, Leipzig, 1916.—Suess' *Das Antlitz der Erde* stands as one of the greatest syntheses of geological science.

SWEDENBORG, Emmanuel. 1688–1772. Nathorst, A. G. *Emmanuel Swedenborg as a Geologist.* Vol. 1, Sect. 1 of *Emmanuel Swedenborg as a Scientist.* Afterbolodets Tryckeri, *Stockholm,* 1908.—Swedenborg's fame as philosopher and theologian obscures his extensive and prescient work in geology, paleontology, mineralogy and metallurgy. Excellent brief study with references.

SYMONDS, William Samuel. 1818–1887. La Touche, J. D. *William S. Symonds, Rector of Pendock. Sketch of his life.* Gloucester, n.d.—Another of the devoted band of amateur geologists who flourished in England in the 19th Century, friend of Murchison and Lyell, and student of the geology of the Malvern Hills.

TENISON-WOODS, Julian Edmund. 1832–1889. O'Neill, G. *Life of the Reverend Julian Edmund Tenison-Woods.* Pellegrini, Sydney, Melbourne, & Brisbane, 416 p., 1929.—Dedicated and indefatigable Jesuit missionary-priest of Australia, who first made known many of the curious Tertiary fossil corals of that land.

TYRRELL, Joseph Burr. 1858–1957. Loudon, W. J. *A Canadian Geologist.* Macmillan Company of Canada, Toronto, 1930.—Tyrrell was survey geologist, Arctic explorer, Klondike mine manager and famous consulting mining geologist. His life spans much of development of Canadian geology.

Tyrrell, J. W. *Across the Subarctics of Canada, A Journey of* 3,200 *Miles by Canoe and Snowshoe Through the Barren Lands.* T. Fisher Unwin, London n.d. (ca. 1900?).—Matter-of-fact report of an exciting geological journey by J. B. Tyrrell and his brother.

WARD, Henry Augustus. 1834–1906. Ward, R. *Henry A. Ward, Museum Builder To America.* Rochester Historical Society, Rochester, 1948.—A grandson expertly writes the story of the great collector who founded Ward's Natural Science Establishment.

WARD, Lester Frank. 1841–1913. Chugarman, S. *Lester Ward, the American Aristotle.* Duke Univ. Press, Durham, N. C., 1939.—This pretentious book on Ward's sociological diversions barely mentions this self-made, indomitable and indefatigable polyhistor's extensive work in geology and paleobotany.

Cape, E. P. *Lester F. Ward, A Personal Sketch.* Putnam's, New York, 1922.—Completely ignores Ward's work in paleobotany, useless from a scientific standpoint, hardly readable otherwise.

WERNER, Abraham Gottlob. 1749–1817. Beck, C. R. *Abraham Gottlob Werner, eine kritische Wurdigung des Begrunders der modernen Geologie, su seinem hundertjahrigen Todestage.*—Werner was the great neptunist. A great teacher, classifier and mineralogist, his attempts to explain all geology on the basis of a small area in Germany actually retarded geological advances for many years.

Frisch, S. G. *Lebensbeschreibung Abraham Gottlob Werner.* Leipzig, 1825.

WESTON, Thomas C. 1832–1910. Weston, T. C. *Reminiscences Among The Rocks in Connection With the Geological Survey of Canada.* For the author by Warwick Bros. and Butler, Toronto, 1898.—Autobiography of a Canadian geologist who was not always in agreement with his more famous contemporaries. Iconoclastic.

WHEWELL, William. 1794–1866. Douglas, Mrs. Stair. *The Life and Selections from the correspondence of William Whewell, D. D., late Master of Trinity College, Cambridge.* C. Kegan Paul & Co., London, 1881.—Professor of mineralogy at Cambridge; active in geology, mathematics and philosophy; academic administrator who knew all the scientific greats of the time. Carefully selected correspondence and a biography which tells far too little of Whewell's scientific work.

WILLIAMS, George Huntingdon. 1856–1894. Williams, T. *Biography of George H. Williams.* New York, 150 p., 1896.—Professor of geology at Johns Hopkins, mineralogist, petrographer, and student of metamorphic geology.

WRIGHT, George Frederick. 1838–1921. Wright, G. F. *Story of my Life and Work.* Bibliotheca Sacra, Oberlin, Ohio. 1916.—Not very geologically revealing autobiography of a clergyman and professor of theology at Oberlin who worked extensively and independently in glacial geology, but not always in agreement with all others.

WHITNEY, Josiah Dwight. 1819–1896. Brewster, E. T. *Life and Letters of Josiah Dwight Whitney.* Houghton Mfflin, New York, 1909.—Whitney was primarily an economic geologist known for his work in New England, the Lake Superior region, Iowa and California. Excellent biography relating the man to his times and fellow geologists.

HISTORY OF GLACIAL GEOLOGY

EARLY DESCRIPTION AND EXPLANATION OF KETTLE HOLES

CHARLES WHITTLESEY

(1808-1886)

Reprinted from *Journal of Glaciology*, Vol. 5, No. 37, February 1964

EARLY DISCOVERERS
XIX
EARLY DESCRIPTION AND EXPLANATION OF KETTLE HOLES
CHARLES WHITTLESEY (1808-1886)

By George W. White
(Department of Geology, University of Illinois, Urbana, Illinois, U.S.A.)

Charles Whittlesey was a graduate of West Point, an army officer, engineer, geologist, lawyer, historian, archeologist, member of state and national geological surveys and a voluminous writer on a variety of subjects. During his own lifetime he felt that his geological talents and contributions were not sufficiently recognized—an opinion in which he was probably correct. The purpose of the present note is to call attention to his descriptions of kettle holes and his analysis of their origin, which appear to be the earliest in America.

Whittlesey was one of the first in North America to perceive the difference between ice-laid unsorted, unstratified material, stratified outwash and lacustrine material and to recognize that all of these originated from ice action. As early as 1838, Whittlesey (1838, pl., p. 56) had noticed several stratigraphic units of the drift near Cleveland, Ohio, and understood that the drift covering the bedrock in northeastern Ohio varied markedly in thickness. All of Whittlesey's many reports after 1838 stress stratigraphic division of the drift, a general recognition of which came many decades later. He even noted that the gravel and sand "diluvial hills" (kames) in northeastern Ohio have "an external coating of clay to the depth of 10, 20, or 30 feet [3–9 m.]" (1838, p. 56), an observation that was lost sight of for over 100 years (White, 1953, p. 36).

In an early description of drift in Wisconsin, Whittlesey (1852) described in detail stratigraphic sections, with diagrams, of interbedded boulder clay, sand and gravel, and lacustrine clay. In his sections of stratigraphy and structure he showed the relation of the irregular and discontinuous overlying drift to the bedrock.

Whittlesey's first speculations about the origin of kettle holes were in a report before the American Association for the Advancement of Science in 1859 (Whittlesey, 1860), in which he stated that in Wisconsin "there are numberless crater-like depressions in the drift material which are called by the people 'potash kettles'". He traced the area of kettles for 150 miles (240 km.) and thus early recognized the "kettle moraine" of Wisconsin. He had noticed these kettle holes ten years before—"while exploring in 1849 it occurred to me that these cavities could not be explained by the usual and well-known examples of aqueous deposits". His description and discussion (Whittlesey, 1860) is as follows:

> "Along the summits or dividing ridge between the waters of Rock river and those of Lake Michigan, there are numberless crater-like depressions in the drift materials, which are called by the people 'potash kettles'.
>
> "They are in the form of cavities, sunk below the general surface, ten, fifteen, and even one hundred feet, their outline rudely circular, and their sides as steep as the earth will stand. . . .
>
> "The materials in which they are found are the coarse drift, such as everywhere in the west occupies the summit of the country. . . .
>
> "To get an idea of the appearance of the 'potash kettle' country, imagine a region of drift moraines inverted. Instead of a surface thickly set with rounded hillocks, suppose it to be occupied by cavities of irregular size and depth.

"If the grinder of a mastodon is reversed and impressed upon a piece of wax, the depressions which result, will represent the drift cavities as contrasted with drift elevations. In travelling through such a region the explorer frequently finds them so near together, that he no sooner rises out of one than he is obliged immediately to descend into another, the diameter of which may not be more than twice or thrice its depth.

"There is very seldom any water in the bottom, owing to the loose and porous character of the gravel drift. Boulders are seen at the bottom, on the sides, and on the surface around them.

"When they are thickly set, as at the source of the Oconto river, and no hillocks between, the rim or edge between them is sometimes so narrow that large boulders have not base enough to rest upon, and tumble down the sides.

"The internal slope is frequently straight like a funnel or inverted cone, but more often cup-shaped or curved in a manner correctly expressed by the form of a kettle. . . .

"At the foot of the Alps, moraines are formed mechanically by the movements of glaciers, carrying forward earth and stones, that are finally left in rounded heaps on the more level country. Masses of ice become entangled with the loose materials which in due time melt away and disappear.

"Without entering at large into a discussion of the drift force, I assume for the present purpose that, in the early periods of the drift epoch, it was *glacier ice*. Nothing else seems to be equal in energy to the results we observe. A strong objection to this view has been removed by the observations of Dr. H. H. [I. I.] Hayes, of the Kane Arctic Expedition.

"On the north-west coast of Greenland, which is a vast glacier, the ice was found to be progressing toward the coast over a country that was comparatively level. It had a movement not only down inclined surfaces, such as the slopes of mountains, but along flat land, and even *up* inclinations that were opposed to its progress. If the temperature of Greenland or the Arctic Circle was brought down to latitude 40° north, glaciers would exist. Regarding the explanations of Agassiz and Desor as to the cause of the motion of glaciers to be correct, that it is not wholly due to gravity, there is nothing improbable in such a movement over a level country. . . .

"Admitting the probability of such a state of things, it follows that along the southern edge of this all-pervading glacier, fragments and masses of ice would be enclosed in and buried beneath the drift materials. . . .

"It is reported that in Patagonia huge piles of stones and ice are seen mingled together for years.

"My first impression on viewing the cavities at the head of the Oconto, was that they are due to *subsidence*. In the cases just cited, if the mixed mass consisted more of ice than of earth and stones, the surface should be one of pits and depressions. Hillocks or moraines could only occur in such materials where the earthy and imperishable parts are in excess. When the proportions are about equal, there would be both cavities and moraines.

"In the southerly part of Wisconsin both forms are combined, but as we proceed northerly the sunken species increase, and the hillocks and ozars diminish.

"As we proceed northerly there is in them less of stratification and a closer approach to the true glacial moraines. . . ."

Figures to illustrate the kettle holes did not accompany the 1860 report but were included in a report a few years later in which Whittlesey (1866) summarized his observations of over 25 years on the glacial drift of the Great Lakes states and included a map and many sections, some of them in color. His illustrations of kettle holes reproduced here as Figures 1 and 2, with their original captions, are from that report.

Whittlesey was one of the first observers in America to compare the deposits reported around the Greenland Ice Sheet with those of the United States. He recognized the entire role played by glacier ice at a time when the origin of the drift in America was still usually ascribed to icebergs and water. Whittlesey did not fall into the trap—which would have been readily excusable—of burying icebergs to make kettle holes. He unequivocally buried glacial ice. Some of his other observations and explanations of drift sequences, glacial boundaries, raised beaches, and engineering effects of different drift deposits are equally important and deserve to be included in a book-length biographical study.

Fig. 1.

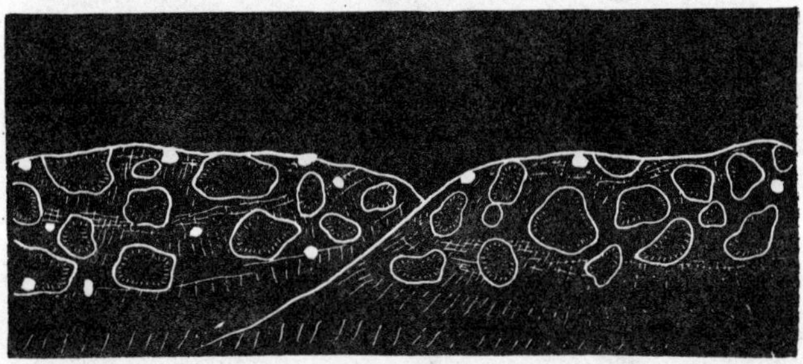

Drift Cavities, or "Potash Kettles," near Greenbush, Wisconsin. Range of drift hills looking west.
○ ○ ○ Boulders of Northern rocks—base 150 feet above Lake Michigan.

Fig. 2.

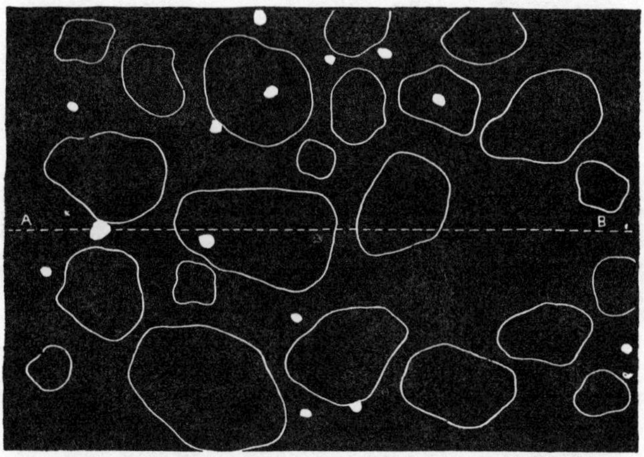

Horizontal Projection of drift cavities 15 to 60 feet deep, head waters of Oconto river, Wisconsin.
○ ○ ○ Large Boulders of Sienite—350 feet above Lake Michigan.

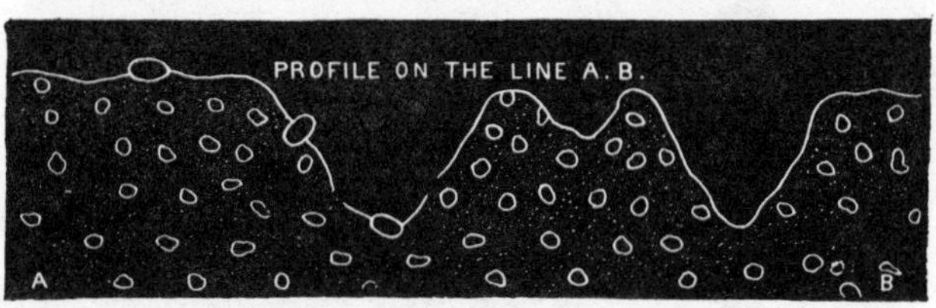

REFERENCES

White, G. W. 1953. Geology and water-bearing characteristics of the unconsolidated deposits of Cuyahoga County. (*In* Winslow, J. D., *and others*. The water resources of Cuyahoga County, Ohio, by J. D. Winslow, G. W. White and E. E. Webber. *Ohio. Department of Natural Resources. Division of Water. Bulletin* 26, p. 36–41.)

Whittlesey, C. 1838. Mr. Whittlesey's report. (*In* Mather, W. W. *Second annual report of the Geological Survey of the State of Ohio*. Columbus, Samuel Redary, p. 41–71.)

Whittlesey, C. 1852. Description of part of Wisconsin, south of Lake Superior. (*In* Owen, D. D. *Report of a geological survey of Wisconsin, Iowa, and Minnesota* . . . . Philadelphia, Grambo and Co., p. 425–73.)

Whittlesey, C. 1860. On the drift cavities, or "potash kettles" of Wisconsin. *Proceedings of the American Association for the Advancement of Science, thirteenth meeting 1859*, p. 297–301.

Whittlesey, C. 1866. On the fresh-water glacial drift of the north-western states. *Smithsonian Contributions to Knowledge*, No. 197, vii, 32 p.

[*It is known* (*Flint, R. F.* [*1957*.] Glacial and Pleistocene geology. *New York, John Wiley and Sons, Inc.; London, Chapman and Hall Ltd., p. 140, 149, 151, 152*) *that kettle holes may originate in one of several ways: by the partial burial of the stagnant remnants of a glacier, by the complete burial of the remnants of a glacier, or by the burial of "floated-in or dropped-in ice masses". The first method produces the largest and deepest kettle holes, whereas the "thinner, buried ice masses create shallow kettles". The prominent steep-sided kettle holes which caught Whittlesey's attention appear to belong to the first kind.* Ed.]

The First Appearance in Ohio of the Theory of Continental Glaciation

GEORGE W. WHITE

THE FIRST APPEARANCE IN OHIO OF THE THEORY OF CONTINENTAL GLACIATION[1]

GEORGE W. WHITE

Department of Geology, University of Illinois, Urbana 61801

ABSTRACT

Louis Agassiz published his "glacial theory" in *Etudes sur les glaciers* in October, 1840. Edward Hitchcock wrote approvingly of the theory in 1841 and reproduced some of the Agassiz plates. Professor Samuel St. John of Western Reserve College reproduced one of the Hitchcock glacier plates and wrote favorably of the theory and described glacial drift in general in the first geology textbook published in Ohio, in 1851 at Hudson. St. John's influence was particularly important in the career of his student, John Strong Newberry, the famous geologist. Newberry's interest in glacial deposits, and especially in the origin of kames, may be traced to St. John, his geology teacher.

The first statement in Ohio of the hypothesis that "drift" had been deposited by an ice sheet was presented by Samuel St. John in what is probably the first Ohio textbook on geology. It was published at Hudson, Ohio, in 1851 (fig. 1). This book and St. John's classroom discussions contributed to the early interest in glacial phenomena so well exhibited in Ohio and soon to be so extensively described in publications by J. S. Newberry and other geologists of the Second Geological Survey of Ohio. This early Ohio explanation of the glacial theory was not derived from Louis Agassiz directly, but through the works of Edward Hitchcock of Amherst College.

St. John's pleasant and important book appears to have been the first Ohio textbook dealing exclusively with geology. Hamilton L. Smith, A. M., wrote on geology in 1848, but his book, published in Cleveland and illustrated by J. Brainerd, was on both astronomy and geology. And, although Smith's book shows reasonable acquaintance with the English writers on geology, especially Lyell, it pays little attention to American geology and geologists and says nothing about glaciers. Wm. W. Mather, first state geologist of Ohio, published a little textbook of geology in 1833, but it was not published in Ohio and Mather had not come to Ohio at that time.

Glaciers had been described in more or less detail and their relation to certain deposits noted long before 1840, but it was the exposition of the theory of continental glaciation in the book and atlas of Louis Agassiz in 1840, *Etudes sur les glaciers*, that immediately caught the attention, both favorable and unfavorable, of geologists and other scientists. The development of this theory and the events leading up to the 1840 publication have been recently reported by Carozzi (1967). The impact of the theory in Great Britain has been described by North (1943). The development of the glacial hypothesis in America has been treated at length by Merrill (1924, 1964, p. 615–642), but he did not include treatment of the critical year 1841 and the reproduction of any of the Agassiz plates in America. It was the copying of these plates by Hitchcock that was so important in the evolution of the Ohio text.

On December 1, 1839, Edward Hitchcock, who had been appointed state geologist of Massachusetts in 1830, presented to the governor of that state the "Final Report" of the geological survey (Hitchcock, 1841a, p. iv, vii). The printing of the work progressed slowly and the report did not appear until after April 1, 1841. Hitchcock took advantage of the delay to ". . . discover many facts since it was first presented to the Government. These I have not hesitated

[1]Manuscript received March 25, 1966.

to incorporate into the work . . ." These "many facts" were presented in a "Postscript," which preceded the regular text. The "Postscript" is devoted mainly to "Glacio-aqueous Action between the Tertiary and Historic Period, denominated in my Report, Diluvial Action" (pp. 3a–11a).

The progress of American recognition of glacial deposits and their origin is well illustrated by the publications of Edward Hitchcock. As he had a keen eye for physiographic forms and was always cognizant of relation of surface form

ELEMENTS

OF

GEOLOGY,

INTENDED FOR THE USE OF STUDENTS.

BY

SAMUEL ST. JOHN,

PROFESSOR OF CHEMISTRY AND GEOLOGY IN WESTERN RESERVE COLLEGE.

Chimborazo.

"Geology, in the magnitude and sublimity of the objects of which it treats, undoubtedly ranks next to Astronomy in the scale of the Sciences."

Sir J. F. W. Herschel.

HUDSON, OHIO:

1851.

FIGURE 1. Title page of St. John's textbook.

to internal composition and structure, he wrote accurate and extensive descriptions of glacial features long before he realized their glacial origin. It is instructive to compare his early description and assumed flood origin of "Diluvial" features in his Massachusetts reports of 1832 (p. 6), 1833 (p. 141–172), 1835 (p. 148–178), and in the first edition of his textbook of 1840, with his sudden recognition in early 1841 that Agassiz' glacial theory "explains satisfactorily" so many hitherto puzzling features.

Very late in 1840, or very early in 1841, Hitchcock (1841a, p. 3a) had "been favoured, through the kindness of Professor Silliman of Yale College, with the perusal of a recent work by Professor Agassiz on Glaciers and Glacial action, entitled *Etudes sur les Glaciers*." He also had abstracts of three papers "read last autumn before the London Geological Society, by Agassiz, Buckland, and Lyell . . . The whole subject of diluvium has been made to assume an aspect so new and interesting, that I am unwilling my Report should go out of my hands unaccompanied by a brief review of the facts and inferences concerning it." He did this by means of ". . . an extract from an Address recently published, [given] before the Association of American Geologists at Philadelphia in April, 1841." It is the "Postscript", rather than the "Address" (Hitchcock, 1841b), which is important in the present connection, because the former included illustrations and was even more positive about glacial action than was the "Address."

After describing the appearance, advance, and retreat of glaciers, Hitchcock (1841a, p. 4) asserted that the theory "explains satisfactorily" the origin of deposits of gravel and boulders "which we meet with almost everywhere in the northern parts of our country." He now favored a glacial origin for "lateral moraines [kame terraces] . . . the terminal and medial moraines; . . . smoothing, polishing, and furrowing of the rocks; . . . transportation of boulders; . . . clay and sand above the drift;" and especially for "furrows" which pass over mountains without losing their parallelism.

In this "Postscript" are reproduced (at a little less than half original size) ". . . a few cuts, copied on a reduced scale from the splendid drawings accompanying the *Etudes sur les Glaciers* by Agassiz." He actually reproduced Agassiz' plates 9, 10, 12, and a part of Plate 18, of the total of 18 plates and 14 overlays making up the atlas accompanying Agassiz' text. Figure 2 is a reproduction of Hitchcock's Fig 277, a very faithful copy of Agassiz' Plate 10, "Glacier de Viesch." (It is interesting to note that these same illustrations, in even better impressions, are included in the 31st edition of Hitchcock's textbook (1860). The frontispiece of this edition of the text is of the "Humboldt Glacier, Greenland." A study of the 31 editions of that text, that appeared throughout a 20-year interval, would give much insight into the development and progress in America of the glacial theory).

Hitchcock's reports were widely disseminated and were well known to Samuel St. John, professor of chemistry, mineralogy, and geology in Western Reserve College, then located at Hudson, Ohio. Dr. St. John, who was born in New Canaan, Connecticut, on March 29, 1813, graduated from Yale College in 1834 as valedictorian. He studied law, chemistry, and medicine for five years more, partly in Paris and London, but mainly at New Haven with Benjamin Silliman. He may have known Hitchcock personally. St. John was admitted to the bar, but never practiced. He was elected professor of chemistry, mineralogy, and geology at Western Reserve in 1838, one year after this post had been established. For part of the following year, he was an assistant on the New York Geological Survey. From 1844 to 1856 he was also professor in the Medical Department in Cleveland (Waite, 1946, p. 94–97). He was prominent in the northern Ohio scientific community, and was the first corresponding secretary of the Cleveland Academy of Natural Science, which was founded in 1845 (Meisel, 1926, vol. 2, p. 564; 1929, vol. 3, p. 7). In 1856 St. John was elected professor in the College of Physicians and Surgeons in New York (later affiliated with Columbia), a post he occupied until his death in 1876. (I am indebted to Mrs. D. E. Helmuth, Archivist of Western Reserve University, for information about St. John and for a copy of the biographical reference in Waite).

St. John soon brought the new ideas about glaciers and glacial deposits to his students, among whom was John Strong Newberry, who lived in nearby Cuyahoga Falls and whose family owned mines in adjacent Tallmadge. It is recorded that St. John was the instructor who most impressed the young Newberry and who influenced him to take up a career in science (Stevenson, 1893). Newberry received the degrees of A.B. from Western Reserve in 1846 and M.D. from Cleve-

Fig. 277.

Glacier of Viesch in the Alps.

FIGURE 2. Hitchcock's Fig. 277, a copy of Agassiz' plate. Original size 95 x 139 mm.

land Medical School in 1848. After three years study of medicine, botany, and zoology in France, he returned to northern Ohio to practice medicine, but soon gave more and more of his attention to science, and especially to geology, and became one of the outstanding geologists of the nineteenth century. Newberry's writings, almost from the outset, showed a keen discernment for physiography, and especially for morphology and composition of glacial deposits. It was from Professor Samuel St. John that Newberry learned of the glacial theory, only two or three years after Agassiz had published his explanation in 1840 and Hitchcock had given it such an approving promulgation in 1841. No doubt professor and student saw together the morainic knolls and the till and gravel deposits in the Cuyahoga Falls-Tallmadge-Hudson region, and looked at the striations on the ledges of Sharon Conglomerate in Boston, just west of Hudson, and in Twinsburg to the north (Newberry, 1873, p. 206; Leverett, 1902, pl. 13 and p. 422–425; White, 1953, pl. 6).

Fig. 12.

Glacier of Viesch, with medial and lateral moraines.

FIGURE 3. St. John's Fig. 12, a copy engraved by J. Brainerd from Hitchcock's figure. Original size 75 x 103 mm.

St. John must have been engaged for some time, probably several years, in the writing of his book (fig. 1). It therefore provides a summary of what he had been teaching for the several years before 1851. The material in his book was in part derived from ". . . Lyell, Murchison, Buckland, Ansted, Agassiz, Hitchcock, Dana, and the State Geological Surveys, particularly that of Professor Hall of New York." The book contains 175 figures and a final folding plate of a "jaw of a fossil fish found at Delaware, Ohio." These were ". . . executed by Professor Brainerd, and in some instances subjects were sketched from nature by him." Although it is not the present purpose to analyze St. John's whole book, it may be noted that its organization is quite similar to any modern general elementary text that includes both physical and historical geology. The stratigraphy is modern in terminology. A long chapter on "Practical Geology" includes material on what today is called "environmental geology." The brief chapter on "History of Geology" is an early American summary of the subject. The concluding chapter is on regional geology of the world.

St. John described and discussed the work of glaciers (p. 25–28) and included illustrations of a glacier, striations, and an iceberg. The "Glacier of Viesch with medial and lateral moraines," shown here as figure 3, is a three-quarter-size copy of Hitchcock's copy (fig. 2) of Agassiz' original plate 10. A comparison of figure 3 with figure 2 will show how faithfully the engraver Brainerd copied Hitchcock's cut. Agassiz' original text and atlas are now rare, and possessed by only a few fortunate libraries, but excellent photographic reproductions of the plates are now available in the new critical translation of the *Etudes* (Carozzi, 1967).

St. John's figure 13, "Striae of Glaciers," was probably taken from an actual Ohio specimen by Brainerd, as it does not correspond to any of the striated rocks illustrated by Hitchcock.

The application of the glacial theory to the explanation of drift is both implicit and explicit in the long Chapter IX (pp. 223–251), "Rocks of the Quaternary Period." This chapter is divided into "The Drift Period" (pp. 223–240) and "The Alluvium" (pp. 240–251). The running head for the "Drift Period" is "The Drift or Pleistocene Period," surely a very early use of this time term in America. In this section the author discussed the limit of drift; direction of drift movement; large boulders; grooves and striations; *roches moutonnees;* "altered drift" (stratified drift); marine terraces, with an illustration of the "Parallel Roads and River Terraces of Glen Roy"; caves with fossils; and large vertebrate fossils, especially mastodons and particularly those from Big Bone Lick.

"Theories of the Drift" are discussed on pages 253 to 256. The "iceberg theory" was examined in a closely reasoned argument and found insufficient. The "elevation theory," so popular at the time (and later), was likewise found wanting. His statement of the "glacier theory" is as follows (pp. 255–256):

> 322. The *glacier theory* supposes that the climate, which in the Tertiary period had been so warm as to allow the palms to grow within the temperate zones, became much colder, causing enormous sheets of ice—polar glaciers—to advance far beyond their previous limits, moving along the surface by alternate advance and retreat, rounding, polishing and striating the rocks, and afterward when melted depositing their loads of boulders and detritus, where the drift is now found. In Europe the center of expansion is supposed to have been the Scandinavian mountains, and in North America in the polar regions, from which the glaciers advanced southerly.
>
> The advocates of this theory contend that the phenomena of glaciers as witnessed in the Alps (§ 32,) are perfect miniature representations of the drift—its striae, furrows, boulders and moraines; that the elevation of extensive regions in high latitudes, like those of the Cordilleras in Mexico, and the high plains of Central Asia, would produce such a reduction of temperature as to cause immense glaciers, even thousands of feet in thickness. This theory is advocated by Prof. Agassiz.
>
> The principal objection to the glacial theory is that glaciers are at present entirely confined to valleys, and the origin of such an enormous sheet of ice as it contemplates is altogether hypothetical.
>
> 323. Neither of these theories is deemed quite satisfactory; the proximate cause of the phenomena is very generally supposed to have been the joint action of ice and currents of water, but their origin and exact modes of operation are not determined.

In 1851 little was known of the Greenland ice cap and practically nothing of the Antarctic ice. Soon afterward, however, information about these ice sheets became known to geologists (Whittlesey, 1860), and thus the glacial theory with the "enormous sheet of ice as it contemplates" became less and less hypothetical.

The relation of ice-laid drift (till) to water-laid drift (outwash) was a puzzle to the early proponents of the glacial theory and they assumed that nonglacial water action was required to explain the stratified deposits. Agassiz himself understood the marine and lacustrine origin of some stratified drift deposits, but did not understand the intimate relation of till and fluvioglacial material in some deposits, especially those of kames. He never understood the role of glacial meltwater (Agassiz, 1850, p. 410) and his keenly observing one-time associate, Edward Desor, similarly required lakes or seas to explain outwash deposits (Desor, *In* Foster and Whitney, 1850, p. 217). Indeed it was sometime after 1874 that the precise role of meltwater in forming stratified drift became generally understood. It is eminently fitting that a student of Samuel St. John, John Strong Newberry (1874, p. 41–54), contributed greatly to the understanding of the puzzle. His excellent descriptions of Ohio kames and his illustrations of their forms and internal structure contributed the data that were required for final understanding. Although Newberry himself, in 1874 (p. 47), still believed that the Stark County kames required standing water and icebergs for their formation, he put kame deposition closer and closer to an actual glacier front. At any rate, he precisely recorded the data that would soon lead to an ice-front and meltwater origin for kames, and for the meltwater-stream origin of valley trains.

Samuel St. John enters the history of glacial geology in 1851. He plays an important part through the first book published west of the mountains to record a clear statement of the theory of continental glaciation. He published nothing else on glacial geology. No other geological books or regular journal articles of his are recorded. His single book must have influenced many students and his influence passed on through the prestigious Newberry was great. As yet we do not know the extent of his influence on others, especially on the "Cleveland group" of Charles Whittlesey (White, 1964) and others, but it almost certainly was significant with them also. Professor Samuel St. John has a clear claim to the title of the first expositor of the theory of continental glaciation in Ohio.

REFERENCES

Agassiz, Louis. 1840. Etudes sur les glaciers; ouvrage accompagne d'un atlas de 32 planches. Aux frais l'auteur. Neuchatel. 347 p. and atlas of 32 plates.

———. 1850. Lake Superior: its physical character, vegetation, and animals. Gould, Kendall and Lincoln. Boston. 428 p.

Carozzi, A. V. (translator and editor). 1967. Agassiz' studies on glaciers. Hafner Publishing Co., New York.

Foster, J. W. and **J. D. Whitney.** 1850. Report on the geology and topography of a portion of the Lake Superior Land District, in the State of Michigan. Part 1, Copper lands. H. R. Ex. Doc. No. 69, Washington, 224 p.

Hitchcock, Edward. 1832. Report on the geology of Massachusetts, examined under the direction of the Government of that State during the years 1830 and 1831. Am. Jour. Science **22**: 1–70.

———. 1833. Report on the geology, mineralogy, botany, and zoology of Massachusetts. J. S. and C. Adams, Amherst. 700 p.

———. 1835. Report on the geology of Massachusetts. Second edition revised and enlarged. J. S. and C. Adams, Amherst. 702 p.

———. 1840. Elementary geology. J. S. and C. Adams, Amherst. 329 p.

———. 1841a. Final report on the geology of Massachusetts. J. H. Butler, Northampton. 831 p.

———. 1841b. First anniversary address before the Association of American Geologists at their Second Annual Meeting in Philadelphia, April 5, 1841. B. A. Hamlen, New Haven. (Also Am. Jour. Science 41: 232–275).

——— and **C. H. Hitchcock.** 1860. Elementary geology. 31st Edition. Ivison, Phinney and Co., New York. 430 p.

Leverett, Frank. 1902. Glacial formations and drainage features of the Erie and Ohio basins. U. S. Geol. Survey Monograph 41. 802 p.

Mather, W. W. 1833. Elements of geology, for the use of schools. William Lester, Jr., Norwich. 139 p.

Meisel, Max. 1924, 1926, 1929. A bibliography of American natural history. 3 vols. Premier Publishing Co., Brooklyn. 244, 741, 749 p.

Merrill, G. P. 1924. The first one hundred years of American geology. Yale University Press, New Haven. 773 p. (Also 1964 facsimile, New York, Hafner).

Newberry, J. S. 1873. Report on the geology of Summit County. Ohio Geol. Survey, Vol. 1, Part 1. p. 201–222.

———. 1874. Surface geology of Ohio. Ohio Geol. Survey Vol. 2, Part 1. p. 1–80.

North, F. J. 1943. Centenary of the glacial theory. Proc. Geol. Assoc. 54 Part 1: 1–28.

St. John, Samuel. 1851. Elements of geology, intended for the use of students. Pentagon Press, Hudson, Ohio. 334 p.

Smith, H. L. 1848. The world: or first lessons in astronomy and geology, in connection with the present and past condition of our globe. M. C. Younglove and Company, Cleveland. 324 p.

Stevenson, J. J. 1893. John Strong Newberry. Am. Geol. 12: 1–26.

Waite, F. C. 1946. Centennial history of the school of medicine. Western Reserve University Press, Cleveland.

White, G. W. 1953. Geology and water-bearing properties of the unconsolidated deposits; p. 18–27 *in* Smith, R. C. The ground-water resources of Summit County, Ohio. Ohio Div. Water Bull. 27. 130 p.

———. 1963. Glacial geology, p. 118–156 *In* DeLong, R. M. and G. W. White, Geology of Stark County. Ohio Div. Geol. Survey Bull. 61. 209 p.

———. 1964. Early description and explanation of kettle holes: Charles Whittlesey (1808–1886). Jour. Glaciology 5: 119–122.

Whittlesey, Charles. 1860. On the drift cavities, or "potash kettles" of Wisconsin. Proc. Am. Assoc. Adv. Science, thirteenth meeting 1859. p. 297–301.

ANNOUNCEMENT OF GLACIATION IN SCOTLAND

WILLIAM BUCKLAND

(1784-1856)

Journal of Glaciology, Vol. 9, No. 55, 1970

EARLY DISCOVERERS
XXVII

ANNOUNCEMENT OF GLACIATION IN SCOTLAND
WILLIAM BUCKLAND (1784–1856)

By GEORGE W. WHITE

(Department of Geology, University of Illinois, Urbana, Illinois 61801, U.S.A.)

BEFORE 1840 Louis Agassiz had twice visited Great Britain upon invitations to discuss fossil fish, the subject on which his fame then rested. By the autumn of 1840 his great work, *Études sur les glaciers* (Agassiz, 1840[a]) was safely in final press, and in September and October 1840 he traveled with William Buckland through Scotland, where he discovered traces of glaciation and demonstrated them to Buckland. Buckland was President of the Geological Society of London and Professor of Geology at Oxford, and in 1845 he became Dean of Westminster. His route and discoveries have been determined in great detail by Davies (1968).

On 4 November 1840, Agassiz read a paper before the Geological Society of London "On glaciers, and the evidence of their having once existed in Scotland, Ireland, and England" (L. Agassiz, 1840[b]; E. C. Agassiz, 1885, vol. 1, p. 309; Marcou, 1896, vol. 1, p. 168–74; North, 1943, p. 22). Buckland had written on 15 October 1840 to Agassiz telling of Lyell's acceptance of the glacial theory (E. C. Agassiz, 1885, vol. 1, p. 309).

A letter written by Buckland on 4 October 1840, here reproduced, is his announcement of the discovery. It was sent the day after Agassiz had written to Robert Jameson about the glacial origin of the famous parallel roads of Glen Roy (Davies, 1968, p. 137). As far as I know the letter has, except for one sentence, never been published. The letter had been loaned to the biographer of Buckland, who barely mentioned the announcement as "1840" and quoted only one sentence of the letter (Gordon, 1894, p. 14; Davies, 1968, p. 137). The letter is not mentioned in Marcou's or in Mme Agassiz's biographies of Agassiz, because it was never known to them. The letter also provides some details of the routes taken by Dr and Mrs Buckland and Agassiz during 3–7 October 1840, and it strongly suggests that Agassiz then traveled by the most direct route from Aberdeen to Glasgow. The exact routes for these days have been less certain than those for the remainder of Agassiz's 1840 tour (Davies, 1968, p. 138–39).

The letter is on one folded sheet 11 by 17 cm, apparently enclosed in a small envelope 8.5 by 5.5 cm, which was presumably discarded. Fortunately, a pencil note establishes that Fleming was the recipient. The Rev. Professor John Fleming (1785–1857) was Professor of Natural Philosophy at the University of Aberdeen, 1834–45, and in 1845 he became Professor of Natural Science at the Free University in Edinburgh, having moved there as a result of his adherence to the Free Church at the time of the Disruption in 1843 (Duns, 1859). The note on the front of the letter is in the handwriting of Ralph Richardson, who was secretary of the Geological Society of Edinburgh and the husband of the granddaughter of Fleming. It was found inserted at page lxxv of the copy of Duns' memoir of Fleming, which appears to have belonged to Richardson, as this copy has other letters inserted and many penciled notes.

This book itself is of glacial interest because the second half (p. 1–102), consisting of Fleming's *Lithology*, is almost entirely on surficial materials, with detailed sketches and sections showing the anatomy of the glacial deposits. However, Fleming was uncertain of the origin of the deposits. Although he had entertained the two chief early proponents of the glacial theory almost at the very moment of their discovery of glacial action in Scotland, Fleming (always a stubborn man) did not really accept the glacial origins of the deposits. He asserted

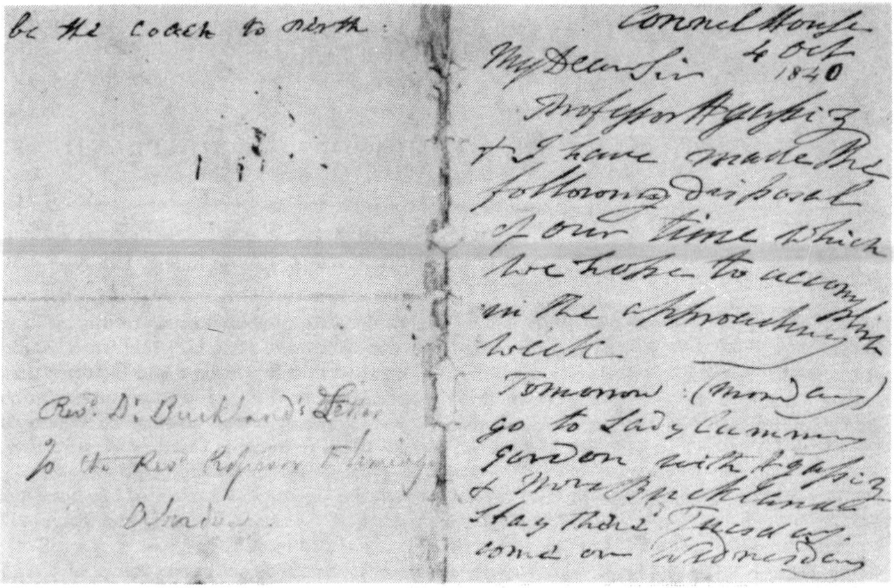

be the coach to Perth:

Revd Dr Buckland's Letter
to the Revd Professor Fleming
Aberdeen

Connel House
4 Oct
1840

My Dear Sir

Professor Agassiz
& I have made the
following disposal
of our time which
we hope to accomplish
in the approaching
week

Tomorrow (Monday)
go to Lady Cumming
Gordon with Agassiz
& Mrs Buckland
stay there Tuesday
come on Wednesday

Page 4

Page 1

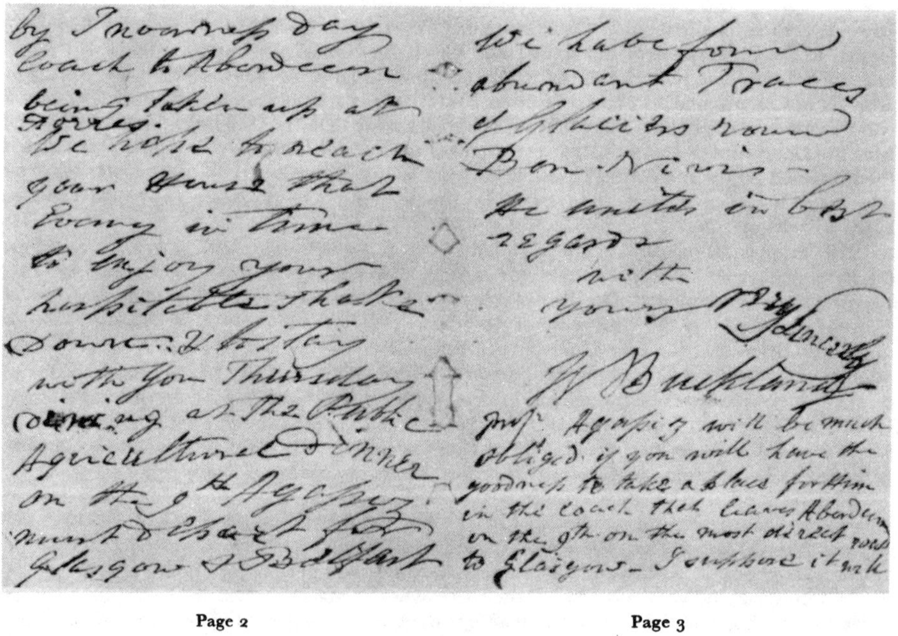

by Inverness day
coach to Aberdeen
being taken up at
Forres.
We hope to reach
your House that
Evening in time
to enjoy your
hospitable shake
down & to stay
with you Thursday
dining at the Public
Agricultural Dinner
on the 9th Agassiz
must depart for
Glasgow & Belfast

We have found
abundant Traces
of Glaciers round
Ben Nevis –
We unite in best
regards
with
yours very
sincerely
W Buckland

Prof. Agassiz will be much
obliged if you will have the
goodness to take a place for him
in the coach that leaves Aberdeen
on the 9th on the most direct road
to Glasgow – I suppose it will

Page 2

Page 3

(Duns, 1859, p. 61–62) that "The *droppings of melting icebergs* and the *moraines of glaciers* . . . may account for a small portion of these facts. . . . That a *debacle* of a singular kind had taken place, will scarcely admit of a doubt. . . . We cannot explain the phenomena, but we trust the reader will give us credit for having indicated the facts of the case. . . ." He did indeed set forth most carefully the "facts of the case" and was thus one of the very first anatomists of glacial deposits. It is fortunate that he preserved Buckland's letter, even if he could not accept Buckland's and Agassiz's glacial theory.

Transcript of Buckland's letter to Fleming

[In pencil on back of letter: "Rev^d^. D^r^. Buckland's Letter to the Rev^d^. Professor Fleming Aberdeen"]

Connel House
4 Oct
1840

My Dear Sir

Professor Agassiz & I have made the following disposal of our time which we hope to accomplish in the approaching week.

Tomorrow: (Monday) go to Lady Cumming Gordon with Agassiz and M^rs^. Buckland. Stay there Tuesday, come on Wednesday by Inverness day coach to Aberdeen being taken up at Forres.

We hope to reach your House that Evening in time to enjoy your hospitable shake down & to stay with you Thursday dining at the Public Agricultural Dinner. On the 9th Agassiz must depart for Glasgow & Belfast.

We have found abundant Traces of Glaciers round Ben Nevis—He unites in best regards

with

yours very Sincerely
W. Buckland

Prof^r^. Agassiz will be much obliged if you will have the goodness to take a place for Him in the coach that leaves Aberdeen on the 9th. on the most direct road to Glasgow—I suppose it will be the coach to Perth.

MS. received 3 September 1969

REFERENCES

Agassiz, E. C., *ed.* 1885. *Louis Agassiz, his life and correspondence.* Boston, Houghton, Mifflin and Co. 2 vols.

Agassiz, L. 1840[a]. *Études sur les glaciers.* Neuchâtel, [privately printed]. 2 vols.

Agassiz, L. 1840[b]. On glaciers, and the evidence of their having once existed in Scotland, Ireland, and England. *Proceedings of the Geological Society of London*, Vol. 3, Pt. 2, No. 72, p. 327–32.

Davies, G. L. 1968. The tour of the British Isles made by Louis Agassiz in 1840. *Annals of Science*, Vol. 24, No. 2, p. 131–46.

Duns, J., *ed.* 1859. *The lithology of Edinburgh, by the late Rev. John Fleming, D.D., F.R.S.E., with a memoir by the Rev. John Duns.* Edinburgh, William P. Kennedy.

Gordon, E. O. 1894. *The life and correspondence of William Buckland, D.D., F.R.S., sometime Dean of Westminster, twice President of the Geological Society, and first President of the British Association.* London, John Murray.

Marcou, J. 1896. *Life, letters and works of Louis Agassiz.* New York, Macmillan and Co. 2 vols.

North, F. J. 1943. Centenary of the glacial theory. *Proceedings of the Geologists' Association*, Vol. 54, Pt. 1, p. 1–28.

HISTORY OF INVESTIGATION AND CLASSIFICATION OF WISCONSINAN DRIFT IN NORTH-CENTRAL UNITED STATES

GEOLOGICAL SOCIETY OF AMERICA
MEMOIR 136

History of Investigation and Classification of Wisconsinan Drift in North-Central United States

GEORGE W. WHITE

Department of Geology, University of Illinois, Urbana, Illinois 61801

ABSTRACT

Outwash deposits in the Ohio Valley that were thought to be alluvium were illustrated by Volney in 1803. Drake (1815) proposed an iceberg origin for Wisconsinan erratics in Ohio. Hitchcock (1841a) in Massachusetts reviewed the glacial theory of Agassiz favorably, and in Ohio, St. John (1851) wrote in detail on the subject. Ice sheet origin of drift was widely and generally accepted by 1865.

Multiple glacial advances, many of which later turned out to be Wisconsinan, had been noted by Lyell (1849), Whittlesey (1866), Worthen (1868), Orton (1870), Winchell (1873), and Newberry (1874). Chamberlin (1878) gave the first detailed description and analysis of the different ages of surface drifts in the Kettle Moraine region. After Chamberlin introduced the term Wisconsin (1894, 1895), Leverett (1899) divided the Wisconsinan into early, middle, and late, and Leighton (1931) assigned the Iowan to the Wisconsin and gave the names Tazewell, Cary, and Mankato to the early, middle, and late substages of Leverett. Rock- and time-stratigraphic terminology of the Lake Michigan Lobe was defined by Frye and Willman (1960), and the terms Altonian, Farmdalian, Woodfordian, and Twocreekan were introduced. Other geologists have recently used this classification to differentiate the drift of other lobes.

INTRODUCTION

The classification of glacial deposits in the central part of North America has progressed from the recognition of surficial material as "diluvium" or "drift," to recognition that drift was of glacial origin, and finally to the realization that the drift was deposited by more than one episode of glaciation. There then followed more detailed and refined divisions of the Pleistocene deposits, particularly of the later, mainly Wisconsinan deposits, which have the widest surface outcrop, are best preserved, and can be dated in actual years by isotopic methods.

Evidence for multiple glacial advances in the Pleistocene and in the Wisconsinan in North America is best exhibited and most studied in the Midwest. The stratigraphy and terminology are developed mainly from this region, and it is now the scene of active studies. For lack of space the history of classification of deposits of the Ontario Basin in the United States and Canada and of the Canadian part of the St. Lawrence Basin is not included in this paper. The classification in southern Ontario has been summarized by Goldthwait and others (1965) and is discussed by Dreimanis and Goldthwait (this volume, p. 71-105).

Very Early Notices of Drift

The earliest diagrams of Quaternary deposits in the Middle West, and indeed in America, were published by Volney in France in 1803 and in America in 1804. He described the material in the Ohio River terraces at Louisville (1803, p. 62, Pl. 1; 1804, p. 48, Pl. 1) and the terraces and their material at Cincinnati (1803, p. 90, Pl. 1; 1804, p. 68-69, Pl. 1). He recognized that these materials (Fig. 1) had been carried down the Ohio Valley from the northeast, but he did not know that they were glacial outwash.

The first suggestion in America of the influence of ice on origin of drift, and especially of erratics, was by Drake (1815, p. 75). In 1817 he described erratics in greater detail as "foreign and adventitious debris . . . consisting of granite . . ." (Drake, 1825, p. 133-135) and gave the first description (in the United States) of till, which he called "geest" (p. 135-137). He explained the origin of the erratics and geest as being from "large fields of ice in a region far beyond the lakes and floated hither by the same inundations that brought down and spread over the surface of this country the *geest* in which they are imbedded" (1825, p. 137). His diagram (p. 138) shows the "alluvial formation" in the Ohio Valley and the geest on the uplands. However, Drake's ice origin of these materials was too radical for general acceptance at the time, even by such an astute observer as Thomas (1819, p. 248).

Maclure, in a brief but important paper on American geomorphology, noted the presence of erratics in the Central Interior. He suggested that those in Ohio had been transported by icebergs in a once far more extensive Lake Erie (Maclure, 1823, p. 102).

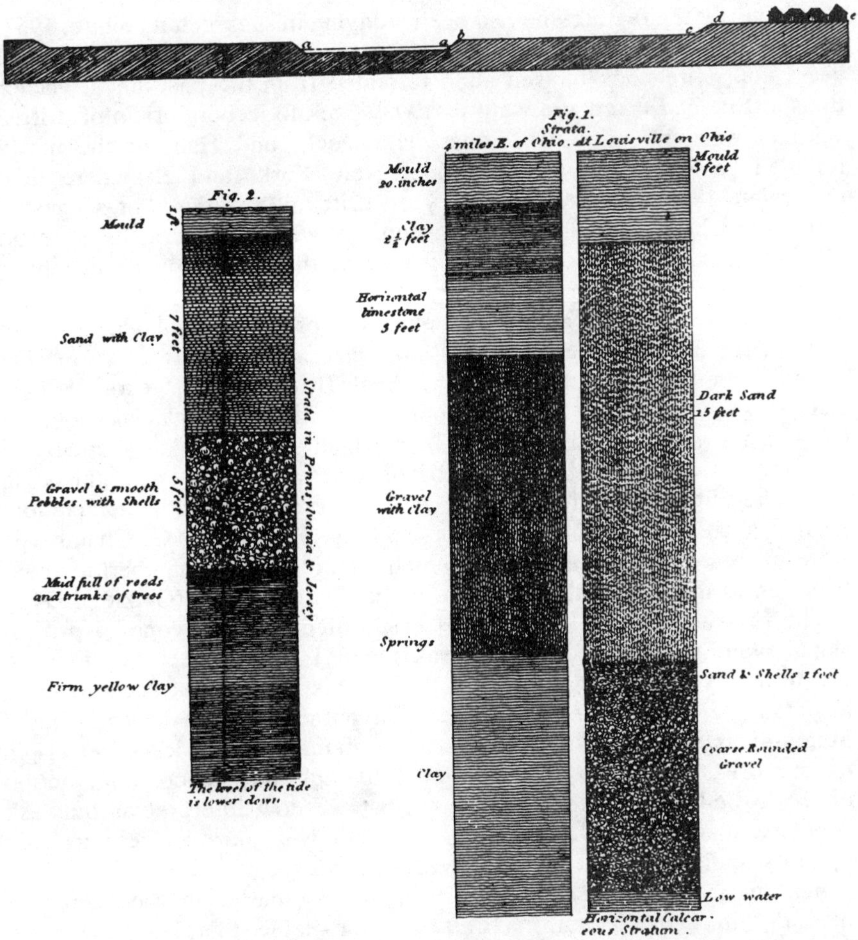

Figure 1. Volney's sections. Upper figure is of the Ohio River terraces which he observed at Cincinnati in August 1796, in "the history of which three distinct periods may be traced" (Volney, 1804, p. 69; Pl. 1). The two columnar sections at the right are of deposits in the Ohio River Valley and are thus the first sections of American Pleistocene deposits. The section on the left is from the Coastal Plain in the Philadelphia region.

Ice Sheet Origin of Drift

The ice sheet origin of drift in America was first seriously proposed and supported by close reasoning from field data by Hitchcock[1] (1841a, 1841b). Earlier (Hitchcock, 1832) he had extensively described drift; but upon reading Agassiz's *Etudes sur les Glaciers,* which had been published in October 1840, he immediately recognized that this theory explained

[1]Conrad (1839, p. 241-242) thought that "freezing of enormous lakes . . . converting them into immense glaciers" might explain erratics and striations in New York, but he did not go into detailed and extended explanations as did Hitchcock.

satisfactorily the features he had been studying in such detail (White, 1967, p. 212). The ice sheet concept was not enthusiastically accepted for several years, and Hitchcock himself soon retreated from the position of glacier transport he had taken and went part way back to iceberg origin of drift.

Many workers, such as Mather, Hitchcock, and Hall, in the newly founded geological surveys of Ohio, New York, and elsewhere, had in the late 1830s written extensively on drift and erratics, but suggested iceberg origin of the drift. These men, as well as others, as a result of Hitchcock's bringing Agassiz's theory to their attention so cogently, began to consider the glacial theory. However, for about 20 years after Agassiz and Buckland announced evidence of continental glaciation in Great Britain, glacial geology did not generate much interest in that country nor in America. Then, in Great Britain, the "years 1861 to 1865 saw a very remarkable development of interest in glacial geology" (Newbiggin and Flett, 1917, p. 157) in which James Geikie participated and soon became the leader in Britain and was widely recognized in the United States. Soon thereafter, as a proponent of multiple glaciation, he had a great effect on American workers. He and T. C. Chamberlin were in frequent communication, which culminated in Chamberlin's contribution to the classic third edition of the *Great Ice Age* (Geikie, 1894).

The first acceptance of ice sheet origin of drift by anyone of professional standing in the Midwest was in 1851, by Samuel St. John of Western Reserve College (White, 1967). This is not the place to trace in detail the acceptance of the glacial hypothesis in the central United States—it is sufficient to realize that by about 1870 the ice sheet origin of the drift was generally accepted, although the concomitant action of ice in bodies of water to explain what we now interpret as outwash deposits, made by water from melting ice, was more or less insisted upon by some until late in the nineteenth century.

In America, studies of the drift rapidly increased in the 1860s, as they did in Great Britain, and it was inevitable that, as field work became more detailed, the layers of the drift and the varying degrees of erosion and weathering on different surface drifts would be recognized.

PREVIOUS HISTORICAL STUDIES

One of the earliest reviews of the literature, with a comprehensive bibliography citing 500 earlier reports up to 1899, is that of Leverett (1901, p. 29-49). This important historical study deals mainly with the Erie Lobe and adjacent regions and includes some references to divisions of the post-Illinoian drift. He compiled another exhaustive bibliography (Leverett and Taylor, 1915, p. 33-54) which included scores of references on classification. Still later, Leverett (1932, p. 209) presented another, briefer history and bibliography.

In his review of the "Development of the glacial hypothesis in North

America,'' Merrill (1906, 1924) touched very briefly on the early recognition of multiple glaciation (his references were usually very sketchy), but had nothing to say about the Wisconsinan and its duration.

The paper by Thwaites (1927), "The development of the theory of multiple glaciation in North America,'' traces year by year from 1873 to 1926 the appearance of reports on this subject. It emphasizes the "Central District,'' but also includes historical summaries of the East, the western mountains, and Canada. It deals mainly with the whole Pleistocene classification, but does include some notices of reports on Wisconsinan divisions up to 1926. This fundamental study with its accurate references has been most useful in the preparation of this paper.[2]

Leighton (1964) summarized some of the history of discoveries of multiple drifts in midwestern North America and gave detailed consideration and references to his most recent classification of Wisconsinan deposits (p. 124–133).

The brief but excellent paper by Flint (1965) deals with the history of classification of the Pleistocene but not with the divisions of the Wisconsinan. His Table II is an accurate summary of the origin of stage names in the central United States.

Charts in recent reports by Frye and Willman (1963, Fig. 1) and by Frye and others (1968, p. E4) show the evolution of Wisconsinan classification. A report on a part of the Lake Michigan Lobe by Frye and his associates (1969, p. 2–5) discusses the history of classifying and naming of Pleistocene units (mainly late Pleistocene) in Illinois. The references and maps are valuable parts of this report. An even more extensive bibliography was included by Willman and Frye (1970) in their definitive bulletin on Pleistocene of Illinois.

The history of discovery and classification of Wisconsinan drifts in Wisconsin is included in a series of papers in elaborate guides prepared for the Pleistocene field trips before and after the 1970 Annual Meeting of The Geological Society of America in Milwaukee. The historical summaries are particularly detailed for southeastern Wisconsin by Laska (*in* Black and others, 1970, p. B3–B7, C3–C14, E3–E13) and by both Black and Bleuer for south-central Wisconsin (*in* Black and others, 1970, p. G1-G11, H5, J6–J8). The history of the Two Creeks buried forest is summarized by Black (1970, p. 21).

THE INQUA VOLUME—1965

The great Quaternary volume published in 1965 on the occasion of the Seventh Congress of the International Association for Quaternary

[2]The paper by Thwaites (1927) is not readily accessible and in our opinion should be reprinted to be available to interested researchers. It might be noted that Keyes (1928) made lengthy and strident objections to Thwaites's paper, mainly on the basis that multiple glaciations were doubtful and had not been proved, and that the merits of the proponents of a "lone Great Ice Age'' had not been recognized.

Research (Wright and Frey, 1965) contains chapters descriptive of the Pleistocene geology of various parts of North America. Most of them emphasize Wisconsinan deposits; all the chapters have some historical references, but some have many more than others.

In the northern Great Plains—Montana, North Dakota, and South Dakota—Lemke and others (1965, p. 20, 21-25) summarized the evidence for "six distinct and separate glacial advances during Wisconsin time," related them to moraines, shown by a useful map, but did not name the tills. Advance 1 is the Altonian; the others are Woodfordian. Wright and Ruhe (1965) summarized the classification and extent of the complex Wisconsinan drifts in Minnesota and Iowa from four major ice lobes and included a useful series of maps. (In this volume, Wright and others discuss the current classification.)

In Illinois and Wisconsin the Wisconsinan divisions were described by Frye and others (1965) following the classification introduced earlier by Frye and Willman (1960). Elaboration of this classification has been recently presented by Frye and others (1968, 1969), and Willman and Frye (1970). Wayne and Zumberge (1965) presented a table showing the Wisconsinan divisions in Indiana after Wayne (1963) and the geologic-climatic units after Gooding (1963).

The Wisconsinan deposits of the Huron, Erie, and Ontario Lobes, long differentiated into substages called early, middle, and late Wisconsin, and later given rock-stratigraphic names by a number of workers, are reviewed and summarized by Wayne and Zumberge (1965), Goldthwait and others (1965), and Muller (1965). The recognition at Sidney, Ohio, of a buried soil more than 22,000 years old below three tills and above at least two earlier tills (Forsyth, 1965, Fig. 2) has provided the basis for elaboration of Wisconsinan classification in the eastern part of the Miami Lobe and hopefully for correlation with deposits of the Scioto Lobe. The term Sidney (?) Interval has been used for the time represented by the formation of the Sidney paleosol (Goldthwait and others, 1965, p. 91).

The rock units of the Erie Lobe and their age assignments are dealt with in detail in this volume by Dreimanis and Goldthwait.

It should be pointed out that ice sheets in much of the Central Lowlands advanced into low regions and generally downslope. In contrast, in the Allegheny Plateau in eastern Ohio, northwestern Pennsylvania, and New York, the Wisconsinan ice sheets, as well as the earlier ones, advanced upslope and, therefore, to not nearly as great a distance. Consequently, in the plateau the various tills are exposed at the surface only in narrow belts, some of which are less than 5 mi wide, causing severe mapping problems. However, drifts in the subsurface can now be correlated with surface drifts in these narrow belts so that classification is more certain (White, 1969; White and others, 1969). Also in the plateau, organic deposits that can be used for C-14 dating are much more scarce than in the Central Lowlands. Fewer than six localities in the plateau, aside from

Figure 2. A very early (earliest?) map of glacial boundary in North America constructed by Charles Whittlesey in 1864. It shows "Southern limits of boulders and coarse drift" in a line extending from New Jersey to Iowa. Original map is 8 in. wide (Whittlesey, 1866).

flood plain and postglacial peat-bog materials, have so far provided organic material for dates.

MULTIPLE DRIFTS 1849–1878

Before 1868 there had been a few notices of differences in ages of the drift, but in the years 1868 to 1878 came the more general recognition of multiple drift deposits and the realization that they could only be explained by one or more glaciations that followed one or more interglacial times. Criteria other than buried organic layers came to be recognized by 1880.

Charles Lyell in his second visit to the United States in 1845–1846 had sensed that "the northern drift, however, is by no means all of the same age" (Lyell, 1849, vol. 2, p. 265). However, it was not until much later that buried organic layers were reported.

In one of the most important early papers, Whittlesey (1866) described at length the morphology of the drift and its different units. His catalog of the many sections that showed "buried timber" (p. 13–15) is the earliest precise description of interglacial (or interstadial) material. Whittlesey's paper also contains the earliest map (dated December 1864) to show the drift boundary in the United States (Fig. 2). The map also shows "striae," and "Moraine Knolls and Cavities."[3]

Another very early notice of buried organic deposits that separated drift sheets was by Worthen in Illinois (1868, p. 75–87). He described buried organic deposits below upper drift. Although realizing their importance, he did not speculate on their origin, saying that "wherever these beds are penetrated in sinking wells or are otherwise exposed, a careful examination should be made . . . for any organic remains . . . as these would no doubt throw some light on their true origin, and the conditions under which they have been deposited" (p. 87). Some of these organic deposits may have been Sangamonian in age, but all those checked have turned out to be Farmdalian, and thus separated two substages of Wisconsinan drift as now recognized (Willman, 1971, oral commun.).

Although the forest bed in southeastern Ohio was known before 1870 and was described in detail by Orton (1870), the explicit recognition in the Midwest of multiple glacial and interglacial deposits that required more than one episode of glaciation to explain their origin was in 1873. Orton (1873, p. 430) first used the term "the *interglacial stage*" in America in his description of the "forest bed" in southwestern Ohio and southeastern Indiana. At the location he studied, the forest bed lies between outwash deposits that he recognized as requiring two glacial stages (Fig. 3). This is the earliest illustration of a deposit specifically labelled "interglacial." It is quite likely an interstadial deposit of Wisconsinan-Farmdalian age and hence records the first division of Wisconsinan glacial materials.

Winchell (1873, p. 61) suspected that some thin weathered material in southeastern Minnesota was "the remains of a previous glacial sheet." He suggested "vegetable growth" had been removed "toward the north" by later ice advance, but that in the southeast at places it was buried "beneath the debris from the hillsides."[4]

Newberry (1874, p. 3) recognized a widespread forest bed in Ohio

[3] Stewart and MacClintock (1969, p. 55–56) have summarized early observations of evidence of multiple glaciations in Vermont.

Figure 3. Section of Pleistocene deposits located 15 mi west of Cincinnati, Ohio, as measured in 1870 by Edward Orton. This may be the earliest diagram that includes an interglacial or interstadial deposit, the "forest bed" (Orton, 1873, p. 427).

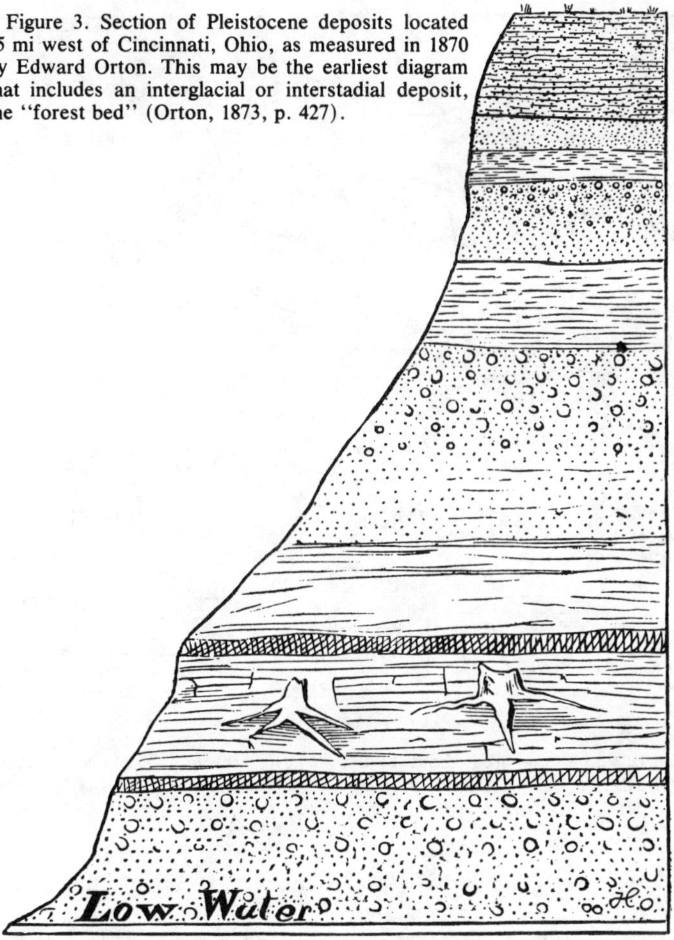

resting upon "boulder clay" and overlain by drift, mainly "water transported material" (p. 7). Although Whittlesey's map of 1864 is the first to show the drift border, Newberry's (Fig. 4) is the earliest map to show the drift border in a more detailed way. In his earlier papers (Newberry, 1870), there is reference to drift margins and to different kinds of drift.

McGee (1878) referred to Newberry's report on the forest bed and asserted that a similar bed is to be "observed at many points in Illinois, all through southwestern Indiana, and at many localities in Wisconsin, throughout northeastern Iowa, in Canada, and in many other places" (p. 339). He gave illustrative sections and stated that "the forest

[4]Winchell's small but significant five-page paper included distinctions between meltwater deposits in valleys sloping away from the ice and those sloping toward the ice, and also an analysis of superglacial deposits let down when the glacier melted.

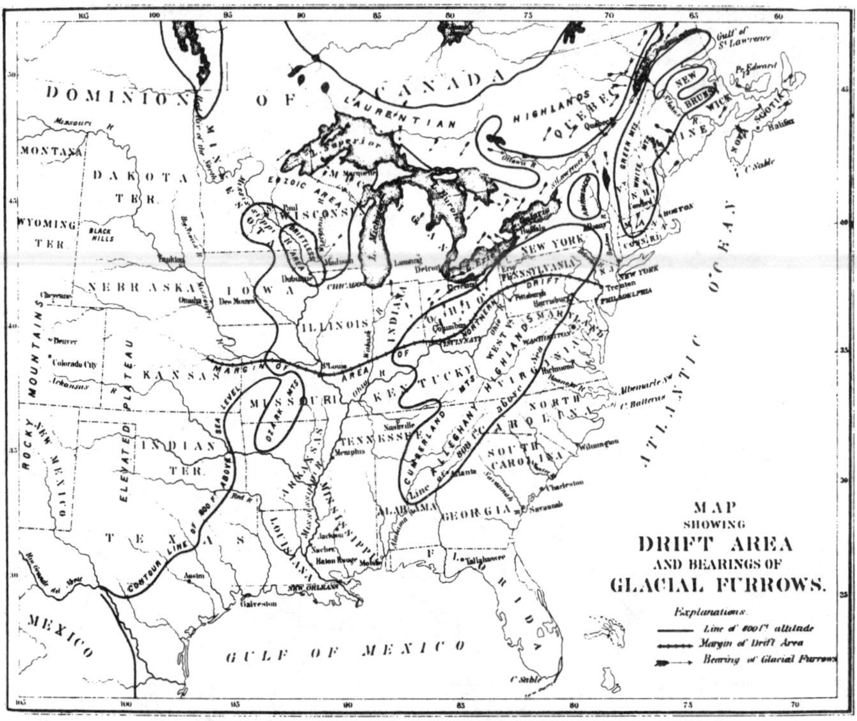

Figure 4. Newberry's map prepared about 1870 of "Drift area and bearings of glacial furrows." Note "Driftless Area." Original map is 9-1/4 in. wide (Newberry, 1874, opposite p. 76).

bed is overlaid by true glacial drift, and hence must be of interglacial age" (p. 341).

Winchell (1875, p. 185) thought that buried peat in Mower County, Minnesota, "seems to mark a period of interglacial conditions." He said that patches of much weathered drift in Fillmore County, on the margin of the Driftless Area, "are believed to belong to a glacial epoch that preceded the epoch that produced the great drift sheet of the Northwest," and an "interglacial epoch separated them." In 1878 he reported on wood deposits "between the two drift periods" in Goodhue County (Winchell, 1878, p. 44).

Chamberlin, who was beginning his work in Wisconsin about the same time, was familiar with the discoveries in Ohio and Minnesota. In his chapter on the Quaternary deposits in *The Geology of Eastern Wisconsin* (1877, p. 196–246), he ascribed the multiple layers of drift to "alternating advance and retreat of the ice-mass" (p. 214). Many of his diagrams illustrate multiple layers of drift, especially those in his Plate 8. The relations of ground-water resources to the drift layers were well illustrated in a series of diagrams. A separate paper that elaborated on these ideas, and which was Chamberlin's first paper entirely on glacial subjects, was his report on the Kettle Moraine (1878), a revision of a paper he had presented orally in 1875. In this he clearly distinguished between the drift in the Kettle Moraine and the drift outside on the basis of different development of drainage and of weathering rather than on stratigraphic grounds. However, the drift outside the Kettle Moraine also is largely Wisconsinan. After describing the Kettle Moraine and tracing it (Fig. 5), both far to the west and far to the east, he summarized his ideas about the different ages of material within and outside the moraine: "*outside* of the Altamont Moraine" is a "fairly established drainage system," in contrast to "the much less established drainage inside the moraine" (p. 16). He recognized moraines buried beneath later deposits (p. 17, 19, 29). He assumed "the moraine was, therefore, formed *after the retreat of the glacier had commenced and marks a certain stage of its subsequent history*" (p. 30). The moraine was ascribed not to a halt, but to an "advance of the ice-mass" (p. 31). It is "*a definite historical datum line* . . . separating the formations on either hand by a chronological barrier" (p. 33-34). Thus, almost 100 years ago, as a result of rigorous analysis of field data, the divisions within the last glacial stage were recognized, and a map was presented to show different ages of drift which are Wisconsinan in age.

It is significant that Newberry, Winchell, Orton, and Gilbert had all worked in Ohio in the first part of the 1870s and had published reports in the Ohio Survey volumes. They were familiar with each other's discoveries, and those of Winchell in Minnesota and McGee in Iowa strongly supported the Ohio findings. Chamberlin had seen the drift in Ohio and correlated the Ohio findings with his own studies in Wisconsin.

These men supposed that the drift above and below the "interglacial deposits" was derived from separate ice advances, but they did not distinguish between stages and substages. They were, however, the first to recognize multiple deposits in the late Pleistocene, that is, within what we now call Wisconsinan.

MULTIPLE DRIFTS AND THEIR CLASSIFICATION, 1878-1894

After the definitive paper on the Kettle Moraine by Chamberlin (1878), an increasing number of reports of interglacial deposits and of multiple tills appeared, as recorded by Thwaites (1927). Only a few can be selected for mention. Intertill deposits, some interglacial and others interstadial, were reported by McGee and Call (1882) in Iowa and by Upham (1883, 1884) in Minnesota. They recognized both stratigraphic evidence and degrees of weathering, and varying amounts of valley cutting where different drifts were exposed at the surface.

McGee and Call (1882) also recognized loess between the drift sheets and stated that this proved readvance of the ice sheet "not during an

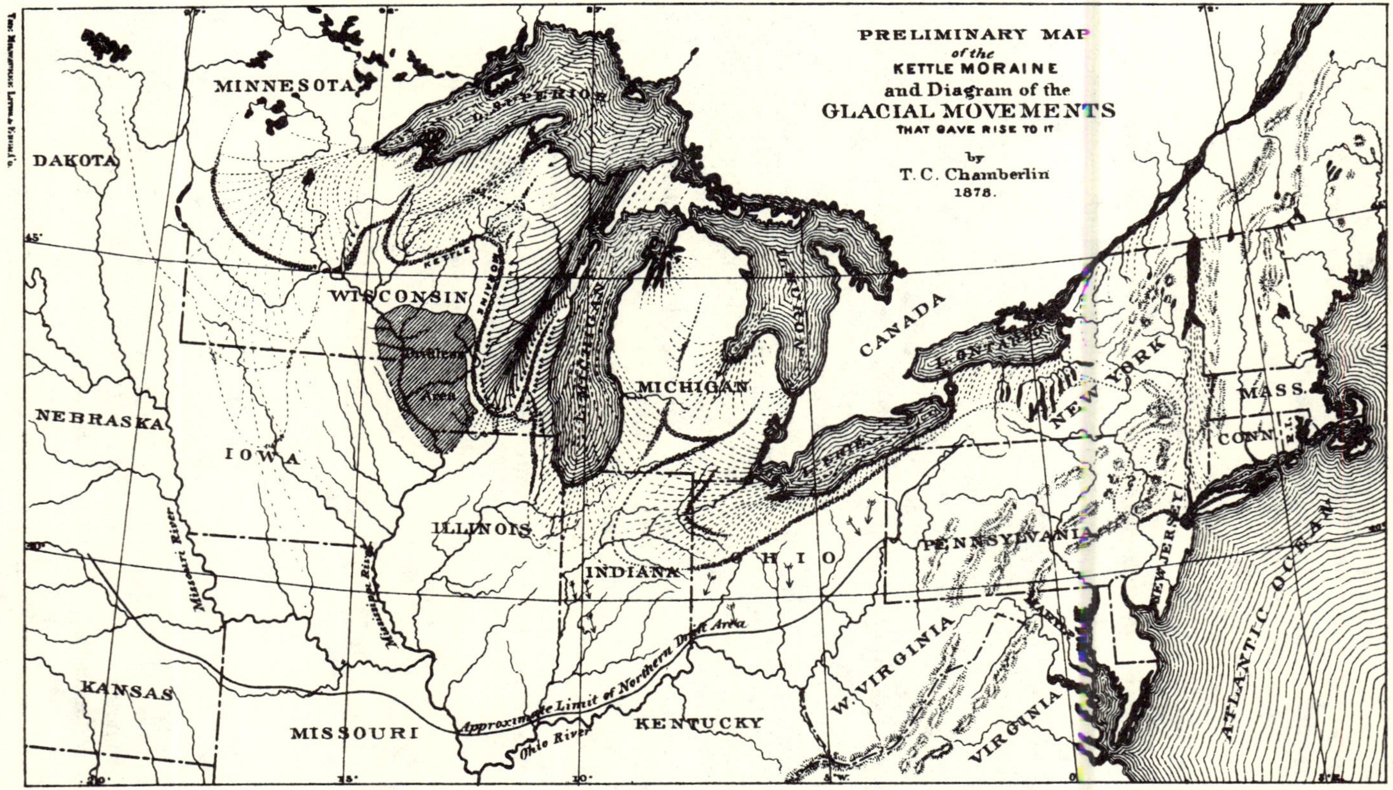

Figure 5. Chamberlin's map of 1878, showing "Approximate limit of northern drift area," the "Driftless Area," and "Kettle Moraine." This is the first map to show explicitly drift of more than one age, because the Kettle Moraine formed the margin of drift definitely younger than that between it (Kettle Moraine) and the drift limit. Original map is 7-1/2 in. wide (Chamberlin, 1878, p. 10).

independent ice-period, but during a temporary halt and slight readvance of the slowly retreating ice-sheet." Here was an early recognition of stratigraphic divisions of the drift of the last glacial stage.[5] In a classic report McGee (1891, p. 472-547) described "upper till" separated from "lower till" by a forest bed or other organic material, and included many diagrams and maps. However, by this time he had interpreted the forest bed as interglacial (p. 576-577).

Chamberlin was becoming more certain about two glacial advances in Wisconsin, both of which are now known to be of Wisconsinan age. He developed this concept in increasing detail and certainty in a series of papers. Chamberlin (1882, p. 93) recorded the "extensive range of terminal moraines" extending from Pennsylvania to Dakota, beyond which was earlier drift.[6] "*The moraine marks a second glacial advance separated from the former by a considerable interval of time*" (p. 94). This was an elaboration of the extramorainal drift he had first described in 1878.

The classic paper by Chamberlin on "The terminal moraine of the second glacial epoch" (1883) is one of the most important for the advancement of Pleistocene classification. It also contains a history of the study of glacial deposits in North America up to that time, and the references to notices of moraines and drift are very valuable. Chamberlin traced the great series of moraines from the Atlantic Ocean to Dakota (p. 310) and stated that they do not lie at the drift margin (p. 314). The extramorainic drift was recognized as older on the basis of more erosion and of more advanced development of the drainage, greater weathering, and thinner and less continuous drift (p. 347). He considered that in places an earlier moraine had been overridden by a later one (p. 347).

Another more elaborate classification of the Quaternary epochs was that of Chamberlin and Salisbury (1885, p. 212). In this classification "the Earlier glacial epoch" was divided into two sub-epochs separated by an "interglacial sub-epoch." The following "Later glacial epoch," which followed the "Chief interglacial epoch" [Sangamonian] was divided into four episodes or stages. Curiously, the oldest was marked by the "Kettle or Altamont" Moraine, the next by the Gary Moraine, then the Antelope Moraine, followed by still "later stages." It is interesting to note that they correlated the Altamont Moraine with that of Nantucket, and the Gary with the Cape Cod Moraine (p. 215). This was long-range correlation indeed!

By 1884, Newberry, who had held out longer than most for iceberg origin for much of the drift, finally understood that a great ice sheet "possessing a peculiar plasticity," and not icebergs floating in a great

---

[5]This paper also contains much of interest concerning superglacial erosion and deposition, some "odd" ideas on the secondary origin of the "blue color" of buried till, and references on the latter subject not easily found elsewhere.

[6]This is essentially the break within the Woodfordian at the front of the Valparaiso Moraine, and not the margin of Woodfordian drift as now understood (Frye and Willman, 1960).

sea, had deposited the till of the northern United States, and adduced evidence to disprove the iceberg theory. He now asserted that "the ice advanced with many arrests and retreats" (1884, p. 86), and that "*striated pavement*" boulder pavements were formed by readvances of the ice (p. 88). He cited contorted drift as evidence of ice readvance and stated that "two boulder clays separated by . . . beds of peat" were indications of "*interglacial warm periods*" (p. 90). He assumed many episodes of advance and retreat and disappearance of the ice, as he was now strongly influenced by Croll's (1875) explanation of glacial episodes caused by variations in the eccentricity of the earth's orbit.

In a closely reasoned paper, Salisbury (1893) discussed the criteria for the recognition for distinct glacial epochs. He listed 12 criteria, which included various kinds of organic and inorganic intertill beds, weathering zones, varying amounts of erosion, and different compositions of tills. His paper was a significant one and is still an important reference for study of multiple glacial deposits.

James Geikie invited Chamberlin, his long-time friend and correspondent, to contribute the section "Glacial phenomena of North America" to the third edition of *The Great Ice Age* (Geikie, 1894). It was thus that the first classification using geographic names, on which our present classification is based, came to be published outside the United States. The text and legend of the colored map shows "Kansan Formation," "East Iowan Formation," and "East Wisconsin Formation." The East Wisconsin extended to the Kettle and Altamont Moraines in the west and the Shelbyville and "Main Morainic System" in the east. The East Iowan was shown in eastern Iowa and in an area in central northern Illinois, extending into southern Wisconsin (Chamberlin, 1894, p. 724–775; Pl. 15).

At Chamberlin's invitation, Geikie (1895) summarized at length the classification of European deposits as a basis for possible correlation with American deposits. In a following paper, Chamberlin (1895) made tentative suggestions about the correlation of American deposits with the European. He "cheerfully accepted Upham's suggested changes"[7] and introduced the term "Wisconsin Formation" instead of East Wisconsin, as he had proposed the previous year, and "Iowan" in place of East Iowan. He recognized that there were stages of halts and "possibly advances" in the latest stage of glaciation (p. 276).

FOREIGN RELATIONS

In the late 1870s and through the 1880s (and later), the glacial geologists of America, Great Britain, and the European continent, communicated with each other through personal visits, exchange of publications, and

[7]In a review of Geikie's book, Upham (1895, p. 56) suggested changing "East Iowan" to "Iowan" and "East Wisconsin" to "Wisconsin."

correspondence. The observations and reports of the great triumvirate of James Giekie of Scotland, T. C. Chamberlin of America, and Otto Torrell of Sweden, with conclusions based in part on the theories of James Croll, were received, studied, taught, and expanded by most American geologists. Those few who were more conservative and reluctant to accept multiple glaciation, and particularly to accept several interglacial periods, were in such minority that their views were hardly regarded as respectable. They tended to be ignored by the mainstream of workers, although from time to time they were graciously allowed to present their conflicting views.

The validity of multiple glaciation and interglacial episodes, and thus the theoretical basis for seeking evidence for these in the deposits were given impetus by the many papers of Croll, beginning in 1864 and culminating in his book *Climate and Time in Their Geological Relation; A Theory of Secular Changes of the Earth's Climate* (1875). The central part of Croll's theory was the change in eccentricity of the earth's orbit, which had produced (partly indirectly) changes in climate throughout geologic time and especially during the glacial period. Croll's book deserves attention today, not only for the theories presented, but also for the voluminous geological data adduced in support of the theories. His arguments were well received by many American geologists, and his ideas were as influential in the last quarter of the nineteenth century as those of Milankovich in the middle part of the twentieth century.

Otto M. Torrell, Swedish professor of geology and later head of the Swedish Geological Survey, whose work in glacial geology in Europe was highly regarded, visited America in 1877. His views were welcomed, inasmuch as they confirmed the findings of the Americans. He supported the concept of advance and retreat of the ice and in a paper summarizing his views states "the retrograde movement of a glacier during the period of melting is characterized by deposits formed . . . by local backward and forward movements of a glacier during successive intervals of time" (Torrell, 1877, p. 77).

James Geikie first came to the United States from Edinburgh in 1884. He was able to see and study glacial and interglacial deposits throughout the Middle West under the guidance of the men who were working in the various states (Newbiggin and Flett, 1917, p. 106–110). Always an enthusiastic proponent of multiple glaciation, he encouraged the Americans to recognize not only one interglacial time, but a multiplicity of them. Some of these are within the Wisconsinan Stage as we now know it. His ideas were brought to the notice of Americans not only in the three editions of his book, *The Great Ice Age,* but also in a paper in the newly established *Journal of Geology* (Geikie, 1895), which was followed by a paper of Chamberlin's (1895). It is interesting to note that Geikie chose for the frontispiece of the third edition of his book (1894) a magnificently reproduced photograph of an exposure "near Williamsport, Indiana," which can now be interpreted as showing five tills, the upper ones being of Wisconsinan age. The development of the concept of multiple glaciation and interglacial intervals in Europe and James Geikie's *avant garde* place in these are discussed at length by Newbiggin and Flett (1917, p. 149-210).

WORK OF LEVERETT

Frank Leverett began to work with Chamberlin in 1886 and was closely associated with him for many years. Leverett's work resulted in a series of papers which were published in the first of his U.S. Geological Survey monographs, *The Illinois Glacial Lobe,* in 1899. Anyone working in territory described by Leverett must begin by a detailed study of his various publications.[8]

In his 1899 monograph Leverett stated that he studied "several thousand well sections" and thus was the first to use subsurface records in detail, although it is to be noted that he relied mainly on "records" and not on sample study, as in present-day methods. McGee (1891) studied well records, but not as extensively as did Leverett.

In this 1899 monograph Leverett distinguished sharply between the "Iowan of Illinois" and the "Iowan of Iowa" (Pl. 6 and p. 131-152). He demonstrated that the Iowan of Illinois was younger than Illinoian, but older than Wisconsin (see also Leverett, 1897, 1898a, 1898b). The extreme northern part of his Iowan of Illinois is now called Altonian, the earliest division of the Wisconsinan. The Wisconsin was divided into "early" and "late"; the "early Wisconsin" was divided into Shelbyville, Champaign, Bloomington, and Marseilles till sheets (Pl. 6, p. 191-379). These are all within the Woodfordian as presently understood.

Leverett's *U.S. Geological Survey Monograph 41* on the Erie and Ohio Basins appeared in 1901. It covered Indiana, Ohio, Kentucky, much of Pennsylvania and western New York, and part of Ontario. In this monograph Leverett for the first time had the advantage of extensive topographic map coverage that brought a new dimension into the study of glacial geology. He extended the use of "geographic names" (p. 50), but mainly for the moraines and their tills. He retained the Iowan Stage and Peorian interval and classified the Wisconsin as "early," to include the Shelbyville, Champaign, Bloomington, and Marseilles Morainic Systems, followed by a "fifth interval of recession (unnamed), shown by shifting of ice lobes." This was followed by the "late Wisconsin drift sheet" with Valparaiso, St. John's or Salamonie, Mississinawa, Wabash, and Fort Wayne Morainic Systems and the lake beaches and their "correlative moraines." The Iowan, he stated "appears not to extend beyond the limits of the Wisconsin drift" (p. 51). He wrote "This outline

[8]Leverett appears to have made some changes in some of his monographs either during printing or after they had been approved for printing, so that in places conflicting views appear and modifications of earlier views are inserted.

is essentially the same as the writer's outline presented in Monograph XXXVIII, the only important modification being in the subdivisions of the late Wisconsin glacial stage. These are more complete in the Ohio District than in the Illinois, and require a corresponding elaboration of the outline."

Leverett studied well records mainly for drift thickness. Unless obvious organic layers were recorded, he did little interpretation of the stratigraphy. Following Chamberlin's earlier preference, he depended mainly on the surface character of the different drifts for his classification (Leverett, 1909). His first love was morphology and he described in great detail the morainic areas, their extent, appearance, and at times in a most detailed way, the rounded cone of a single two-acre kame.

Leverett classified the drift beyond the "Wisconsin moraine" on the east side of the Grand River Lobe in Pennsylvania as "Kansan," but later (1934) decided this outer drift was "Illinoian" and "pre-Illinoian." The more westerly part of the "Illinoian" is now known to be about 40,000 yrs old and is therefore Wisconsinan (Altonian) in age (White and others, 1969, p. 23-32). Apparently Leverett had second thoughts about this "fringe drift" and—obviously out of place—inserted a page (Leverett, 1901, p. 351) of exact and excellent description leading to the statement that a portion "nowhere more than 10 miles" is too eroded to be late Wisconsin and too fresh to be pre-Wisconsin. It is curious that he did not speculate that this might be his "Iowan." He sensed (p. 452) the difference in composition of the till between the coarse, sandy till (Kent) of the outer part of the Grand River Lobe and the clayey till (Lavery and Hiram) of the inner part, but did not carry the subject far enough to indicate a considerable time difference of retreat and then readvance of the ice to account for the difference.[9]

Leverett's "late Wisconsin" drift of the east side of the Scioto Lobe is, in present terms, Woodfordian as far north as Utica. North of Utica, in Licking and Knox Counties, Leverett (1901, Pl. 13) recognized a northeastern extension of the Wisconsinan drift to Danville and Jelloway. Because this drift is more eroded it was called "Illinoian" by White (1934) and by Forsyth (1961). Now that an Altonian Stage of the Wisconsinan is recognized, the possibility must be considered that Leverett was right in calling this Wisconsin, with a narrow and possibly discon-

---

[9]The history of the "fringe drift" (Wright, G. F., 1884, 1889) of the Grand River Lobe in Ohio outside of what was earlier regarded as the outermost Wisconsinan moraine has already been discussed by the author (White, 1951). It has been variously called "Wisconsin" (Leverett, 1901), "Illinoian" and "pre-Illinoian" (Leverett, 1934), "Illinoian" (White, 1951), and "Kansan" (Stout, 1943). Narrow as it is, it is now interpreted as being of two ages—the larger part, called "Inner Illinoian" by Shepps and others (1959, p. 21) is 40,000 years old and hence Wisconsinan-Altonian (White and others, 1969, p. 23-32). A narrow and discontinuous outer part (Shepps and others, 1959, p. 20) is pre-Wisconsinan and probably Illinoian. Thus, the Grand River Lobe has, at the surface and in the subsurface, Woodfordian and Altonian Wisconsinan drifts, which now have rock-stratigraphic names (White and others, 1969, p. 9).

tinuous, outer part of this tract, being of Illinoian age.[10] Leverett's "early Wisconsin drift" of the Miami Lobe (1901, p. 304, Pl. 11) may possibly be early Wisconsinan (Altonian) in the present sense. The "Illinoian drift" beyond is Illinoian in the present sense and no part is Wisconsinan-Altonian. He discussed the difference between the early Wisconsin drift and that to the north in some detail (p. 352–353).

U.S. Geological Survey Monograph 53, The Pleistocene of Indiana and Michigan and the History of the Great Lakes, by Leverett and Taylor (1915) contains an extensive bibliography which is important for historical purposes. The folding chart (p. 30), "Approximate time relations of ice lobes, morainic systems, and moraines of the Wisconsin stage of glaciation in Indiana and southern Michigan" also shows the Illinois, Miami, Scioto, Lake Michigan, Saginaw, and Huron-Erie Lobes. The map (Pl. 5), which was later reprinted without change (Leverett, 1931, Pl. 18), shows the Shelbyville as the oldest and the Port Huron as the youngest Wisconsinan drifts in the region. This is a key map to show his concept of the divisions in the Wisconsinan.

It is interesting to note that "Illinoian and Iowan Drifts" are at one place combined (Leverett and Taylor, 1915, p. 26), but on the following page the "post-Sangamon and main loess (Iowan?)" indicates Leverett's growing belief that the Iowan was more closely related to the Wisconsin as he understood it than to the Illinoian. His uncertainty about the age assignment of post-Illinoian, pre-Wisconsinan (Woodfordian) drift and the realization that it is closely allied to the Wisconsinan Stage is well illustrated by his later statement that "it may be more consistent to regard the Illinoian, Iowan, and Wisconsin as a triple glaciation. . . . The data at hand do not seem to be decisive in this matter" (Leverett, 1930, p. 194).

LEIGHTON AND THE EXTENSION AND CLASSIFICATION OF THE WISCONSIN—1931-1956

In 1913, M. M. Leighton published his first paper on Pleistocene deposits in the Mississippi Valley. His work was to continue for more than 50 years.[11] After more than 15 years of field investigation Leighton proposed

[10] Leverett actually spent less time in this "re-entrant angle" tract of Knox and southern Ashland Counties than was usual in his studies (personal commun. in a conference with Leverett at Ann Arbor, 1930). He was a master at dealing with questionable areas or interpretations. Note his statement that part of the fringe drift of the Grand River Lobe may be Wisconsin (1901, p. 351). His detailed description and analysis as he proceeds from place to place are convincing and masterful. When, after scores of pages of detailed descriptions mile by mile, he jumps 20 or more miles, as from Utica to Mansfield, one must follow a detailed map to see that for the intervening space there are no real data!

[11] It is interesting to note the long time span covered by the work of several other geologists working on the glacial geology of the Mississippi Valley. A. C. Trowbridge published his first glacial paper in 1912, and thus his publication record reaches to almost 60 years. W. C. Alden has a 54-year record. Frank Leverett just reached 50 years, and G. F. Wright, F. T. Thwaites, Charles Whittlesey, James Walter Goldthwait, and Paul MacClintock closely approached 50-year records.

a classification that soon achieved wide usage. He concluded that "the Iowan drift is the first deposit of the last glacial stage and it must be known as the early Wisconsin or else the older terms, early Wisconsin, middle Wisconsin, and late Wisconsin must be dropped" (Leighton, 1931, p. 51). He proposed the terms "Manitoban Substage" for the Iowan, "Quebecan Substage" for the early and middle Wisconsin, and "Hudsonian Substage" for the late Wisconsin. However, two years later Leighton (1933, p. 168) proposed that the term "Iowan" be retained for the oldest subdivision of the Wisconsin, and that "Tazewell," "Cary," and "Mankato" be used for the early, middle, and late subdivisions, as previously differentiated by Leverett. These names came into general use for more than 30 years and still have some currency. When radiocarbon dating showed that the Mankato drift in Minnesota was older, not younger, than the Two Creeks forest bed, Leighton (1957) modified his classification by adopting the name "Valders" (Thwaites, 1943) for a substage younger than the Mankato.

Kay and Leighton (1933, p. 673) used the term "Eldoran Epoch," which had been introduced by Kay (1931) for the Wisconsin and Holocene Stages, and other new names for each of the three earliest ages and their following interglacial times. These names for combined glacial and interglacial stages never achieved general usage.

A major advance in elucidation of Wisconsinan history and in classification was the recognition by Leighton that a major glacial advance had taken place in post-Sangamonian time before the deposition of Shelbyville drift (earliest "classical Wisconsin"). Leighton (1926) observed that the loess which he had called "Late Sangamon Loess" rested on the Sangamon Soil and was only moderately weathered before a loess, then correlated with the Iowan glaciation, was deposited on it. He named the early Wisconsin loess "Farmdale" (*in* Wascher and others, 1948), and the Farmdale was made the first substage of the Wisconsin Stage (Leighton and Willman, 1950, p. 602). This Farmdale Loess was interpreted as silt blown from valley trains coming from Farmdale ice an unknown distance away. Later, Shaffer (1954, 1956) identified till in northwestern Illinois as the deposit of this ice and named the till "Farmdale."

The Farmdale Loess was later (Frye and Willman, 1960) subdivided into a lower unit, largely loess, named "Roxana Silt," and Farmdale was restricted to an upper organic silt and peat which had been widely dated. The name "Winnebago" was applied to the till in northern Illinois that was the source of the Roxana Silt, and the name "Altonian" was introduced for the earliest Wisconsinan glacial substage which included the Winnebago Till and the Roxana Silt.

Although the Wisconsinan classification of 1931 was modified by Frye and Willman (1960) in developing a multiple classification scheme, Leighton (1960, 1964) defended the previous classification and introduced the terms "Farm Creek," "Gardena," "St. Charles," "Bowmanville," and "Two Creeks" for intraglacial substages between the glacial substages.

IOWAN DRIFT

The Iowan drift, originally described as "East Iowan" (Chamberlin, 1894) and soon renamed "Iowan" (Chamberlin, 1895), was originally set off as a separate stage[12] older than the Wisconsinan ("classical Wisconsin"-Woodfordian). As its affinities in the type area of eastern Iowa were thought to be much closer to the Wisconsinan than to the older drifts, it was eventually classified as the earliest Wisconsinan substage (Leighton, 1933).

A problem with the Iowan, not present with any other stage, is that the term "Iowan" has been used for widely different features. This history has been summarized by Ruhe (1969, p. 93–95) and his map (Pl. 1) shows the "Iowan of Iowa" not as a till but as an "Iowan erosion surface" developed largely on the Kansas till. The till is overlain by "loam sediments," except on the southwest where it is covered by "Wisconsinan loess." At the extreme eastern edge the surface cuts across Illinoian till. Leverett (1909, p. 368) came to the same conclusion, but later was not so sure. Ruhe proposed the same explanation for the western part of "Iowan drift" in northwestern Iowa (Ruhe, 1969, p. 104).

The "Iowan of Illinois" (Leverett, 1899, Pl. 6) is a "real" till of Altonian age in extreme northern Illinois and southern Wisconsin, but farther south it is a till now interpreted as earliest Woodfordian (Frye and others, 1969, Figs. 1-1899, 1-1969; Fig. 2). Leverett also used the term "Iowan" in Illinois for the loess outside the Shelbyville (Woodfordian) drift and for the loess underlying the drift, but Alden and Leighton (1917) substituted the term "Peorian" for the loess, and Kay and Leighton (1933) restricted Peorian to the loess outside the Shelbyville and reintroduced Iowan for the loess beneath the Shelbyville. This was the Iowan of Illinois until Frye and Willman (1960) dropped Iowan as a glacial substage and introduced the term "Morton" for the loess previously called Iowan.

As the term "Iowan" was used in Iowa for a till, now believed to be the Kansas till, it is proper that it now be dropped. In northern Illinois this "real" till is now given the rock-stratigraphic name of Winnebago and is referred to the Altonian Substage of the Wisconsinan (Frye and others, 1969).

It may be noted that Carman (1917) explained the "Iowan drift" in northwestern Iowa as eroded Kansas till. Like Ruhe's explanation for eastern Iowa 50 years later, he suggested that erosion had truncated drift of two different ages. The idea was so revolutionary that it was not allowed to be printed by the Iowa State Geologist until it was considerably modified (Kay and Apfel, 1929, p. 114). Later, Carman's ideas were

[12] Much of Chamberlin's description was based on the detailed descriptions, diagrams, and illustrations of McGee, who, however, did not apply names to these units, but described them as "upper till" and "lower till" (McGee, 1891, p. 472–542).

more acceptable, particularly about the two ages of the drift, and were printed in a revision of his earlier bulletin (Carman, 1931).

INTRODUCTION OF ROCK-STRATIGRAPHIC TERMINOLOGY AND DISTINCTION BETWEEN ROCK AND TIME TERMS

Until about 1960 glacial deposits were referred to by names which carried both time and rock connotation.[13] For example, the term "Tazewell" for early "classical" Wisconsinan was used from the eastern United States to the western. By 1960 Pleistocene deposits began to be regarded as rock strata which not only could, but should, be described in rock-stratigraphic terms according to the Stratigraphic Code. Time and space does not permit a review of the exploding use of rock-stratigraphic terminology in the decade 1960 to 1970, but two examples will be used.

In Illinois the work of Frye and Willman (1960) on the drift of the Lake Michigan Lobe led to the time-stratigraphic classification of the Wisconsinan Stage into the Altonian Substage, in which were deposited the Winnebago Drift and Roxana Silt; the Farmdalian Substage, in which were deposited silt and peat; the Woodfordian Substage, in which were deposited tills and loesses; the Twocreekan Substage; and the Valderan Substage. Further elaboration of time-stratigraphic and rock-stratigraphic classifications was presented by Frye and others (1968, 1969), and by Willman and Frye (1970). The time terms "Altonian" and "Woodfordian" as Wisconsinan substages have now been used by some workers for deposits at least as far east as Pennsylvania (White and others, 1969, p. 9) and as far west as Saskatchewan (Christiansen, 1968, p. 335).

The author and his associates (Shepps and others, 1959) used rock-stratigraphic terms for Wisconsinan tills in northwestern Pennsylvania in 1959 and used "Cary" and "Tazewell" as time terms. In 1960 and later, Wisconsin tills in northeastern Ohio were defined as rock-stratigraphic units according to rigid Code specifications (White, 1960), again using "Tazewell" and "Cary" as time terms, but in 1966 (p. 20) "uncertain" was used in place of "Tazewell." A few years later, White (1969) and White and others (1969), were using the same rock-stratigraphic terms in Pennsylvania and Ohio, but the time terms now were "Wisconsinan Stage" with "Altonian," "Farmdalian," and "Woodfordian" Substages.

CONCLUSION—SUMMARY OF STAGES IN INVESTIGATION OF WISCONSINAN DEPOSITS

The surface material of the Central Interior is mainly late Pleistocene (Wisconsinan) in age. During the past 130 years the investigation of this

[13] Multiple classification, to the degree of time-stratigraphy and rock-stratigraphy, was introduced for the Pleistocene of Kansas by Frye and Leonard (1952).

"drift" and the classification of its parts were based on the materials themselves, the varying amount of surface weathering and dissection, the intertill materials, and other features. These investigations progressed through a series of stages, each roughly 20 to 30 years in duration.

Recognition of Drift and its Possible Iceberg Origin—1815-1840

Erratics and the matrix in which they were embedded were observed and an iceberg origin suggested by Drake as early as 1815, and soon thereafter by others, including Maclure in 1823. The observations increased rapidly in the 1830s as state surveys were organized. It was in this period that the concept of origin of the drift from icebergs floating in a continent-covering sea was increasingly common.

Recognition of Glacial Origin of the Drift—1841-1860

In this period the belief in iceberg origin of the drift slowly gave way to ice sheet origin. The activity of state surveys decreased from the level of the 1830s, but independent geologists continued field observations.

Recognition of Multiple Glaciation—1865-1894

In the early part of this period state surveys became more active. In the later part of the period the United States Geological Survey was established and soon supported work on glacial deposits under the leadership of T. C. Chamberlin. Interglacial deposits were recognized, and it was realized that more than one episode of glaciation was required to explain the sequence of deposits. By 1875 Chamberlin determined that the last glacial episode was complex, and in it there had been several advances and retreats of the ice sheets. In 1894 and 1895 he proposed the name "Wisconsin" and stated it had several divisions.

Beginning of Detailed Field Study—1895-1915

Topographic maps for parts of the Central Interior began to be available by 1895 and glacial studies took a quantum jump. Led by Frank Leverett, detailed field study, in some cases square mile by square mile was inaugurated. The four encyclopedic monographs of Leverett provided the basic knowledge of glacial geology of the Central Interior. Leverett's divisions of early, middle, and late Wisconsin, and the named moraines and morainal systems became a classification to be used for 40 years; many of his moraine names are still used. During this period he had varying views on the place of the Iowan.

Elaboration of Detailed Field Study—1915–1940

During this period topographic maps became available for most of the Central Interior. More and more detailed studies could be made, and moraine mapping was highly developed. Each moraine was believed to mark an important stand of the ice front and to record an ice advance to that position. Discordant moraines that cut off others at an angle were supposed to mark more important episodes of retreat, "regrouping," readvance, and a change of direction of advance of the continental glacier. It was assumed that each moraine was made of the till of the last ice advance to that position, and that this till extended northward from the moraine to the margin of the next younger moraine. It was in this period that Leighton proposed his widely used classification of the Wisconsinan substages—Iowan, Tazewell, Cary, and Mankato.

Air Photos Come into Use—1940–1950

From about 1940 air photos began to supplement maps as tools for observation and as bases for recording field data. By the use of these for observing topography, drainage, and soil types, much more refined mapping of surface drift, moraines, and other features became possible. Moraine mapping reached its highest degree of detail.

C-14 Dating—1950-

Soon after 1950[14] enough C-14 dates began to appear to make it possible for the first time to attach ages in years[15] to different drift units. At first the ages were regarded as shockingly low, but mental adjustment was soon made to the chronology they imposed (Horberg, 1955). The earliest Wisconsinan drift—"classical Wisconsin" or Tazewell—was only about 24,000 years old! More and more dates between 30,000 and 40,000 years appeared for "pre-Wisconsin" deposits. It was increasingly recognized that a drift, earlier than classical Wisconsin, but much younger than Illinoian, was widespread; and that the weathered or organic material, or both, in the subsurface was not Sangamonian. After stillborn proposals of "Ohioan," "Rockian," and other terms (but curiously not of "Iowan"), the name "Altonian" began to achieve wider and wider usage for this early Wisconsinan substage.

---

[14]The history of the discovery and use of C-14 dating in Pleistocene geology would require a lengthy paper in itself—parts have appeared in various places, but a connected study would produce a long thesis in geology or in history of geology.

[15]The term "absolute age" is sometimes used, but the term "radiocarbon age" is to be preferred, because radiocarbon years are not calendar or solar years, and are known to vary through time. Where the term "years," referring to age of deposit, occurs in this paper, it should be taken as "radiocarbon years."

Stratigraphic and Subsurface Study—1960-

The great increase in deep excavations that started in the late 1950s for superhighways, subways, strip mining, quarrying, and large building foundations was continuing at an accelerated rate, laying bare an almost embarrassment of riches in the form of drift anatomy that showed what was "really" below the surface over wide areas and not just in an occasional narrow ravine or in a stream cutbank.[16] Drill cores and samples from engineering investigations became available—indeed, more of them than there were Pleistocene geologists to study all of them. Pleistocene geology made a rapid right-angle turn from mainly study of surface morphology and surface drift (including soil study) to stratigraphic study of the whole drift column. It was discovered that—although there are some exceptions—the surface drift is usually only a few feet thick, and that the "real story" was below the surface. It was also learned that till sheet margins do not always coincide with moraines. Moraine mapping became less "interesting," for it became apparent in many cases that the surface drift did not make the bulk of a moraine, but was only a veneer over older material. While moraines remain important geomorphic features, each one must be critically re-examined to determine if indeed it does mark the edge of the last ice sheet in the region.[17]

Through extensive subsurface study of drift units, then tracing them to surface outcrop, it became more and more evident that these were really rock-stratigraphic units that could be treated according to the Stratigraphic Code. Although the distinctiveness of some Wisconsinan tills was known as early as 1930, or even before, it was only in this period that rock-stratigraphic names began to be applied to Wisconsinan drifts, and assigned to time-stratigraphic substages. These units began to be studied in the laboratory as rocks and it was discovered that over wide areas different tills maintained distinctive color, texture, lithology, and content of heavy, light, and clay minerals.

Exploratory Drilling—1960-

Drilling specifically for subsurface Pleistocene study has been done in several states, especially Illinois, Iowa, and Saskatchewan, and is now

[16] "There are extensive areas, for instance in southern Ontario, where 100–300 feet of Pleistocene stratigraphy are well visible in natural exposures, and therefore the 'real story' has been deciphered even without deep drilling . . . during the last 20 years" (Dreimanis, 1970, written commun.).

[17] The reader must be warned that the author's statements here are based on "conditioning" from working mainly on the drifts on the eastern sublobes of the Erie Lobe. Here the ice advanced upslope out of the basin and did not go far, so that the pulsating glacier piled thin sheet on thin sheet. In the Lake Michigan Lobe, where till sheets may indeed be thick, there was a rapid downslope flow of the ice, followed by a pulsing retreat with several major readvances, producing the situation described above under the heading "Elaboration of detailed field study—1915–1940."

being stepped up in intensity.[18] In many states samples from state highway and ground-water investigations have been available and have been studied in varying degrees of detail. Saskatchewan, where drift is more than 1,000 ft thick over wide areas, leads the way in exploratory drilling—each Pleistocene geologist is issued a powerful drilling rig, with two shifts of geologically skilled crews for collection of samples, cores, and electric logs, plus a field laboratory, and a headquarters chemical laboratory for more detailed mineral and chemical analyses. Investigations to a depth of 1,100 ft are routine.

Geophysical Methods—1970 (?)-

For more than 20 years it has been possible to distinguish some drift units by electrical and seismic methods. Electric logging of bore holes, routine in petroleum geology, has been almost unused in Pleistocene studies, except in Saskatchewan. The whole field of Pleistocene geophysics is still essentially uncultivated but will be explored and used widely in the future.

Paleontology

The history of study of Pleistocene pollen, grass and other seeds, insects, gastropods and other invertebrates, and vertebrates, almost entirely found in Holocene or interglacial and interstadial deposits, would make a multivolume monograph. The stratigraphic usefulness of these materials has not been great to this time, because most geologists are not comfortable in the biological field, and the biologists are uncertain in the geological field. More and more Pleistocene workers are becoming able to operate in the two worlds. Great results are already beginning to appear.

Another paleontological field is that of macro- and micro-fossils in till derived from fossiliferous pre-Pleistocene rocks. The fossil content of till—*Tasmanites*, bryozoa, conodonts, and other forms—may be characteristic of a given till. Studies in Holland and in Great Britain already give indication of this happy situation.

The Next Stage—1970-

Pleistocene geology in the Central Interior for the past 130 years has moved in 25- to 35-year cycles. The first period, to 1870, was of recognition of ice sheet origin, followed by the Chamberlin period to 1895, the Leverett period to 1930, the Leighton period to 1955, and the present

[18] The 1969 recommendation of the Quaternary Subcommittee of the National Advisory Committee [of Canada] on Research in the Geological Sciences has been accepted in the following wording: "The mapping should include stratigraphic investigations and test drilling to a greater extent than in the past" (Dreimanis, 1970, written commun.).

period dominated by those heretics, now become respectable, who treat Wisconsinan deposits as rock-stratigraphic units.

The next period, soon to be entered—for geological periods become exponentially shorter—will certainly be one in which subsurface geology, the study of the whole drift column, will predominate. Anything less will not be "respectable." Pleistocene units are rock units, and will be treated strictly as such. Geophysical methods, both in boreholes and from the surface, will be as routine as in petroleum geology, and will enable the geologist to look into the drift as he now can look into cores by radiography. Paleontology in its many branches will be extensively used. New isotopic dating methods will sooner or later be found to extend the record to 100,000 years and hopefully far beyond. The study of deep-sea Pleistocene sediments will be more and more correlated with terrestrial and continental shelf Pleistocene deposits. All this will require teamwork. The day of Leverett with his little trowel, or of Leighton (and many others!) with only a Pleistocene pick and an acid bottle, is over. The team will be composed of a geological drilling crew, mineralogist-petrographer, geochemist, geophysicist, geochronologist, and stratigrapher, led by a Pleistocene geologist. The cost, although high in dollars, will be moderate in comparison with the cost per day of marine geology and oceanographic ship-time or of North Slope petroleum exploration. After all, over half the people in the United States and all those of Canada live on, and more and more in, Pleistocene (mainly Wisconsinan) deposits that constitute a major part of their environment. These deposits are a key factor in that now magic term "environmental geology."

ACKNOWLEDGMENTS

An earlier draft of this paper was sent to other participants in this Symposium; their useful comments are much appreciated. I am particularly indebted to J. C. Frye and H. B. Willman for additional references, and especially to the latter for suggestions which saved me from some errors and which led to extensive revisions; and to R. P. Goldthwait for many suggestions and especially for gently tempering some of my heretical enthusiasms in the latter part of the paper. He has almost convinced me that "heresy by itself, however, is no token of truth."

BIBLIOGRAPHY

Agassiz, Louis, 1840, Etudes sur les glaciers: Ouvrage accompagne d'un atlas de 32 planches; aux frais l'auteur: Neuchatel, 347 p. and atlas of 32 pls.

Alden, W. C., and Leighton, M. M., 1917, The Iowan drift, a review of the evidences of the Iowan Stage of glaciation: Iowa Geol. Survey Ann. Rept. 1915, v. 26, p. 49-212.

Black, R. F., 1970, Glacial geology of Two Creeks Forest Bed, Valderan type locality, and northern Kettle Moraine State Forest: Wisconsin Geol. and Nat. History Survey Inf. Circ. 13, 40 p.

Black, R. F., Bleuer, N. K., Hole, F. D., Lasca, N. P., Maher, L. J., 1970, Pleistocene geology of southern Wisconsin, a field trip guide: Wisconsin Geol. and Nat. History Survey Inf. Circ. 15, 175 p.

Carman, J. E., 1917, The Pleistocene geology of northwestern Iowa: Iowa Geol. Survey, v. 26, p. 233-445.

____ 1931, Further studies on the Pleistocene geology of northwestern Iowa: Iowa Geol. Survey, v. 35, p. 15-193.

Chamberlin, T. C., 1877, Geology of eastern Wisconsin: Geology of Wisconsin, Survey of 1873-1877: Wisconsin Geol. Survey, v. 2, p. 91-405.

____ 1878, On the extent and significance of the Wisconsin Kettle Moraine; n.p., n.d. [1878], 36 p.: a separate publication of a paper in Wisconsin Acad. Sci., Arts and Letters Trans., v. 4, p. 201-234.

____ 1882, The bearing of some recent determinations on the correlation of the eastern and western terminal moraines: Am. Jour. Sci., v. 124, p. 93-97.

____ 1883, Preliminary paper on the terminal moraine of the second glacial epoch: U.S. Geol. Survey, 3d Ann. Rept., p. 291-402.

____ 1894, Glacial phenomena in North America, *in* Geikie, J., The Great Ice Age (3d ed.): London, p. 724-775.

____ 1895, The classification of American glacial deposits: Jour. Geology, v. 3, p. 270-277.

____ 1896, Nomenclature of glacial formations: Jour. Geology, v. 4, p. 872-876.

Chamberlin, T. C., and Salisbury, R. D., 1885, Preliminary paper on the Driftless Area of the upper Mississippi Valley: U.S. Geol. Survey, 6th Ann. Rept., p. 199-322.

Christiansen, E. A., 1968, A thin till in west-central Saskatchewan, Canada: Canadian Jour. Earth Sci., v. 5, p. 329-336.

Conrad, T. A., 1839, Notes on American geology; observations on characteristic fossils, and upon a fall of temperature in different geological epochs: Am. Jour. Sci., v. 35, p. 237-251.

Croll, J. A., 1875, Climate and time in their geological relations; a theory of secular changes of the earth's climate: London, Daldy, Isbister & Co., 577 p., 8 pls.

Drake, Daniel, 1815, Natural and statistical view or picture of Cincinnati and the Miami country: Cincinnati, Looker & Wallace, 251 p, 2 maps.

____ 1825, Geological account of the Valley of the Ohio: Am. Philos. Soc. Trans., v. 2, p. 124-139.

Dreimanis, A., and Goldthwait, R. P., 1973, Wisconsin glaciations in the Huron, Erie, and Ontario Lobes: Geol. Soc. America Mem. 136, p. 71-105.

Flint, R. F., 1965, Introduction: Historical perspectives, *in* Wright, H. E., Jr., and Frey, D. G., eds., The Quaternary of the United States: Princeton, N. J., Princeton Univ. Press., p. 3-11.

Forsyth, J. L., 1961, Pleistocene geology, *in* Root, S. I., Rodriguez, Joaquin, and Forsyth, J. L., Geology of Knox County: Ohio Div. Geol. Survey Bull. 59, p. 107-138, pl. 5.

____ 1965, Age of the buried soil in the Sidney, Ohio, area: Am. Jour. Sci., v. 263, p. 571-597.

Frye, J. C., and Leonard, A. B., 1952, Pleistocene geology of Kansas: Kansas Geol. Survey Bull. 99, 230 p.

Frye, J. C., and Willman, H. B., 1960, Classification of the Wisconsinan Stage in the Lake Michigan glacial lobe: Illinois Geol. Survey Circ. 285, 16 p.

—— 1963, Development of Wisconsinan classification in Illinois related to radiocarbon chronology: Geol. Soc. America Bull., v. 74, p. 501-506.

Frye, J. C., Willman, H. B., and Black, R. F., 1965, Outline of glacial geology of Illinois and Wisconsin, *in* Wright, H. E., Jr., and Frey, D. G., eds., The Quaternary of the United States: Princeton, N.J., Princeton Univ. Press, p. 43-61.

Frye, J. C., Willman, H. B., Rubin, M., and Black, R. F., 1968, Definition of Wisconsinan Stage: U.S. Geol. Survey Bull. 1274-E, 22 p.

Frye, J. C., Glass, H. D., Kempton, J. P., and Willman, H. B., 1969, Glacial tills of northwestern Illinois: Illinois Geol. Survey Circ. 437, 45 p.

Geikie, James, 1894, The Great Ice Age (3d ed.): London, 850 p. (Also see review by Salisbury, R. D., 1894, Jour. Geology, v. 2, p. 730-747.)

—— 1895, Classification of European glacial deposits: Jour. Geology, v. 3, p. 241-269.

Goldthwait, R. P., Dreimanis, Aleksis, Forsyth, J. L., Karrow, P. F., and White, G. W., 1965, Pleistocene deposits of the Erie Lobe, *in* Wright, H. E., Jr., and Frey, D. G., eds., The Quaternary of the United States: Princeton, N.J., Princeton Univ. Press, p. 85-97.

Gooding, Ansel, 1963, Illinoian and Wisconsin glaciations in the Whitewater Basin, southeastern Indiana, and adjacent areas: Jour. Geology, v. 71, p. 665-682.

Hitchcock, Edward, 1832, Report on the geology of Massachusetts, examined under the direction of the government of that state during the years 1830 and 1831: Am. Jour. Sci., v. 22, p. 1-70.

—— 1841a, Final report on the geology of Massachusetts: Northampton, Mass., J. H. Butler, 831 p.

—— 1841b, First anniversary address before the Association of American Geologists at their second annual meeting in Philadelphia, April 5, 1841: reprint New Haven, Conn., B. A. Hamlen, 48 p.; Am. Jour. Sci., v. 41, p. 232-275.

Horberg, C. L., 1955, Radiocarbon dates and Pleistocene chronological problems in the Mississippi Valley region: Jour. Geology, v. 63, p. 278-286.

Kay, G. F., 1931, Classification and duration of the Pleistocene period: Geol. Soc. America Bull., v. 42, p. 425-466.

Kay, G. F., and Apfel, E. T., 1929, The pre-Illinoian Pleistocene geology of Iowa: Iowa Geol. Survey, 304 p., 3 pls.

Kay, G. F., and Leighton, M. M., 1933, Eldoran epoch of the Pleistocene period: Geol. Soc. America Bull., v. 49, p. 669-674.

Keyes, C., 1928, Theory of multiple glaciations: Pan-Am. Geologist, v. 50, p. 131-144.

Leighton, M. M., 1913, An exposure showing post-Kansan glaciation near Iowa City, Iowa: Jour. Geology, v. 21, p. 431-435.

—— 1926, A notable type Pleistocene section; the Farm Creek exposure near Peoria, Illinois: Jour. Geology, v. 34, p. 167-174.

—— 1931, The Peorian loess and the classification of the glacial drift sheets of the Mississippi Valley: Jour. Geology, v. 39, p. 45-53.

—— 1933, The naming of the subdivisions of the Wisconsin glacial age: Science, v. 77, p. 168.

—— 1957, The Cary-Mankato-Valders problem: Jour. Geology, v. 65, p. 108-111.

——— 1960, The classification of the Wisconsin glacial stage of the north-central United States: Jour. Geology, v. 68, p. 529-552.

——— 1964, Elements in the classification of the late glacial Quaternary of midwestern North America: Advancing frontiers in geology and geophysics: Hyderabad, India, Osmania Univ. Press, p. 115-133.

Leighton, M. M., and Willman, H. B., 1950, Loess formations of the Mississippi Valley: Jour. Geology, v. 58, p. 599-623.

Lemke, R. W., Laird, W. M., Tipton, M. J., and Lindvall, R. M., 1965, Quaternary geology of the northern Great Plains, *in* Wright, H. E., Jr., and Frey, D. G., eds., The Quaternary of the United States: Princeton, N.J., Princeton Univ. Press, p. 15-27.

Leverett, F., 1897, The Pleistocene features and deposits of the Chicago area: Chicago Acad. Sci. Geol. and Nat. History Survey Bull. 2, 86 p.

——— 1898a, The weathered zone (Sangamon) between the Iowan loess and Illinoian till sheet: Jour. Geology, v. 6, p. 171-181.

——— 1898b, The Peorian soil and weathered zone (the Toronto Formation?): Jour. Geology, v. 6, p. 244-249.

——— 1899, The Illinois glacial lobe: U.S. Geol. Survey Mon. 38, 818 p., 24 pls.

——— 1901, Glacial formations and drainage features of the Erie and Ohio Basins: U.S. Geol. Survey Mon. 41, 802 p., 26 pls.

——— 1909, Weathering and erosion as time measures: Am. Jour. Sci., v. 177, p. 349-368.

——— 1929, Moraines and shore lines of the Lake Superior region: U.S. Geol. Survey Prof. Paper 154, 72 p.

——— 1930, Relative length of Pleistocene glacial and interglacial stages: Science, v. 72, p. 193-195.

——— 1931, Quaternary system, *in* Cushing, H. P., Leverett, Frank, and Van Horn, F. R., Geology and mineral resources of the Cleveland district, Ohio: U.S. Geol. Survey Bull. 818, p. 57-81.

——— (with contributions by F. W. Sardeson), 1932, Quaternary geology of Minnesota and parts of adjacent states: U.S. Geol. Survey Prof. Paper 161, 149 p., 5 pls.

——— 1934, Glacial deposits outside the Wisconsin terminal moraine in Pennsylvania: Pennsylvania Geol. Survey Bull., v. 7, 123 p.

Leverett, Frank, and Taylor, F. B., 1915, The Pleistocene of Indiana and Michigan and the history of the Great Lakes: U.S. Geol. Survey Mon. 53, 529 p., 32 pls.

Lyell, Charles, 1849, A second visit to the United States of North America: London, John Murray, 2 vols., 273, 287 p.

Maclure, William, 1823, Some speculative conjectures on the probable changes that may have taken place in the geology of the continent of North America: Am. Jour. Sci., v. 6, p. 98-102.

McGee, W. J., 1878, On the relative positions of the forest bed and associated drift formations in northeastern Iowa: Am. Jour. Sci., v. 15, p. 339-341.

——— 1891, The Pleistocene history of northeastern Iowa: U.S. Geol. Survey, 11th Ann. Rept., pt. 1, p. 189-577.

McGee, W. J., and Call, R. E., 1882, On the loess and associated deposits of Des Moines: Am. Jour. Sci., v. 124, p. 202-233.

Merrill, G. P., 1906, The development of the glacial hypothesis in America:

Popular Sci. Monthly, v. 68, p. 300-322.

Merrill, G. P., 1924, The first one hundred years of American geology: New Haven, Conn., Yale Univ. Press, 773 p. (Reprint, 1964, New York, Hafner Publishing Company, Inc., 773 p.)

Muller, E. H., 1965, Quaternary geology of New York, *in* Wright, H. E., Jr., and Frey, D. G., eds., The Quaternary of the United States: Princeton, N. J., Princeton Univ. Press, p. 99-112.

Newberry, J. S., 1870, On the surface geology of the basin of the Great Lakes, and the valley of the Mississippi: Annals Lyceum Nat. History New York, v. 9, p. 213-234.

____ 1874, Geology of Ohio; Surface geology: Geol. Survey Ohio, Rept. 2, pt. 1, p. 1-80, maps.

____ 1884, The drift deposits of Indiana: Indiana Geol. Survey, 14th Ann. Rept., p. 85-97.

Newbiggin, M. I., and Flett, J. S., 1917, James Geikie, the man and the geologist: Edinburgh, Oliver and Boyd, 227 p.

Orton, Edward, 1870, On the occurrence of a peat bed beneath deposits of drift in southwestern Ohio: Am. Jour. Sci., v. 50, p. 54-57, 293.

____ 1873, Report on the third geological district; geology of the Cincinnati group; Hamilton, Clermont, Clarke Cos.: Ohio Geol. Survey, Rept. 1, pt. 1, p. 365-480.

Ruhe, R. H., 1969, Quaternary landscapes in Iowa: Ames, Iowa, State Univ. Press, 255 p., colored map.

St. John, S., 1851, Elements of geology, intended for the use of students: Hudson, Ohio, Pentagon Press, 334 p.

Salisbury, R. D., 1893, Distinct glacial epochs and the criteria for their recognition: Jour. Geology, v. 1, p. 61-84.

Shaffer, P. R., 1954, Farmdale drift: Science, v. 119, p. 693-694.

____ 1956, Farmdale drift of northwestern Illinois: Illinois Geol. Survey Rept. Inv. 198, 25 p.

Shepps, V. C., White, G. W., Droste, J. B., and Sitler, R. F., 1959, Glacial geology of northwestern Pennsylvania: Pennsylvania Geol. Survey Bull. G 32, 54 p., colored map.

Stewart, D. P., and MacClintock, Paul, 1969, The surficial geology and Pleistocene history of Vermont: Vermont Geol. Survey Bull. 31, 251 p.

Stout, Wilber, 1943, Glacial geology, *in* Stout, Wilber, Ver Steeg, K., and Lamb, G. F., Geology of water in Ohio: Ohio Geol. Survey (4th ser.), Bull. 44, p. 17-50, pls. 1-3, 5, 7.

Thomas, David, 1819, Travels through the western country in the summer of 1816, including notices of the natural history . . .: Auburn, N.Y., David Rumsey, 320 p.

Thwaites, F. T., 1927, The development of the theory of multiple glaciation in North America: Wisconsin Acad. Sci. Trans., v. 23, p. 41-164.

____ 1943, Pleistocene of part of northeastern Wisconsin: Geol. Soc. America Bull., v. 54, p. 87-144.

Torrell, O., 1877, On the glacial phenomena of North America: Am. Jour. Sci. (3d ser.), v. 13, p. 76-79.

Trowbridge, A. C., 1912, Geology and geography of the Wheaton quadrangle: Illinois Geol. Survey Bull. 19, 79 p.

Upham, W., 1883, The Minnesota Valley in the Ice Age: Am. Jour. Sci., v. 127, p. 34-42, 104-111.

____ 1884, The geology of Faribault County (and other counties): Minnesota Geol. Survey, v. 1, p. 404–532.

____ 1895, Review of the Great Ice Age and its relation to the antiquity of man, by James Geikie, 3d ed.: Am. Geologist, v. 15, p. 52–57.

Volney, C. F., 1803, Tableau du climat et du sol des États Unis d'Amérique. Suivi d'eclaircissements sur la Floride, sur la colonie Française au Scioto, sur quelques colonies Canadiennes et sur les sauvages. Enrichi de quatre planches gravées, dont deux cartes géographiques et une coupe figurée de la chûte de Niagara: Paris, Courcier, Dentu, An XII-1803, 2 v., xvi, 300 and 532 p., 2 folding maps, 2 folding plates.

____ 1804, A view of the soil and climate of the United States of America: C. B. Brown, trans., Philadelphia, J. Conrad & Co. 446 p., maps (Reprinted, with introduction by G. W. White, New York, Hafner Publishing Company, Inc., 1968).

Wascher, H. L., Humbert, R. P., and Cady, J. G., 1948, Loess in the southern Mississippi Valley—Identification and distribution of loess sheets: Soil Sci. Soc. America Proc. 1947, v. 12, p. 389–399.

Wayne, W. J., 1963, Pleistocene formations of Indiana: Indiana Geol. Survey Bull. 25, 85 p.

Wayne, W. J., and Zumberge, J. H., 1965, Pleistocene geology of Indiana and Michigan, *in* Wright, H. E., Jr., and Frey, D. B., eds., The Quaternary of the United States: Princeton, N.J., Princeton Univ. Press, p. 63–84.

White, G. W., 1934, Illinoian drift region of northeast central Ohio: Ohio Jour. Sci., v. 37, p. 1–19.

____ 1951, Illinoian and Wisconsin drift of the southern part of the Grand River Lobe in eastern Ohio: Geol Soc. America Bull., v. 62, p. 967–977.

____ 1960, Classification of glacial deposits in northeastern Ohio: U.S. Geol. Survey Bull. 1121-A, 12 p.

____ 1966, Glacial geology, *in* Winslow, J. D., and White, G. W., Geology and ground-water resources of Portage County, Ohio: U.S. Geol. Survey Prof. Paper 511, p. 17–38.

____ 1967, The first appearance in Ohio of the theory of continental glaciation: Ohio Jour. Sci., v. 67, p. 210–217.

____ 1969, Pleistocene deposits of the north-western Allegheny Plateau, U.S.A.: Geol. Soc. London Quart. Jour., v. 124, p. 131–151.

White, G. W., Totten, S. M., and Gross, D. L., 1969, Pleistocene stratigraphy of northwestern Pennsylvania: Pennsylvania Geol. Survey Bull. G-55, 88 p., 43 figs.

Whittlesey, Charles, 1866, On the fresh-water glacial drift of the northwestern states: Smithsonian Contr. Knowledge 197 (also in v. 15, 1867), 32 p.

Willman, H. B., and Frye, J. C., 1970, Pleistocene of Illinois: Illinois Geol. Survey Bull. 94, 204 p.

Winchell, N. H., 1873, The surface geology: Minnesota Geol. and Nat. History Survey, 1st Ann. Rept., p. 61–62.

____ 1875, The geology of Mower County: Minnesota Geol. and Nat. History Survey, 3d Ann. Rept., p. 20–36.

____ 1878, The geology of Rock and Pipestone Counties: Minnesota Geol. and Nat. History Survey, 6th Ann. Rept., p. 93–111.

Worthen, A. H., 1868, Geology and paleontology: Illinois Geol. Survey, v. 3, 574 p., 20 pls.

Wright, G. F., 1884, The glacial boundary in Ohio, Indiana and Kentucky: Cleveland, Ohio, Western Reserve Historical Co., Tract No. 60, p. 199–263.

—— 1889, The ice age in North America and its bearing on the antiquity of man: New York, D. Appleton and Company, 622 p.

Wright, H. E., Jr., and Frey, D. G. (eds.), 1965, The Quaternary of the United States, a review volume for the VII Congress of the International Association for Quaternary Research: Princeton, N. J., Princeton Univ. Press, 922 p.

Wright, H. E., Jr., and Ruhe, R. V., 1965, Glaciation of Minnesota and Iowa, *in* Wright, H. E., Jr., and Frey, D. G., eds., The Quaternary of the United States: Princeton, N.J., Princeton Univ. Press, p. 29–41.

Manuscript Received by The Society December 27, 1971

Printed in the United States of America

GLACIAL GEOLOGY OF OHIO
IN 1874
THE END OF AN ERA

THE OHIO JOURNAL OF SCIENCE 76(2): 51, March, 1976

GLACIAL GEOLOGY OF OHIO IN 1874—THE END OF AN ERA[1, 2]

GEORGE W. WHITE, Department of Geology, University of Illinois, Urbana 61801

Abstract. The Second Geological Survey of Ohio in 1869–73 produced more or less detailed reports and maps of the 88 counties of which 83 were published in 1873, 1874 and 1878. Each report included a shorter or longer section on "Surface Geology," all organized in a similar fashion, as directed by J. S. Newberry, State Geologist. In 1874 Newberry summarized, in an 80-page essay, the knowledge of glacial deposits in Ohio and the current theories of their origin. The synthesis was developed from his own work and that of M. C. Read in northern and northeastern Ohio, G. K. Gilbert and N. H. Winchell in northwestern Ohio, and Edward Orton in southwestern Ohio. It was recognized that the "bowlder clay" was an ice sheet deposit. It was believed that erractics were concentrated on the surface and had been transported by icebergs in an inland lake, which was later drained. An extensive "Forest Bed" recorded a period of warmer climate, named by Orton an "Interglacial Period", the first use of this term. The detailed descriptions, maps and sections are valuable records of the observations of these outstanding geologists as are the explanations they proposed for the Quaternary history of Ohio.

OHIO J. SCI. 76(2): 51, 1976.

Although glacial features and deposits have been observed in Ohio for more than 200 years, the earliest scientific observations were by C. F. Volney, who published in 1803 a diagram of outwash deposits in Cincinnati. He recognized the sediments as stream deposits, but did not associate them with glacial melt-water (White, 1973). Observations of till and erratics by Drake in 1815, followed by reports of other workers who recognized widespread polishing, striation, and grooving of rock surfaces were initially attributed to the action of icebergs in an inland sea, but later related to ice sheets (Leverett, 1902, p. 24–29; White, 1973). The theory of glaciation by an ice sheet proposed by Agassiz in 1840 was embraced in the United States by Hitchcock in 1841, and was widely accepted in Ohio by 1851, as shown by the extensive treatment of ice-sheet glaciation in the geological textbook of Samuel St. John, published at Hudson, Ohio (White, 1967).

The first Geological Survey of Ohio (the "Mather Survey") was established in 1837 and discontinued in 1839. This left Ohio without a survey until the Second Survey (the "Newberry Survey") was established in 1869. The extensive publications of the Second Survey provided an unparalleled exhibit of the knowledge of the time regarding the origin and effects of glacial deposits throughout Ohio.

The publications of the Newberry Survey deserve attention because they mark the end of an era in glacial studies. Immediately after their appearance, new concepts based on moraine studies led to views of much greater complexity and provided a broader framework for understanding the origin of glacial deposits. Leverett's great monograph (1902) marked the height of the "new dispensation," which is still congenial to glacial geologists 75 years later. This new framework is not the subject of this paper, as it has been partly explored elsewhere (White, 1973, p. 13–16, 18–20).

In the Second Survey, the State was divided into four districts, and the reports for 83 of the 88 counties were pre-

[1]Manuscript received May 30, 1975 (#75-37).

[2]In 1926, George White published his first paper in The Ohio Journal of Science, "The Limestone Caves and Caverns of Ohio," Ohio J. Sci. 26(2): 73–116. Dr. White asked us to hold the paper for publication until the March 1976 issue so that we could celebrate his 50th year of contributions to the Science of Geology together. We extend our heartiest congratulations to Dr. White.

pared by 1873. For some reason, Adams, Scioto, Lawrence, Jackson, and Vinton Counties, all unglaciated, were not published. The State Geologist, Dr. J. S. Newberry, a distinguished physician, explorer, paleontologist, and stratigrapher, issued to all his assistants elaborate and detailed "directions for observing and collecting." He directed them to collect information on surface geology—"special materials (clay, sand, gravel and so forth)." He instructed them to record whether these materials were of local or foreign origin, note if they were stratified, and tell their thicknesses and what fossils were present. Was the bedrock surface planed, scratched, or furrowed, and in what direction? What was the composition, extent, and altitude of terraces and lake ridges? What was the extent of peat bogs and marl beds, determined by boring? Were there fossils of elephant, mastodon, or other animals? What was the depth to rock, including that in valleys? (Newberry, 1870, p. 11).

These detailed instructions were taken seriously by all the geologists. The various county reports follow the same general pattern, and all of them include descriptions, some of them very extensive, of the surface geology. The detailed sections and diagrams continue to be useful today. The county reports, all of which seem to have been written by 1873, were published in three volumes in 1873, 1874, and 1878. Those for the northeastern counties were prepared by Newberry and M. C. Read; those for the northwestern counties by G. K. Gilbert and N. H. Winchell; for the southwest by Edward Orton, and for the southeast (mostly unglaciated) by E. B. Andrews. Several counties were reported on by others (see "Contents" for each volume). The staff was notably able and four of the men had already, or were about to achieve, national eminence and were to become members of the prestigious National Academy of Science.

NEWBERRY'S "FACTS"

A summary and synthesis of the county reports was prepared by Newberry (1874). He noted that "Drift deposits" cover so much of Ohio that "the Quaternary system deserves . . . as full and thorough an exposition as our . . . space will permit." He had before him the reports of his staff for all the counties of the state, and from these and his own extensive observations, he was able to synthesize the great mass of data, and to relate the Ohio material to what he and others knew of other parts of the Continent.

Newberry summarized the "important facts" in straightforward statements of observations in the first four pages of his essay. In the next four pages, he set forth the "history deduced from facts cited." He then devoted 70 pages to a detailed consideration of all the "facts" he had listed earlier. He included many maps and sections, mainly using those already published in the various county reports by himself and others, but without giving exact citations to their first appearance. These "facts" and "history" will be discussed in the order in which Newberry listed them.

1. Ice had ground down the surface "as low as the fortieth parallel of latitude . . . The track of a glacier is as unmistakable as that of a man or a bear." The scratches are generally north-south, but locally conform "in a rude way to the present topography." His map of the "Drift Area" is the best depiction to that time of the glacial boundary from New Jersey to Kansas (White, 1973, p. 12, Fig. 4).

2. Some of the valleys have been excavated during the "ice period, or an earlier epoch" far below the present lakes and stream levels.

3. "Bowlder clay," a tough, blue, unstratified clay, with stones, lies upon the striated bedrock.

4. At certain places the bowlder clay is overlain by fine laminated clay without pebbles, the "Saugeen clay." This is the upper part of the "Erie clay" of William Logan: the lower part of the Erie clay is the bowlder clay: both are "parts of one formation."

5. Over the surface of the bowlder clay is found "over a large area in Ohio and other Western States, a layer of carbonaceous matter, with logs, stumps, and sometimes upright trees . . . the Forest Bed, *. . . apparently an ancient soil . . . on the area previously occupied by the ice-sheets." In places in "southern Ohio this horizon is marked by deposits of peat now deeply buried under the later-formed deposits of the Drift". This sometimes contains mastodon and other bones.*

6. "Above the old soil . . . we find a series of stratified deposits . . . evidently the product of a 'submergence'." These are "sometimes of considerable thickness." In southern Ohio, there are at places "white clays . . . containing bowlders." In west-

ern Ohio and farther west occurs the "Loess or Bluff formation," which is called the "Lacustrine Drift."

7. Bowlders of northern rocks may be very large and mark the margin "of the great ice-sheet . . . but most of the bowlders . . appear to have been deposited by another agency, at a much later date . . . they must have floated *to their present resting places. The evidence is conclusive that they were transported by icebergs, and hence I have called these the* 'Iceberg Drift'."

8. The "hills, ridges, and banks of well rounded gravel and sand, with some bowlders . . . correspond closely with the 'Kames' and 'Eskers' of the Old-World Drift. These peculiar accumulations of drifted material were evidently produced by special and local causes . . ."

9. Most recent of the features are the "Lake Ridges" from 100 to 250 feet above the present level of Lake Erie.

NEWBERRY'S GLACIAL HISTORY

Based on the "facts" he had enumerated, Newberry divided the history of the "Quaternary or Glacial Epoch" into several divisions, which he listed in order as:

1. Period of Arctic climate, formation of glaciers, first local, then general, to cover the whole area, then again local in the waning stage.
2. The movements of the ice were governed by topography and basins and certain valleys were excavated to great depths to form the Great Lakes basins.
3. The continent stood "several hundred feet higher than now," and deep valleys were excavated.
4. A period of glacier retreat was followed by a "water period," and by subsidence, and as ice "retreated northward it thrust out and left behind it a succession of heaps of bowlder clay, which now form a nearly continuous sheet over the glaciated surface."
5. "When the retreating ice-sheet had passed the great watershed of Ohio, basins of water began to form along its margin, and in these . . . were deposited laminated clays . . . (to) form the upper subdivision of the Erie clay." This is "usually stratified in thin leaves . . . and contains no bowlders." [By "great watershed" Newberry meant the highland extending east-west across Columbiana, Stark, Holmes, Ashland and Richland counties, which was indeed a very early divide through which the streams mentioned later by Newberry do flow southward.]
6. After the retreat of the ice, forests grew, for perhaps thousands of years, to form the "Forest Bed."
7. Submergence of the land so that the Gulf of Mexico extended to cover all of the lower half of our State. At that time, "the clays, sand, and gravel overlying the peat beds in southern Ohio, the lacustrine clays of northern Ohio, and finally the loess of the Mississippi Valley were deposited. These filled and obliterated many of the valleys of the Forest Bed era, as the Erie clay had done those of pre-glacial date." The Erie clay included both bowlder clay and the lacustrine clay of the Erie basin and of southeastern Ohio.
8. During the submergence, icebergs floated from the north and in melting, erratics were "scattered broadcast over all the submerged area."
9. "In this last submergence, portions of the highlands . . . were low islands and shallows, exposed to the full action of shore waves . . . the drift accumulations were assorted . . . and many of the gravel hills and sand banks (kames) of the summit of the watershed produced."
10. "With the subsidence of the waters . . . lines of drainage were established in the gaps of the watershed . . . through these . . . strong currents of water poured," and vast quantities of gravel deposited in the "great drainage lines" of the Wabash, Miami, Scioto, Muskingum, and the Beaver.
11. "The retirement of the sea at the close of the Drift period took place very gradually, with intervals of rest." The terraces in valleys were formed in this "Terrace epoch, the last chapter in the Drift history."
12. The Ohio Valley was emptied, but the lake basin was still filled with water. Various outlets opened at various places "by the

cutting away of barriers, or the warping of the earth's crust." The stationary intervals are marked by wave cut terraces or by "lake-ridges."

Newberry was not certain if there had been a return of the glaciers, "although in southern Ohio, the sheet of pebbly clay which overlies the Forest Bed seems to indicate a return in that region of something like the conditions in which the first bowlder clay was deposited." (Newberry, 1874, p. 8).

Newberry regarded the history of the Great Lakes important enough to deserve a final section in his "History" which he listed as:

1. Wearing down of high land to the north in Canada.
2. Establishment of drainage lines in valleys.
3. The major drainage lines were to determine the location of the present lakes.
4. Local glaciers began excavation of the basin.
5. The Great Ice Sheet continued its erosion, and eroded the southern margin, spilling over and cutting the Finger Lakes in New York and deepening some north-south valleys in Ohio, such as the Grand River lowland.
6. Rise in temperature caused dissipation of the ice sheet and the return of local glaciers, which continued some further basin erosion. [Rise in temperature caused dissipation of the ice sheet and led to localization of glacial activity, which continued some further erosion within the basins.]
7. Melting of the glacier, sinking of the continent and flooding of the Atlantic Ocean to Lake Champlain, Hudson Valley and probably through the Mohawk and Lake Erie to the Mississippi.
8. Elevation of the continent and draining of the inland sea. The elevation was irregular and several outlets operated at different times through the Mohawk, later through the Niagara and St. Lawrence and through the Wabash to the Mississippi and other trunk river systems; irregularity of ice may also have contributed to uncovering new outlets.

DISCUSSION

From Newberry's essay and from the detailed county reports we can determine what the Ohio geologists thought about the glacial deposits and their origin and attempt to find out why they thought as they did in 1874. It is significant that they recognized the "bowlder clay" as an ice deposit, made during ice retreat. This was a significant change from earlier explanations of its deposition by water. Here and elsewhere, Newberry stressed the topographic control of direction of ice advance. This is almost the final statement on the "Water Period," in which erratic-bearing icebergs floated in a large body of water and in which the upper part of the Erie clay was deposited. It should be noted that soon after his essay appeared, Newberry (1884) gave up this idea and asserted that the erratics were deposited by an ice sheet.

The idea of an inland body of water in the United States and in North America goes back at least to 1750 and was a firm belief of Volney, Mitchill, Maclure, and Drake. Until about 1830, the flat-lying bedrock strata west of the Allegheny Mountains were believed to have been deposited in this lake, and by implication, the "Drift" was the last deposit in the inland body of water. Thus a mechanism for floating icebergs carrying erratics was at hand. The idea of an inland body of water was as firmly fixed in the minds of these workers as the idea of permanent continents and ocean basins was fixed in the minds of all geologists until only a very few years ago. It should be noted, however, that soon after 1830, it was recognized that the sedimentary bedrock was not a deposit in the inland body of water, but was a series of marine deposits belonging to the Paleozoic Era separated by unconformities and thus separated in time and very much older than the "Drift Epoch." A continuous "Primal Lake" was no longer accepted, but Newberry and his associates could see the lacustrine deposits of a higher Lake Erie as a confirmation of the accepted idea of a

Quaternary inland lake, and then could interpret the clayey tills with not very many pebbles as also a lacustrine deposit. If observations indicated that the erratics were generally concentrated on the surface, this would be a further confirmation of the presence of the lake in which icebergs carried erratics across the surface of the lake. Erratics are indeed concentrated on the surface in belts in western Ohio. At the present day, the puzzling presence of erratics far beyond the accepted glacial boundary in Kentucky (Leverett, 1929) is explained by some as floated by icebergs in a lake, which existed in the Ohio Valley.

In northwestern Ohio, Gilbert had determined that the shore of the highest early lake (highest Maumee) was 85 miles southwest of the present shore of Lake Erie, and to the east Newberry saw that the silts and clays of the earliest lake stages extended from 5 to 15 miles south of the present lake shore. Both noted that the lake clays overlie till. South of the upper limit of the highest lake, the Hiram Till and its correlatives are very clayey, contain few pebbles and are easily confused with lacustrine silty clay. This clayey till extends as much as 40 miles south of the highest Maumee shoreline. The concept of a "Water Period" was fully justified on the basis of the facts available in 1873, and indeed it may be necessary to revive it in a modified form a century later, for over considerable areas in Ashtabula, Trumbull, Mahoning, northwestern Columbiana, and northeastern Stark Counties, the Hiram Till does have lacustrine components, where ponding occurred during and after the ice retreat. Beyond the glacial boundary in southeastern Ohio, widespread silts of laustrine origin indicate an early Pleistocene lake of considerable extent (Lessig, 1963; Lessig *et al.*, 1968; White and Totten, in press). The presence of shallow ponds as well as more extensive bodies of water, is again being proposed to explain some of the latest glacial deposits in the Allegheny Plateau in Ohio. Studies in Canada now indicate that over tens of thousands of square miles of deposits were formed in shallow bodies of water upon the waning ice or in front of it to form irregular thin units of reworked secondary materials (Moran, 1969).

The presence of southward-flowing streams in narrow valleys cutting across the ancient divide that extends westward from Columbiana County indicates stream diversions. For each of them, a lake is indeed required to persist until an outlet could be cut across the lowest point in this divide. Newberry believed in a large lake with several outlets, perhaps successive ones, whereas our present-day view is that a number of separate lakes were held in between retreating ice and the high divide.

The lacustrine origin of the loess and of the white clays in southern Ohio was a generally held one at the time and even later. It was not until several decades later that it was generally accepted that the loess was an eolian deposit, and even today it does appear that parts of the loess may have been deposited in local pools and puddles.

The description of the "Forest Bed" in many of the counties is valuable today, as it provides the most extensive information about this unit (which might actually be several units), which Orton (White, 1973) was the first to recognize as indicating an "Interglacial Period."

The final exposition of the theories that had been forming for over 60 years, from Drake to Newberry, carried in itself some of the bases for the next level of glacial theory that was to develop so rapidly after 1874. Gilbert (1873, p. 540–544) and Winchell (1873; 1874) had been the first in the United States to recognize, name, and suggest an origin for the end moraines—those of northwestern Ohio. Winchell had suggested that gravel knolls (kames) in the moraines had been formed by meltwater streams on, in, or at the edge of glacial ice. Newberry barely mentioned the moraines in his summary report, and did not even mention the Winchell explanation of origin of kames. The recognition of the importance of these features was to come very soon, but these were not really a part of the "glacial model" of Newberry in 1874. It was the "new glacial geology" that was to be based on mapping and description of end moraines and upon the deciphering of the history they revealed of repeated ice advances and retreats, with shorter or longer periods of climatic amelioration and weathering episodes between the advances.

CONCLUSION

The work of the geologists of the Second Geological Survey of Ohio must be judged on the basis of what data were available to them, what kind of base maps they had on which to record their observations, and what framework or model ("paradigm") they found or constructed to explain these features and their origin. These men were not in error, nor were they mistaken in terms of the science in 1873 (Westgate, 1942). Indeed, they were in the forefront of the science of their time and their contributions led directly to the further development of glacial geology, which in the next 25 years culminated in the work of Leverett and his associates. To criticize rather than to praise them is to criticize them for not having topographic maps, C^{14} analyses, X-ray devices for mineralogical determination, coring devices, and a whole host of techniques available a hundred years later. Such criticism is called presentism by a technical historian, and geologists with historical interests are well advised to judge the work of another time in terms of that time, and not in terms of the present. On this basis, Newberry and his associates must be given high marks for their keen observations and respectable theories.

LITERATURE CITED

Gilbert, G. K. 1873. Surface geology of the Maumee Valley. Ohio Geol. Survey, vol. 1, part 1, p. 535–590, maps.

Lessig, H. D. 1963. The Calcutta Silt, a very early Pleistocene deposit in the upper Ohio Valley: Geol. Soc. Amer. Bull. 74: 129–140.

Lessig, H. D., Hole, W. F., and Yohn, J. L. 1968. Soil Survey of Columbiana County, Ohio. U. S. Dept. Agric. Soil Conserv. Service., 132 p. 95 maps.

Leverett, Frank. 1902. Glacial formations and drainage features of the Erie and Ohio basins: U. S. Geol. Survey Mono, vol. 41, 802 p., maps.

Leverett, Frank. 1929. Pleistocene of Northern Kentucky: Kentucky Geol. Survey. vol. 31, 80 pp.

Moran, S. R. 1969. Geology of the Hudson Bay Area, Saskatchewan. Univ. of Illinois, Urbana, unpublished Ph.D. thesis. 194 p.

Newberry, J. S. 1870. Report of Progress of the Geological Survey of Ohio in 1869. Ohio Geol. Survey.

Newberry, J. S. 1874. Surface geology. Ohio Geol. Survey 2: 1–80.

Newberry, J. S. 1884. The drift deposits of Indiana: Indiana Geol. Survey, 14th Ann. Rept., p. 85–97.

Ohio Geological Survey. 1873. Vol. 1, pt. 1.

——— 1874. Vol. 2, pt. 1.

——— 1878. Vol. 3, pt. 1.

Westgate, L. G. Newberry on the Ohio Drift: Ohio J. Sci. 42: 215–219.

White, G. W. 1967. The first appearance in Ohio of the theory of continental glaciation. Ohio J. Sci. 4: 210.

White, G. W. 1973. History of investigation and classification of Wisconsinan Drift in north-central United States. Geol. Soc. Amer. Memoir 136. p. 3–34.

White, G. W., Totten, S. M. Glacial geology of Columbiana County, Ohio Geol. Survey of Ohio. (in press)

Winchell, N. H. 1873. The surface geology of northwestern Ohio. Amer. Assoc. Adv. Sci. Proc. 21: 152–186.

Winchell, N. H. 1874. Reports on the geology of Ottawa, Crawford, Morrow, Delaware, Van Wert, Union, Paulding, Hardin, Hancock, Putnam, Allen, Auglaise, Henry, and Defiance Counties. Ohio. Geol. Survey, vol. 2, pt. 1, pp. 227–438.

EARLY ENGRAVINGS OF GLACIERS AND FEATURES WITH BRIEF NOTES

Fig. 277.

Glacier of Viesch in the Alps.

THE GLACIAL THEORY COMES TO AMERICA

The "Glacier of Viesch" is a reproduction of a woodcut (Figure 277) in Edward Hitchcock's *Geology of Massachusetts,* 1841. It was a copy of a lithograph (Plate 10) in the atlas volume of Louis Agassiz's *Etude sur les Glaciers,* published in October, 1840, in Neuchatel. Hitchcock's great report on the geology of Massachusetts, already in press when he received a copy of the Agassiz book, contained elaborate descriptions and illustrations of "Diluvial" features, which at that time were of puzzling and unknown origin. Hitchcock, greatly impressed with Agassiz's glacial explanation, immediately recognized that these "Diluvial" features had been made by a continental glacier. He inserted an eight page "Postscript" on the glacial theory in his report and included copies of several of the Agassiz plates of glaciers and glacial forms. He pointed out that the glacial moraines and ice-scoured bedrock illustrated by Agassiz corresponded to similar forms of New England, and said the glacial theory "certainly explains most of these phenomena in a satisfactory manner." Thus, within a few months of its publication in Switzerland, the glacial theory was presciently presented to a wide American audience.

T. IV. Pl. II.

Vue du Mont-Blanc et de la Route par la quelle on a atteint la Cime,

*A. Cime du Mt. Blanc. B. Dome du Gouté. C. Aiguille du Gouté. D.E. Arve et Vallée de Chamouny. * * Places ou l'on a campé en montant.*

The view of Mont Blanc here reproduced is from de Saussure's Plate II, Vol. IV, and shows his route up the mountain. The view is taken southwest down the fine U-shaped valley of the Arve, with Mont Blanc shown about eight miles to the south. The large glacier, the Bossons, is now much shorter than in 1787. The Taconnaz Glazier is just behind.

THE FIRST ALPINE GEOLOGIST CLIMBS MONT BLANC

Horace Benedict de Saussure (1740-1799), member of an ancient Geneva family, became professor of philosophy at 22 and went on to a brilliant and famous career in science. He spent many decades in exploring the Alpine mountains and in investigating their geology and glaciers. His great work, VOYAGES DANS LES ALPES in four quarto volumes (1779-1796), is a great classic of Alpine geology and travel. It had immediate and wide influence on geologists; James Hutton quoted from it at length, as did his commentator, John Playfair. Unfortunately the French text has never been translated; a critical English translation would be of great benefit to geologists and to historians of science.

De Saussure's book includes stirring reports on climbs on the glaciers and to the peaks. One of the most famous passages is on the climb to the summit of Mont Blanc, at an altitude of 15,781 feet, the highest mountain in Europe (Vol. IV, p. 141-216; D. W. Freshfield, in THE LIFE OF HORACE BENEDICT DE SAUSSURE, London, 1920, p. 197-241, gives some added material derived from original de Saussure notes). Mont Blanc, which had never been climbed, had long presented a challenge to de Saussure. He encouraged the guides of the Chamonix region to explore for a route up the mountain and offered a handsome reward for the first ascent. On August 8, 1786 a guide, Jacques Balmat, accompanied by the local doctor, Michel Gabriel Paccard, was the first to reach the summit. He immediately informed de Saussure that a route had been found. Weather, however, prevented de Saussure's ascent that year.

On August 1, 1787, de Saussure set out from Chamonix with his party consisting of his personal valet and 18 guides, led by Jacques Balmat. The route was from Chamonix (out of sight in the plate, in the Arve Valley on the left), down the valley to the Taconnaz Glacier, thence along the Grand Mullets to the upper Bossons Glacier, where the night was passed (note asterisk at the camp location on the upper end of this rocky ridge). On August 2 the party crossed the heavily crevassed ice fall and the ice and snow of the Grand Plateau. They camped that night not far below the summit. On the morning of August 3, 1787 the party reached the summit, where de Saussure spent several hours in scientific observations. He returned with one overnight camp, to the anxiously waiting arms of his family in Chamonix, who had watched the ascent by telescope.

The engraving here copied is from a drawing by M. T. Bourrit, a skilled artist who published several books on the glaciers of the Alps, and who had some association with de Saussure from time to time.

N°. 8. GLACIER du RHÔNE dessine d'apres nature en 1817 par Mr. Lardy.

THE RHÔNE GLACIER - Then and now

Within the past century and a half the Rhône Glacier has retreated from the position in 1817 well down the valley, as shown here, about to the position of the three prominent rocks below the smooth upper snowfield. The Furka Pass is on the right and the Grimsel Pass off the plate to the left. The present Hotel Belvidere is at about the location of the prominent isolated rock mass in the upper center just to the right of the glacier and above the level of the highest ice pinnacles (seracs). The present highway over the Furka Pass is along the shoulder below the snow field on the right, horizontally along the upper part of the valley to the Belvidere Hotel and thence descends in a series of switchbacks along the valley side at the extreme right where it crosses the Rhône River.

The snow fields above the Rhône Glacier have now almost completely disappeared. The river, which in 1817 rose at the end of the glacier as shown here, now rises at the present end of the glacier, high up the valley. The end moraine known to have been present in 1777 appears to have been overridden by 1817.

The plate here reproduced was published by Jean de Charpentier (1786-1855), Directeur des Mines du Canton de Vaud, et Professeur Honoraire de Géologie à l'Académie de Lausanne, a well known geologist and engineer. He convinced Louis Agassiz of the former greater extent of the Alpine glaciers, which led Agassiz to write his famous ETUDES SUR LES GLACIERS, published in October 1840. Charpentier had been working on his own book ESSAI SUR LES GLACIERS ET SUR LE TERRAIN ERRATIQUE DU BASSIN DU RHÔNE, which he sent to press on October 31, 1840, but which did not appear until early 1841. In this book he dealt at length with evidence of the retreat of Alpine glaciers, especially the Rhône Glacier. This book of 363 pages contained a map, seven plates of great erratics and this plate of the Rhône Glacier. This plate, "drawn from nature by M. Lardy", shows the glacier as it was in 1817.

It is interesting to contrast this plate with that of the artist F. Meyer, which is Plate XII in HISTOIRE NATURELLE DES GLACIERS DE SUISSE by Grouner, 1770 (translated from the German of 1760). Meyer's highly exaggerated depiction may also be contrasted with the almost photographic and far more accurate representation by Besson of "Der Rhone-Gletcher im August 1777", which is reproduced as the frontispiece of GESCHICHTE DE MORÄNENKUNDE by A. Böhm von Böhmersheim, 1901.

Sévestre in lap. — Lith. de Nicolet à Neuchatel.

HOTEL DES NEUCHATELOIS

sur la Mer de glace du Lauter Aar et Finster Aar.

AGASSIZ'S MOBILE HOME ON THE AAR GLACIER--1840

In the summer of 1840 Louis Agassiz, the young professor at the University of Neuchâtel, now famous as the promulgator of the glacial theory, had finished writing his great ETUDES SUR LES GLACIERS, which was just being printed. For relaxation he and several associates went to the Aar Glacier and established headquarters in a hut earlier constructed on the glacier by Fr. J. Hugi, which they facetiously called the HOTEL DES NEUCHÂTELOIS. E. Desor, one of the party, published a report of the expedition*. The hut is shown here on the extreme left of the copy of the lithograph in Desor's report. From this as a base they climbed some of the nearby peaks and studied the glaciers.

The hut is behind and partly underneath the large rock upon which the flagstaff is located. Another view from almost the opposite direction, showing the hut entrance, is in Plate 14 of the ETUDES and a painting of the HOTEL, by an unknown artist, is reproduced as Figure 1 in A. V. Carozzi's translated edition of the ETUDES, STUDIES ON GLACIERS . . ., Hafner, New York, 1967.

It was at this hut that Agassiz read the galley proofs of the ETUDES. With his associates he investigated the structure of the glaciers and measured their rate of motion by observation of lines of stakes set in the ice. The hut had moved down the glacier since their first acquaintance with it at a rate of about 200 feet per year. It was indeed a "mobile home"!

Desor's 1841 little book was later enlarged and published in 1844 in the better known and much longer book of essentially the same title. But it is the 1841 edition that first gives us a report of just how Agassiz set about his glaciological studies and his measurements of glacial motion from his base camp of the "mobile home" on the Aar Glacier.

*Desor. E., EXCURSIONS ET SEJOUR DE M. AGASSIZ SUR LA MER DE GLACE DU LAUTERAAR ET FINSTERAAR, EN SOCIÈTE DE PLUSIEURS NATURALISTES, Tire de la Bibliotheque Universelle de Geneva, Mars et Avril 1841, 103 pp.

Phil. Trans. MDCCCXXXIX. Plate II. p. 39.

Albert Way, delt. Basire, lithog.

View of the Parallel Roads of Glen Roy

THE PLATE FROM DARWIN'S PAPER ON THE PARALLEL ROADS OF GLEN ROY,
A lithograph from the drawing of Albert Way, 1839

THE PARALLEL ROADS OF GLEN ROY

Glen Roy, a tributary valley to the larger Glen Spean, is a fine U-shaped valley in the Highlands of Scotland 100 miles northwest of Edinburgh. The mouth of the glen is easily reached by main highway from Fort William, 10 miles to the southwest. A very passable road leads several miles to the head of the glen. The remarkable feature of Glen Roy is a series of three horizontal, parallel shelves, which are cut into the sides of the valley. They look like grade-lines of superhighways and have been called "roads" from early times. Many books and articles have been written about the glen and the origin of the roads (See James MacFadzen, "Parallel Roads of Glen Roy . . ." Edinburgh, 1883, p. xx; also "Nature", Vol. 22, p. 69).

Louis Agassiz visited the Scottish Highlands with Professor William Buckland in 1840 to look for evidences of glaciation. To their great satisfaction they found abundant signs of former glaciers. Agassiz recognized that the glen was once the site of a glacial lake and that the "roads" were shorelines of the lake, made when it had stood at three different levels (Agassiz, Louis, "On Glaciers", read Nov. 4, 1840, Proc. Geol. Soc. London, Vol. 3, p. 332, 1842; also see Davies, G. L., "The Earth in Decay", Elsevier, N. Y., 1969, p. 279).

The year before Agassiz' discovery, Robert Darwin published a long paper, "Observations on the Parallel Roads of Glen Roy. . . ." in Philosophical Transactions of the Royal Society, 1839, pp. 39-81. Darwin is generally thought of today as a biologist, but he was also a geologist. He was not only a Fellow of the Geological Society of London, but a member of the Council and from 1838-1841 one of the Secretaries of the Society. His several books on geology are much earlier than the "Origin of Species". The "Origin" itself contains much geology and is a necessary book in any geology library. Darwin continued to write on geology as late as 1855.

Darwin had carefully studied the "roads" and wrote in detail about their surface form and internal structure. He concluded that these were indeed shorelines, and that they had been formed when the sea stood at these levels. Later, when the glacial theory provided a mechanism for a dam to form a lake in Glen Roy, Darwin gladly accepted that explanation. His study still remains a testimony of his keen observation of the geological features of the Parallel Roads.

Darwin's article is accompanied by a detailed map and by a view of the glen here reproduced. At the end of his paper he wrote: "I am much indebted to my friend Mr. Albert Way for his kindness in lending me the drawing, from which the accompanying lithographic sketch has been taken. It faithfully represents the general appearance of Glen Roy." It still does.

PETER MARTEL'S DRAWING OF THE ANIMALS HE SAW IN THE ALPS IN 1742

PETER MARTEL'S EARLY OBSERVATIONS ON ALPINE ANIMALS, ON HAY HARVESTING BY MARMOTS, AND ON GLACIERS

Peter Martel was a Swiss engineer who, between August 20 and 26, 1742, travelled with several companions from Geneva to Chamonix along the Arve Valley to the foot of Mt. Blanc. His report was published in London in 1744 as AN ACCOUNT OF THE GLACIERES OR ICE ALPS IN SAVOY . . . ILLUSTRATED WITH A MAP AND TWO VIEWS OF THE PLACE &C (another edition, without plates, was published at Ipswich in 1747). It includes a plate of the Mer de Glace, a map of the valley, a sketch of the valley near Chamonix, and the lively sketch of the animals, here reproduced. The "Bouquetin", at "a" is a form of mountain goat; "b" is the "Chamois"; "c" is the "Marmotte", about like our groundhog or woodchuck. The statement of the animals in the valley and story of the marmots' hay harvesting for winter use is here reproduced in facsimile (with the ancient form of initial and medial "s" which looks like "f"):

there being in this Valley no Beaſt of Prey, though Bears, Wolves, and Foxes abound in the Country all about. Nothing inhabits here but *Chamois, Bouquetins*, who keep in the high Mountains, and a great Quantity of *Marmotes*; this is the Account the Inhabitants gave us of this Animal. They ſleep ſix Months of the Year, that is, all Winter, and in the Summer they provide a warm Couch againſt their Time of Sleeping; for this End they cut Herbs with their Teeth, and in order to carry them to their Holes one of them lays on its Back, and the others load it like a Cart, and then drag it by the Ears to the Hole. They pretend alſo that they provide againſt being ſurprized, by placing Centinels, who give them the Alarm by a whiſtling Noiſe; they eat theſe *Marmotes*, and find them very good, and uſe their Fat to burn in Lamps; there are no Birds of Prey in this Valley, nor Crows, neither are there ever any Swallows.

Martel and his companions travelled on the Mer de Glace (which he called the "Glacier des Bois"). He made important glacial observations, which preceded by several decades the much better known ones of Gruner and of de Saussure. He described feelingly the transverse and oblique "waves" (crevasses) and their danger to travellers. He thought the origin of the ice was because of the low temperature at high levels and the less sunlight on the north-facing slopes, which preserved the snowfall in these locations. He said that "their marks" show the glaciers had once been higher, and the huge stone at Chamonix showed they had once been more extensive. Perhaps his most important statement was that the glacier moves by "sliding" down the valley.

Martel had surely used his time to good advantage, as he not only observed the animals, the plants, the mountains, and the glaciers, but also made impressive sketches. His descriptions of the ice, its movement, extension, retreat, crevasses, and deposits, all made in such a short time, make us wish that he could have followed his own advice at the conclusion of his report:

"that those who later undertake this troublesome and curious journey ought to imploy more time in it." If Martel could produce so much from a few hours observation, we wonder what he could have done had he stayed two or three weeks in the region.

R. Price delin — Vivares Sculpsit

View of ye Ice Valley, & Mountains that Surround it, from Mount Anver

PETER MARTEL'S OBSERVATIONS ON GLACIERS, 1742

Peter Martel was a Swiss engineer who, between August 20 and 26, 1742, travelled with several companions from Geneva to Chamonix along the Arve Valley to the foot of Mt. Blanc. His report was published in London in 1744 as AN ACCOUNT OF THE GLACIERES OR ICE ALPS IN SAVOY . . . ILLUSTRATED WITH A MAP AND TWO VIEWS OF THE PLACE &C (another edition, without plates, was published at Ipswich in 1747). It includes a plate of the Mer de Glace, a map of the valley, a sketch of the valley near Chamonix, and a lively sketch of the animals. This sketch of the animals was reproduced together with the story of hay harvesting by marmots in the "G W W Note" of 1972.

Martel and his companions travelled on the Mer de Glace where he made important glacial observations, which preceded by several decades the much better known ones of Gruner and of de Saussure. He described feelingly the transverse and oblique "waves" (crevasses) and their danger to travellers. He thought the origin of the ice was because of the low temperature at high levels and the less sunlight on the north-facing slopes, which preserved the snowfall in these locations. He said that "their marks" show the glaciers had once been higher, and the huge stone at Chamonix showed they had once been more extensive. Perhaps his most important statement was that the glacier moves by "sliding" down the valley.

The plate "View of ye Ice Valley . . .", drawn by Martel's associate R. Price, here reproduced at a reduced scale, is of the glacier now known as the Mer de Glace. The depiction of the bordering mountains is not to be regarded as "quaint", or "exaggerated"; most views of mountains of this period similarly show jagged peaks. This is the way Martel and his associates saw the peaks, in a way strikingly similar to the "gothic rendering" of the earliest drawing of the Grand Canyon of the Colorado in the 1860's and 1870's. Perhaps two centuries from now our current sketches of mountains will be just as curious to the person of 2195!

Martel had surely used his time to good advantage, as he not only observed the animals, the plants, the mountains, and the glaciers, but also made impressive sketches. His descriptions of the ice, its movement, extension, retreat, crevasses, and deposits, all made in such a short time, make us wish that he could have followed his own advice at the conclusion of his report: "that those who later undertake this troublesome and curious journey ought to imploy more time in it." If Martel could produce so much from a few hours observation, we wonder what he could have done had he stayed two or three weeks in the region.

Gezeichnet nach der Natur von J.L. Aberli. Gegraben von A. Zingg.

Ces Rochers, ces Glaçons qui forment ces Montagnes
Servent a nos Besoins, abreuvent nos Campagnes

THE LAUTERBRUN VALLEY AND FALLS, 1756

The Lauterbrun Valley and its waterfalls is a well-known illustration of Switzerland and is a favorite travel poster subject. One of the earliest illustrations of this valley (perhaps the earliest?) is the frontispiece of the earliest book devoted entirely to glaciers.

Die Eisgebirge des Schweizerlandes, by Gottlieb Sigmund Gruner, 1717-1778, published in Bern in 1760, was the report of his observations in 1756. A French translation, Histoire Naturelle des Glacieres de Suisse, was published in Paris in 1770. Gruner's work is of great importance in the history of the study of glaciers. He understood that erratics and drift were carried by the ice, and reported that bodies of men and animals that had fallen into crevasses in a glacier sometimes appeared a long time later at the glacial margin. His observations on rate of glacier movement are very early, perhaps the earliest. In addition to the frontispiece here reproduced, the book contains 18 magnificent plates of glaciers and glaciated valleys, and two large geological maps, among the very earliest examples of such maps. Gruner made many significant geological, paleontological, and mineralogical observations in this book on glaciers, as well as in many other publications.

The Lauterbrun Valley is in Bernese Oberland about 10 km south of Interlaken, in central Switzerland. The view is south from the village of Lauterbrunnen. The mountains in the distance ("a-e" on the plate) are in the Breithorn-Grosshorn massif. The end of the Breithorn Glacier ("f" on the plate) was much closer to the village in 1756, but has retreated to its present position about 13 km from the village.

The U-shaped valley is a classic example of this form, deeply carved by a glacier. The tributary valleys remain uneroded to form "hanging valleys," from which their streams plunge over the almost vertical cliffs in magnificent waterfalls.

The drawing is by J. L. Oberli, engraved by A. Zingg. We can quite agree with the quotation from Haller that "These rocks, these glaciers which form these mountains fulfill our needs and water our countryside."

GREAT GLACIER OF HUMBOLDT.

From a sketch by Dr Kane.

THE HUMBOLDT GLACIER

The Humboldt Glacier, in far northwest Greenland, was discovered by Dr. Elisha Kent Kane in 1854, who named this 62-mile wide great outlet glacier of the Greenland ice cap for the famous German scientist. At this point Kane had reached 80° 35', the farthest north of any explorer at the time. His illustration of the glacier is one of the plates in his great book, "Arctic Explorations: The Second Grinnell Expedition in Search of Sir John Franklin . . . 1853, '54 '55" (1856).

Elisha Kent Kane (1820-1857) was one of the most scholarly of Arctic explorers. His training was in engineering (including geology and mineralogy) and in medicine. As a naval surgeon he had traveled extensively in South America, Asia, and the Mediterranean and Alpine regions. After notable service in the Mexican War, he was surgeon on the first Grinnell Expedition to the Arctic in 1850-1851 in search of the lost Sir John Franklin. He returned to the Arctic in 1853 as commander of a second Grinnell Expedition. Kane was an artist of considerable ability, and his two books, one on the first expedition, and one on the second, contain hundreds of drawings and diagrams. He was continually impressed by the structure in ice flows, pack ice, icebergs, and the ice cap itself. His analysis of the formation and movement of ice in the Arctic regions was most perceptive. He was especially impressed by the structure of ice flows as they collided, overthrust, or moved past each other to form folds and faults which he compared to similar bedrock structures in the Alps, in Greenland, and elsewhere. His geological and glaciological observations are scattered through his two books, with repeated statements that their brevity was because he was preparing a more extensive monograph on glaciology and geology. Such a monograph was never completed because he died in 1857 at the age of 37, due in part to the great hardships this always frail scholar, physician, scientist, and explorer had suffered on his expeditions.

REVIEWS

ASSAY OF THE WEST

Assay of the West

John Charles Frémont. *The Expeditions of John Charles Frémont.* Volume I: *Travels from 1838 to 1844*, and *Map Portfolio.* Edited by **Donald Jackson** and **Mary Lee Spence.** xliv + 845 pp. 26 plts., index, 5 maps in 11 large sheets, with 16 pp. Commentary by Donald Jackson. Urbana: Univ. Illinois Press, 1970. $22.50 (Map Portfolio only, $10.00).

John Charles Frémont (1813–1890) has a secure reputation as mapmaker-explorer of the western United States,[1] a reputation not yet fully assessed as a scientific explorer of the West, a less secure reputation as the first presidential canditate of the Republican Party, and a very much less than brilliant reputation as a commanding general in two very brief periods in the Civil War. His greatest contribution was made before the age of forty as he explored the West and reported on its geography, geology, and botany. He crossed the plains and the Rockies five times in expeditions to "map and determine the nature of the country." His published reports and maps immediately became the bibles of the travellers and immigrants who came after him. His 1845 report probably had the greatest influence on the final objective

[1] Allan Nevins, *Frémont, Pathmarker of the West* (New York: Appleton-Century, 1939; several later editions).

of the Mormon migration from Nauvoo to Utah. He became a very rich man through land investments in California, lost it all after the Civil War, and was then supported by several minor appointments and by his wife's writings.

Frémont was fortunate in his early associations with men who could both teach him and advance his career: Joel Poinsett, Secretary of the War Department; J. N. Nicollet, the emigré French scientist; F. R. Hassler, head of the Coast and Geodetic Survey; Col. J. J. Abert, head of the Corps of Topographical Engineers; and especially Thomas Hart Benton, Senator from Missouri and Chairman of the Military Affairs Committee. But the greatest influence was Benton's daughter, Jessie, who before her sixteenth birthday decided to marry the twenty-eight-year-old Army officer. Their runaway marriage, just after Jessie's seventeenth birthday, was much against her parents' wishes, but they soon became reconciled and staunchly supported the young Frémonts. Jessie immediately began to act as her husband's literary secretary, and in the next forty-nine years this team first reached heights of justified acclaim and later depths of disappointment that approached classical Greek tragedy in intensity.

Frémont was a brilliant student at the College of Charleston, where he excelled in mathematics and astronomy. He was a drop-out three months before graduation but later was awarded a degree. Through the influence of Joel Poinsett, he was appointed civilian teacher of mathematics to the midshipmen aboard the U.S.S. *Natchez* on a two-year cruise in South American waters. References to Italian volcanoes and geology in his reports indicate that the ship may also have visited the Mediterranean. He later received an appointment as professor of mathematics in the Navy, but declined because of the delay in the official appointment. Surveying in the Cherokee country gave him his first experience in precise surveying in wild regions. Through the influence of Poinsett he was appointed as civilian assistant to J. N. Nicollet and shortly received a commission as second lieutenant in the Corps of Topographical Engineers.

This first volume of a planned three-volume series deals with Frémont's travels with Nicollet in 1838 and 1839 to the Upper Mississippi Valley and with Frémont's own expeditions to the Rocky Mountains in 1842 and to Oregon and California, in 1843–1844. Donald Jackson's earlier publications on the Lewis and Clark and the Zebulon Pike expeditions are well known.[2] His method has been to present the relevant documents connected with an expedition: the orders, the requisition for materials, some of the enabling Congressional acts, correspondence relating to contemporary comments and reviews and, for the Pike report, the complete text and maps of the original journals. In this work on Frémont, in which Jackson is joined by Mary Lee Spence, the editors identify every man in Frémont's parties, every correspondent, and even anyone casually mentioned in letters and reports. The correspondence, the orders, and the financial records and their notations (by Frémont, the auditors, and his commanding officer, Col. Abert) allow the reader to form his own conclusions about the various men involved. These basic documents, as well as the reprinted reports and maps, will now enable historians of science to begin to work on Frémont's contribution and his place in the scientific activity of the time.

Nicollet, the French scientist whose work has recently been described in detail by Bray,[3] explored, surveyed, and mapped the Upper Mississippi Basin in great detail. Under Nicollet, Frémont learned sophisticated methods of geodetic surveying, a considerable amount of geology, and especially how to get along with voyageurs, Indians, fur traders, and other

[2] Donald Jackson, ed., *Letters of the Lewis and Clark Expedition* (Urbana: Univ. Illinois Press, 1962), and *The Journals of Zebulon Montgomery Pike, with Letters and Related Documents*, 2 vols. (Norman: Univ. Oklahoma Press, 1966).

[3] M. C. Bray, "Joseph Nicolas Nicollet, Geologist," *Proceedings of the American Philosophical Society*, 1970, *114*:37–59. M. C. Bray, ed., *The Journals of Joseph N. Nicollet*, trans. André Fèrtey (St. Paul: Minnesota Historical Society, 1970).

characters of the frontier. He helped and perhaps oversaw a good part of the preparation of the great map (Map 1 in the *Portfolio*) in the Nicollet report, and because of the illness of his leader, Frémont himself surveyed, mapped, and then drafted the Des Moines River Valley part of the map.

Frémont's first expedition, after his work with Nicollet, left the junction of the Kansas and Missouri rivers on June 10, 1842, and returned to St. Louis on October 1. He went back to Washington, wrote (with much help from Jessie) the report of the 1842 expedition, prepared for the much more elaborate expedition to follow, and on May 29, 1843, just as the report appeared, set out from St. Louis for Oregon, near enough to the Pacific Coast to connect his surveys with the earlier one of Wilkes. He was away fourteen months, returning through almost the length of California and the southern part of what is now the United States. He immediately wrote the great report of this expedition, which appeared in 1845, and made preparations for another journey to California, which will be dealt with by the authors in a succeeding volume.

A whole essay could be written on the frustrations of Frémont with his auditors. The editors reprint the actual vouchers to January 1, 1843, the later ones are either tabulated or summarized. This painstakingly builds up a picture of the expedition for the periods of fitting out progress, and conclusion—the purchase of instruments, supplies, and horses, the fate of these horses, and the refusal of auditors from time to time to approve certain items. The item for buckskin pants was not allowed: the auditors did not see the necessity for buckskin pants in Washington, so why in the valley of the Platte River or in the Columbia? (They had no conception of what *Artemesia* does to ordinary trousers.) Some of the medicines were contested, particularly brandy, and at one time the purchase of macaroni (a delicacy in Washington?) was disallowed until Frémont explained that this was more easily carried than flour and besides it was available at one of the supply depots when almost nothing else was.

The editors, obviously botanically inclined, have provided extensive notes on the plants described and collected by Frémont and have identified every plant mentioned. Frémont himself gave the Latin names, some obviously from his knowledge at the time and others from identification by John Torrey afterward. The plants were carefully collected, but unhappily on both expeditions many were lost: most of those from the first expedition went down when a primitive rubber boat capsized in the rapids of the North Platte River; many of those collected as far as California on the second expedition were lost when a mule fell into a canyon; and those collected from California to Kansas were badly damaged in a sudden flood of one of the tributaries of the Kansas River, just before the expedition reached civilization. Nevertheless, Torrey was able to determine many of the specimens and described them in a special appendix of the report of the second expedition and in a large monograph afterward.[4] Frémont was proud of his botanical knowledge, and the new species reported in the appendix carried the names Torrey and Frémont.

The year after the larger report appeared, the botany was very favorably reviewed in one of the prominent botanical journals of the day.[5] It was recognized that a region had been visited "where probably the foot of civilized man never before trod." The important parts of the collections were mentioned, a number of new species commented upon, and the relation of plants to soil and rock types was favorably noted. The review closed with the statement that "from explorations like these, even under such disadvantages, in the science of botany alone

[4] John Torrey, "Plantae Fremontianae, or Descriptions of Plants Collected by Colonel J. C. Frémont in California," *Smithsonian Contributions to Knowledge*, 6, Art. 2, 1854.

[5] "Notice of the 'Reports of the Exploring Expedition to the Rocky Mountains, in the year 1842—and to Oregon, and North California, in the years 1843–1844. By Brevet Captain J. C. Frémont, of the Topographical Engineers, &c. &c. Printed by order of the Senate of the United States. Washington, 1845' by X," *Magazine of Horticulture, Botany, and all useful Discoveries in Rural Affairs*, 1846, *12*:86–94.

we can anticipate most interesting and valuable results. . . . Many and signal advantages must occur."

A contemporary review in the *American Journal of Science*[6] dealt mainly with the geographical results, included several pages of excerpts from Frémont's report, and noted: "there was no retinue of science attached to the expedition, yet by personal exertion, in connection with his other arduous duties, Captain Frémont made valuable geological and botanical collections. . . . It is right that this bold explorer of the mountains should have his name inscribed among the flowers of the region and about its loftiest heights as well as upon the honored page of history."

The reports contain abundant material for history of surveying. Frémont had learned navigation at sea, surveying in the Cherokee country, and the more sophisticated geodetic methods under Nicollet. In his reports he published scores of pages of astronomical observations for position and barometric observations for altitude. This is source material for investigation of how the positions were determined on this and similar expeditions. Frémont does not describe his methods, but they were based on those described by Nicollet and Hubbard.[7] Altitudes were determined by mercurial barometer (the elaborate means then necessary for providing reference standards are not realized today when bench marks are usually not far away and instantaneous communication with base camp and headquarters is routine). Frémont established latitude and longitude, not only by noon sights of the sun, but he also corrected his chronometer and took precise observations repeatedly from occulation of fixed stars and from the immersion and emersion of the satellites of Jupiter. (It would be most interesting to compare his positions with those now accepted, which have been determined by methods of exact time—not to be available until the telegraph and, later, radio time signals became available.) We have to remember that Frémont's instruments were being carried first in a cart and then on pack mules, that there were hostile Indians commonly present, and that food was often a problem (sometimes a desperate one). The ruggedness of the country, particularly the crossing of the Sierra Nevada, made the transport of the instruments a feat in itself. In his narrative he mentions only occasionally and very briefly the difficulties of transporting the instruments and never dwells on the problems of late-night observations by himself and his topographer assistant, Charles Preuss, who acted as recorder.

The original material on ethnology is scattered through the reports and forms valuable early—indeed in many cases the first—notices of the different Indian tribes and more or less careful description of their characteristics and appearance.

Contributions to the history of meteorology and climatology are provided by the very extensive tables and the continuous notes in the reports. He was the first to describe and explain so succinctly the great climatic difference between the east and west sides of the Cascade Mountains.

The Frémont reports contain the first geological observations for vast tracts of the western country. He collected specimens of rocks, minerals, fossils, and soils. His soil reports were

[6] "Observations on the Rocky Mountains and Oregon: From the Reports of the Exploring Expeditions of Capt. J. C. Frémont," *American Journal of Science and Arts*, 1847 (2nd Ser.), *3*: 192–202.

[7] J. N. Nicollet, *Report Intended to Illustrate a Map of the Hydrographic Basin of the Upper Mississippi River:* Senate Report 237, 26th Cong., 2nd Sess. Washington, 1843, Serial 380 (also printed as 28th Cong. 2nd Sess. House of Rep. Doc. 52, 1845, with different map), pp. 93–106.

J. S. Hubbard, "Note from Professor Hubbard (of the National Observatory, Washington City), describing the instruments used by J. C. Frémont in making the astronomical observations in his third or last expedition, and the methods followed by Professor Hubbard in reducing them." ". . . A table of latitudes and longitudes . . ." in J. C. Frémont, 1848, *Geographical Memoir upon Upper California in Illustration of his Map of Oregon and California*, 30th Cong. 1st Sess. Senate Misc. No. 148, pp. 45–58. The instruments used on the 1842 expedition are listed on p. 312 of the volume being reviewed.

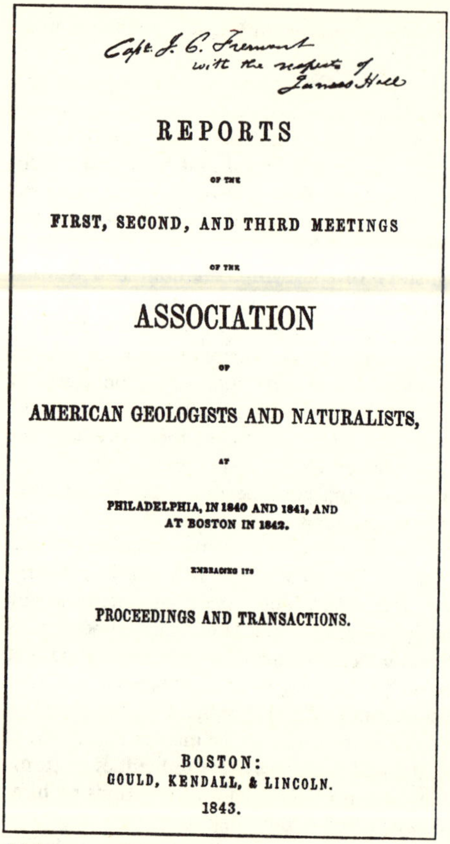

Capt. J. C. Fremont with the respects of James Hall

REPORTS

OF THE

FIRST, SECOND, AND THIRD MEETINGS

OF THE

ASSOCIATION

OF

AMERICAN GEOLOGISTS AND NATURALISTS,

AT

PHILADELPHIA, IN 1840 AND 1841, AND AT BOSTON IN 1842.

EMBRACING ITS

PROCEEDINGS AND TRANSACTIONS.

BOSTON:
GOULD, KENDALL, & LINCOLN.
1843.

Title page of the report presented to Frémont by the geologist James Hall.

certainly the first for the region and documented his assertion that huge areas were suitable for settlement. The second report has many analyses of rocks and minerals in an appendix by James Hall, who from Frémont's notes and specimens attempted a very brief summary of the geology over which the expedition passed. Fremont's association with Hall was friendly enough to result in Hall's presenting him with a report of the Association of American Geologists and Naturalists of 1843.[8]

Frémont was a perceptive observer both of large geological features and of small details. He could identify rocks and minerals and even make some attempt at identifying fossils. In the Columbia Plateau he saw that some of the sedimentary rocks had been metamorphosed by contact with the lavas and that zeolites had formed in the gas cavities of the lava. Frémont and Preuss noticed not only the rocks over which they travelled, but from the heights which they climbed they made some attempt to see their extent far beyond the route of march. The editors have provided abundant and perceptive notes on Frémont's botanical contributions, but they specifically disclaim similar treatment of the geology. (A whole monograph can be written on Frémont's contribution to Western geology.)

Frémont was especially concerned with water quality and composition, and his report contains many analyses of mineral waters. We wonder how, where, and by whom they were made—There is no comment on this by Hall or Frémont. Water samples could hardly have been brought from the Hot Springs of Idaho and elsewhere to Washington or even to St. Louis. This puzzle remains for future historians to unravel.

Frémont collected fresh-water marl deposits in the Columbia Plateau, which he recognized contained microfossils. The samples which he brought back were studied in detail by America's pioneer micropaleontologist, J. W. Bailey, who contributed a plate for the report and compared these diatoms with some that Dana had earlier found (he considered that they were somewhat younger than Dana's specimens).

[8] *Reports of the First, Second and Third Meetings of the Association of American Geologists and Naturalists*, . . . Boston, 1843 (see illustration for complete title; note Hall's inscription to Frémont). The fossil plant specimens which were identified by Hall were long lost sight of but were discovered in 1887, in the collections of the U.S. National Museum, according to G. P. Merrill, *The First One Hundred Years of American Geology* (New Haven: Yale Univ. Press, 1824; reprinted New York: Hafner, 1964), p. 243. (Merrill had an unjustifiably low opinion of Frémont's contribution to geology, but he generally denigrated the geological contributions of explorers who were not professional geologists.)

One of the greatest geological contributions was that of the discovery and naming of the Great Basin as a geologic and geographic entity and the proof that no rivers flowed from it to the sea. This Frémont knew because he had circumnavigated the whole area. His perceptive analysis of why there was no external drainage although large rivers flowed into it, is the first recognition of the great power of evaporation in this region.

His two-page summary on the geology and geography of the rivers and mountains of the West and the comparison with those of the eastern United States deserves wide quotation. It seems to have escaped the observations of those collecting "first" articles for anthologies. His comparison of the Great Basin with the interior of Asia was a very apt one, as was his recognition that the Oregon Territory was geologically different, partly because of the great lava flows and partly because of the Columbia River, the great through-flowing river from the Rocky Mountains to the Pacific. Upon seeing submerged trees at one place in the Columbia River he immediately recognized that their drowning was not the result of the rising of the waters but the result of complicated land-sliding, and upon examination he was able to find the actual shear surfaces on which the slides had occurred.

Frémont repeatedly refers to the regimen of the various streams and the relation of their width and depth to the character of their beds. He refers in more than one place to the origin of the sediments of the Great Plains as materials eroded from the Rocky Mountains and carried far to the east. His explanation of the origin of these widespread Tertiary deposits has been abundantly confirmed.

The great cross-section of the big map in the 1845 report (Map 3 of the *Portfolio*) shows not only a profile of the surface with certain prominent mountains, but also indicates the kind of rocks. From these data Frémont could have constructed a very respectable geological cross-section for the Rocky Mountains and the Columbia Plateau region. Earlier Dr. Edwin James had presented a section from the Colorado Rockies eastward to the Mississippi River,[9] but it was far less sophisticated than the one Frémont could have produced. That Frémont did not collect his geological observations into a connected monograph on the geology is most disappointing.

Frémont was accompanied on the 1842 and the 1843–1844 expeditions by the German topographer and artist Charles Preuss (1803–1854). Preuss's diary, recently found in Germany, provides an illuminating commentary on the expedition.[10] He assisted Frémont in the astronomical observations, but his mathematical training was not sufficient to allow him to assist in the calculations for positions. His role was to map the topography—and later to draft the maps—and to make sketches of important physical features. These sketches, reproduced in the reports, show that Preuss was an outstanding artist of the early West.[11]

The maps are reproduced in a separate *Portfolio*, which includes a reproduction of Nicollet's map of the Upper Mississippi Basin, the 1843 map of the "Country lying between the Missouri River and the Rocky Mountains," the large 1845 map of the "Expedition to the Rocky Mountains in the year 1842 and to Oregon and California in 1843–1844," and the excellent reproduction of Preuss' "Topographical map of the road from Missouri to Oregon" in seven

[9] Edwin James, *Account of an Expedition from Pittsburgh to the Rocky Mountains performed in the years 1819 and '20 . . .*, 2 vols. (Philadelphia, 1823).

[10] Charles Preuss, *Exploring with Frémont. The Private Diaries of Charles Preuss, Cartographer for John C. Frémont on his First, Second, and Fourth Expeditions to the Far West*, trans. and ed. E. G. and E. K. Gudde (Norman: Univ. Oklahoma Press, 1958).

[11] The plates for the reports reproduced in this volume were lithographs made from Preuss' drawings. They exist in different states in the House and Senate issues; however, some copies of the Senate issue of the 1845 report have some plates of each state. These can be distinguished by the presence or absence of borders, varying positions of trees, bushes, men, and animals. If the original Preuss drawings still exist, their exact photographic reproduction would make an exciting additional portfolio for the series.

sections, on which are entered notes from Frémont's report. This last map was used by thousands of immigrants on their travels to Oregon. The maps are introduced by Dr. Jackson in a separate "commentary." The reproduction of the maps and the commentary are in themselves a very important contribution. The original maps are now not only rare and difficult to find, but are often torn or missing from the volumes. The large maps are on fragile paper, and their use subjects them to serious damage. And yet, to follow Frémont's text, it is necessary to have the maps for reference.

Frémont had no scientists attached to his first two expeditions. Although he was sufficiently schooled in botany to make excellent collections and had considerable geologic and mineralogic knowledge, he could not give much time to the scientific part of the expeditions. The wonder is that he was able to do as much collecting as he did. As leader of the expedition he had to take care of the logistics, oversee the mapping, conduct the geodetic surveying, arrange the defense, and keep the records. A geologist or even a physician-naturalist responsible only for the geology could have (except where the Indians were particularly dangerous) ranged from one side of the track to the other and could have mapped a belt of country of considerable width.

We would like to know what books Frémont took along. He certainly had astronomical tables, probably some botanical reference materials, and one or more books on mineralogy. He makes no mention of reference materials or the reports of other travellers. Surely he had knowledge of the Parker and Townsend travels along the Oregon Trail, and it is possible that he had their actual books and maps with him.

In preparation for the study, Jackson followed Frémont's actual tracks, re-covering his camp sites and stops, so that he himself has seen the territory described. Interwoven with Jackson's own notes on the routes are the references to others who have studied parts of the Frémont route, especially over the Sierra and across the Mojave region to Las Vegas, but no one else has studied the routes as a whole. We ardently wish for one more map, constructed from or on present-day road maps to show the exact routes Frémont took.[12] No one other than these editors has that information.

The Frémont documents, reports, and maps are source material for early western science. Drs. Jackson and Spence have mined much of the botanical ore but have left materials for separate studies of the geology, ethnology, meteorology, and geodetic surveying, which we can hope historians of science will now analyze in detail, showing how succeeding workers in these fields built upon the raw data provided by Frémont.

We look forward to the succeeding volumes in this series. We are assured by these eminent western historians that one of the following volumes will include a complete bibliography of the Frémont reports in their various Senate, House, and independently published forms, and also some notice of the contemporary reviews.

George W. White

Department of Geology
University of Illinois
Urbana, Illinois 61801

[12] A page-size map of Frémont's routes is in W. H. Goetzmann, *Army Explorations in the American West 1803–1863* (New Haven: Yale Univ. Press, 1959), p. 71. Goetzmann also reproduces the 1845 Frémont-Preuss map at a reduced scale, map pocket.

REVIEW OF
Philosophy of Geohistory 1785-1970

QUATERNARY RESEARCH 6, 123–124 (1976)

BOOK REVIEWS

Philosophy of Geohistory: 1785–1970 (Benchmark Papers in Geology, Vol. 13). Claude C. Albritton, Jr. (Ed.). Dowden, Hutchinson and Ross, Inc., Stroudsburg, Pennsylvania, xiii + 386 pp. 1975. $24.00.

In 1684 Thomas Burnet, in the "Preface to the Reader" of his *Theory of the Earth: Containing an Account of the Original of the Earth* . . . told who he considered to be unsuited for geophilosophy:

> As every Science requires a peculiar Genius, ſo likewiſe there is a Genius peculiarly improper for every one; and as to Philoſophy, which is the Contemplation of the works of Nature, and the Providence that governs them, there is no temper or Genius, in my mind, ſo improper for it, as that which we call a *mean* and *narrow Spirit*; and which the *Greeks* call *Littleneſs of Soul.* This is a defect in the firſt make of ſome Men's minds, which can ſcarce ever be corrected afterwards, either by Learning or Age. And as Souls that are made little and incapacious cannot enlarge their thoughts to take in any great compaſs of Times or Things; ſo what is beyond their compaſs, or above their reach, they are apt to look upon as Fantaſtical, or at leaſt would willingly have it paſs for ſuch in the World. Now as there is nothing ſo great, ſo large, ſo immenſe, as the works of Nature, and the methods of Providence, men of this complexion muſt needs be very unfit for the contemplation of them. Who would ſet a purblind man at the top of the Maſt to diſcover Land? or upon an high Tower to draw a Landskip of the Country round about? for the ſame reaſon, ſhort-ſighted minds are unfit to make Philoſophers, whoſe proper buſineſs it is to diſcover and deſcribe in comprehenſive Theories the *Phænomena* of the World, and the Cauſes of them.
>
> This original diſeaſe of the Mind is ſeldom cur'd by Learning, which cures many others; 'Tis like a fault in the firſt *Stamina* of the Body, which cannot eaſily be rectified afterwards. 'Tis a great miſtake to think that every ſort of Learning makes a man a competent Judge of Natural Speculations;

There, Gentle Reader, you have it "like it is." If you do not have a "mean and narrow Spirit," a "little and incapacious soul," and if you are a "competent Judge of Natural Speculations" (or would like to be), you will be delighted and stimulated by this book. I think you will be, for Quaternarians are noted for large and capacious souls.

It is fitting indeed that this book be reviewed in this journal. The editor, Dr. Albritton, is a distinguished geologist, whose investigations have included Quaternary studies and who is eminent in geohistory and geophilosophy. Many of the authors included in the collection are well known for their contributions to Quaternary geology.

The editor's selection of material reproduced begins with the philosopher Toulmin's essay on history of concepts of time and closes with the essay on philosophy of these time concepts by Kitts, who is both a professional geologist and a professional philosopher. His strictures on "time-stratigraphy" and on "absolute time" should be carefully considered by all geologists, especially Quaternary ones. In between these essays are essays or excerpts from books by the geologists Hutton, Playfair, Lyell, Whewell, Chamberlin, Gilbert, Davis, Johnson, Mackin, Anderson, Hubbert, Simpson, and the philosopher–historian Hooykaas. The editor's Introduction is a valuable factual and interpretive review of each essay, which also includes notes on the scientific biography of each man.

Read Hutton's Abstract for the succinct and readable statement of his steady state theory. (You will have to read this or another facsimile reproduction, as there is believed to be only one original in North America.) Then read the Playfair excerpts to admire the logical presentation. As you would expect, Lyell is there, and we are grateful to have Whewell to show that catastrophism is not philosophically indecent. The famous "Multiple Working Hypotheses" of Chamberlin, one of America's greatest glacial geologists, is gratifyingly reproduced in the original form, rather than in the better-known, but much shorter, *Journal of Geology* form.

Gilbert's essay is valuable in showing that in 1896 the preferred one of a number of working hypotheses was not the "correct" solution of the origin of Coon Butte (Meteor Crater). Gilbert's choice was indeed the "correct" one for the time in which he wrote—a meteoritic origin would have been "incorrect" at that time, as the editor, himself an expert on impact craters, so delicately points out. (Gilbert's work on Quaternary geology is well known, but it may not be so well known that in his very first professional work, for the Ohio Geological Survey under J. S. Newberry, he, along with N. H. Winchell, was the first in the United States to describe, name, and explain the origin of an end moraine of the continental glacier.)

One of the pieces that this reviewer had not

known (or had forgotten—he is very good at forgetting) is the delightful one by W. M. Davis, "Outrageous Hypotheses." In 1926, Davis said that just maybe you should not laugh too loudly at the ideas of the crazy German. He meant Wegener, not W. Penck, who Davis thought was really far out. But then we all are apt to be tolerant of outrageous hypotheses in fields other than our own!

Dr. Albritton states in his Introduction that the selection of essays is from 1785 to 1970, as "the guiding principles of geohistory were not bound together in a coherent theory, free of the strictures of natural philosophy, until after 1785," although earlier men had been "prophets of historical geology." Dare we hope for another book on benchmarks in "protogeophilosophy?" How did Leibniz, Hooke, Steno, Kircher, Agricola, Bergman, Buffon, Avicenna, al Biruni, David, and others come to their philosophical conclusions in geohistory from the data they had at their time?

Buy this important collection of benchmarks on geophilosophy 1785–1970, read it, and quote from it. Encourage your students to read it in order to "enlarge their thoughts to take in a great compass of Times and Things."

GEORGE W. WHITE
Department of Geology
University of Illinois
Urbana, Illinois 61801

REVIEW OF
Trek of the Oil Finders

Trek of the Oil Finders: A History of Exploration for Petroleum. EDGAR WESLEY OWEN. American Association of Petroleum Geologists Memoir 6. Semicentennial Volume. Pp. xv + 1647. American Association of Petroleum Geologists, 1444 South Boulder, Tulsa, Oklahoma 74101. 1975. Price $38.00.

In celebration of the semicentennial of the American Association of Petroleum Geologists (founded in 1917), several books were authorized. Two are closely related: *The Sourcebook for Petroleum Geology* (1969) "catalogues the ideas or concepts" and *The Trek of the Oil Finders* "tells of the men who conceived the ideas". The latter tells of the men who explored, discovered, drilled, and organized small, large, or eventually integrated giants. It is a history of men's activities woven into the petroleum development throughout the world.

In 1916 Edgar W. Owen, then 20 years old, was hired by the Empire Oil Company at a salary of $90.00 per month. For the almost 60 years since then he has been prominent in the petroleum field and is admired and respected for his many contributions to petroleum geology, to business, to academia, to science, and to history. But Owen did not enter the oil business as a neophyte—his grandfather, Samuel Thorniley, was active in oil affairs in the Appalachian region as early as 1865 and his grandfather, Josiah Owen, was associated with oil companies and for a time was chief geologist of a large company. Owen's own personal knowledge and knowledge through his family therefore goes back to the very beginning of commercial petroleum production in America. This, together with his acquaintance with oil men throughout the world, his worldwide geographic experience, his amazing extensive correspondence, and his own personal archives, makes him uniquely fitted to write this history.

At first one is not enthusiastic about the title, *Trek of the Oil Finders,* which seems an unnecessary and not very descriptive phrase to precede *"A History of Exploration for Petroleum."* But one dictionary definition is "trek—to make one's way arduously". The oil finders did "make their way arduously" into hills, plains, and mountains, and later into deserts and stormy seas. But this is not a book about hardships; they are taken as a matter of course and even as a matter of pride. The emphasis is on the men who were the oil finders: geologists and nongeologists. The early oil finders were not geologists (although Silliman lent his name to the report which led to the Drake well) or at least they were not so trained. There were, indeed, able geologists in the 19th century who wrote perceptively on petroleum geology, but they mostly did so to explain after the discovery rather than to guide discovery in virgin territory. They were pathologists rather than attending physicians. As late as 1931 the discovery well in America's greatest oil field (East Texas) was found by a man and his advisors who were definitely not technically trained. However, Owen is quick to point out that the field would have been soon found on geological analysis, thus dissipating the myth that East Texas had been condemned by geologists.

Owen worships no heroes but neither is he given to debunking accepted reputations. He does admire a well-written report on an oil field or on a prospect before it was a field. Such a report enables him to quote appropriate passages to forward the history. The organization of the book is effective in joining chronologic and geographic treatments. The author has the happy faculty of compressing the essence of his histories into a reasonable compass and in a straightforward and delightful style, which the editors (God bless them) have not ruthlessly stodgified. There are many gems, from the copyright page quotation from Mrs. Weeks whose husband "waded down into rivers and mushy marshes gathering leeches . . . searching for those funny little anticlines", to the one near the end (p. 1550) that may well become famous as the shortest description of the geology of Britain: "Rocks representing every age, except Miocene, from Precambrian to Recent crop out in the United Kingdom. Most of Scotland is a Caledonian hodge podge, and Wales is a messy Paleozoic massif, leaving England as the only region with appreciable petroleum prospects". The way in which Owen refers to his "Grandpa [Thorniley] who seemed to dislike Andrews" after that famous geologist shot down one of grandfather's prospects in 1865, leads us to wish to know more of grandpa. But Owen's history is not at all what historians call "anecdotal" if they wish to be critical or call "oral history" if they wish to be approving.

The 24 chapters of the book can be divided into several groups or "megasections" which could well stand by themselves as separate books. This is especially true of the first six chapters dealing with the history of petroleum geology and oil finders to 1910. "The Earliest Oil Industry" to 1859 (Chapter 1) shows that there was indeed an oil industry before 1859, and some prescient ideas about petroleum had been expressed in Europe and in America (for example, subsurface geology by Hildreth in 1836). Owen very neatly puts the Drake well of 1859 into context; if not Drake then some other would have drilled specifically for oil—its time had come! Chapters 2, 3, and 4 blend history of theories—trends, belts, and anticlines. (Most Appalachian oil was stratigraphically controlled, even if in some places associated with anticlines.) The well-known, as well as the not so well-known, early geologists made significant contributions to theory: Newberry, Winchell, Hunt, Andrews, Rogers, Hitchcock, Lesley, Hilgard, Lyman, and French, German, Russian, and Rumanian scientists. (Here is one phase of geology in which British geologists did not take much part, although later they became prominent.) Chapter 5, which continues the story of "Contributions and Controversies, 1871–1890" in all the producing countries, does not slight Europe and Asia. Here, as throughout the book, Owen's intimate acquaintance with overseas geologists and other oil finders and his detailed knowledge of geologic, economic, and political controls is impressive. The first group closes with "Inception of Professional Practice, 1891–1910" (Chapter 6), when oil becomes really big business and more and more sophisticated methods come into use by the scores of geologists throughout the world, with all of whom Owen deals in more or less detail.

The second group of chapters, 7, 8, and 9, deals in detail with the period 1911–1920 in the Midcontinent, other parts of the United States, and the international scene. Here again it is the story of the men who made the discoveries, how they made them, and what they did with the oil they found. (Gas was not yet very important.)

Chapters 10 through 16 make up the third group, which is concerned with the development in the United States to 1940 ("From Scarcity to Surplus"—East Texas and all the rest), and after 1940, with separate long chapters on western United States and Canada, the Gulf Coast, and the Permian Basin.

The fourth group consists of one long Chapter 17 on "Latin America, after 1920", a chapter mainly written by J. Herbert Sawyer, who tells everything you want to know and were afraid to ask about men, geology, politics, economics, and problems in these countries.

The fifth section is Chapters 18–23 with three chapters on the Middle East, and one each on the U.S.S.R., Africa, and the North Sea. This will be a very important part of the book for both geologists and non-geologists, for nowhere else will one find so much information so well organized on men and how they dealt with geology, economics, politics, and diplomacy. In these chapters Owen used brief contributions from others, but mostly the material is from his own personal knowledge and from his skillful use of the various publications and areal reports of the A.A.P.G. and similar sources. The chapter on the U.S.S.R. should be singled out for special mention: Owen describes the sequence of developments by men and organizations. In an impartial way he describes the differences between the methods of the "Western World" and those of the U.S.S.R. in planning and central control and their effect on geologists and engineers as well as on development and production. Production is generally at a very rapid rate to fulfill "the plan". There are separate Ministries of Gas and of Oil, and Owen tells how this affects development, repressuring, and end use of these products.

Each chapter is subdivided into numbered sections, and an excellent system of cross references is employed throughout the book to enable the reader to move forward or backward to other related numbered parts or to the appropriate map. Very comprehensive bibliographies are at the end of each chapter and of many longer sections within chapters. A system of clear and uncluttered index maps is used to show location of fields and areas. There are almost no geologic cross sections or detailed maps of structures, but the excellent references will lead to the appropriate originals. The separate subject, geographical, and proper name indexes occupy the last 55 pages. The users of the book will be indeed grateful for these elaborate and well-organized indexes.

This encyclopedic work is not only the greatest book to appear on a phase of economic geology but also a book that must be put in the class of other great histories of geology, those of Zittel, Geikie, Adams, Hölder, and Chorley-Beckinsale. No other history of geology so clearly brings out the economic, geographic, and political factors that influenced and even governed the development of a science and of a resource. It will go through many printings and will long remain an indispensable tool for reference and a delight to read.

George W. White

Department of Geology
University of Illinois
Urbana, Illinois 61801
August 7, 1975

MEMORIALS

MEMORIAL
TO
JAMES WALTER GOLDTHWAIT

J.W. Goldthwait

PROCEEDINGS VOLUME OF THE GEOLOGICAL SOCIETY OF AMERICA
ANNUAL REPORT FOR 1948
PP. 171-182, PL. 10 APRIL 1949

MEMORIAL TO JAMES WALTER GOLDTHWAIT

BY GEORGE W. WHITE*

James Walter Goldthwait was born March 22, 1880, in Lynn, Massachusetts, the son of James Wesley Goldthwait and Olive Jane Parker Goldthwait. The earliest ancestor, who spelled his name "Goldthwaite", came to America and settled in Salem, Massachusetts, in 1632. An early Goldthwait established a glass factory north of Boston which was a competitor of the better-known Sandwich Glass Works.

From his early life it was taken for granted by his family that Walter Goldthwait would go to college, for which he prepared at Lynn Classical High School, where he made a brilliant record. He was well trained in Greek and Latin and was facile in French and German. A part of his facility in German may have been because of the family custom of singing songs in that language. He entered Harvard in 1898, where he took a general course, receiving the A.B. degree in 1902. The quality of his work is shown by his election to Phi Beta Kappa. Before graduation he had determined to become a geologist and was a student assistant during his last undergraduate year.

Immediately after graduation Professor W. M. Davis, under whom he had studied, took Walter Goldthwait and Ellsworth Huntington, who had just received his M.A. degree, to southwestern Utah to take part in the Plateau and Basin Range studies there in progress. After one week with them Professor Davis left the two young men to work along the Hurricane fault near Toquerville where they spent much of the summer. Their time was used to advantage. They made a topographic and geologic map of an area of about 200 square miles, worked out the stratigraphic section, and determined the relative ages of the two major displacements along the fault. A descriptive and analytical report was immediately written and appeared in the Journal of Geology in 1903. A larger report was published the following year in the Museum of Comparative Zoology Bulletin.

From 1902 to 1904 Walter Goldthwait was teaching fellow at Harvard and also taught at Radcliffe. He received the M.A. degree in 1903. During the second year he began work in glacial geology, which was to be his life-long major interest, with an investigation of glacial lake Sudbury under the direction of Professor Davis. In the report which appeared in 1905 he mapped and described the deltas and sand plains at various levels and, while favoring postglacial tilting to account for the varying levels, he gave full attention to F. G. Clapp's recently announced "theory of

* The writer gratefully acknowledges the assistance in the preparation of this memorial given by Professor E. D. Elston of Dartmouth College, Professor T. R. Meyers of the University of New Hampshire and chief geologist of the New Hampshire State Planning and Development Commission, and Commissioner F. E. Everett of the New Hampshire Highway Department. He is especially indebted to Professor R. P. Goldthwait, of Ohio State University, son of James Walter Goldthwait, who provided much essential information and made available an unpublished resolution of the Dartmouth faculty and the memorials which have already appeared in the *Dartmouth Alumni Magazine*, February 1948, pages 27-28 and in the *Geographical Review*, number 3, 1948, pages 504-505.

many lakelets" (irregular disappearance of essentially stagnant ice). His statement that "such an explanation whether correct or not, is certainly convenient", was to be thought of again by him, and with greater favor, in much later investigations of ice margin glacial deposits.

In 1904 he went to Northwestern University as assistant professor of geology. He at once made himself familiar with the local geology and soon had his students working with him on detailed mapping of late glacial shore lines in the Evanston region. He extended the raised shoreline investigation in the summer and fall of 1905 northward, where he thoroughly covered the eastern part of Wisconsin on foot, by boat, and by railroad. Many data were gathered from a comparatively small district, and his use of the Y-level for determining the elevation of the old shore lines was probably the first extensive precise survey of such features. He demonstrated the warped Lake Algonquin water planes and made an analysis of outlet sequence and positions. A report of this work was accepted as a doctoral dissertation at Harvard, and he received his Ph.D. degree in 1906. The complete report was published in 1907 as Bulletin 17 of the Wisconsin Geological Survey. The northeastern Illinois part of his glacial, shore line, and geomorphic studies was published in 1908 and 1909 in Bulletins 7 and 9 of the Illinois State Geological Survey. In 1907 he worked with F. B. Taylor in Michigan and immediately presented his results in a journal paper (1908) which correlated the shore lines on both sides of Lake Michigan and was a closely reasoned analysis of the data and extent of tilting and warping of the Algonquin and Nippissing water planes. He had demonstrated his capacity to see a problem, collect the data, analyze them, and promptly present a report, which was to be his working method to the end of his career.

Doctor Goldthwait's knowledge of Great Lake history and present lake conditions led to his being retained as a consultant on the Great Lakes diversion suits and to his preparation of a brief and testimony before the Supreme Court.

In 1908 when Charles H. Hitchcock retired as professor of geology at Dartmouth Doctor Goldthwait was invited to come to Hanover. His appointment was as assistant professor of geology, and in 1911 he was elected Hall Professor of Geology, which position he was to hold for more than 36 years. Although at one time or another Doctor Goldthwait taught most of the various courses offered in geology, his own special field of teaching was glacial geology. He took an active and leading part in the large elementary course, which through the years became one of the best-known and most highly regarded courses in the college. In none of his courses did he use a stereotyped method, but drew upon personal experience to a large degree and presented the material in a stimulating fashion. He used a minimum of highly technical and ponderous terminology even in his advanced courses. At the same time, even in elementary courses, he made use of the latest professional reports, especially those dealing with geological bases of practical problems and their solution. A favorite course in later years, "Application of Geology", was taught almost entirely by the use of original reports and by the assignment of problems in which his students had to determine the geological data from the literature or from field study in order to present a possible solution for a ground-water, construction-material, dam-location, or a similar problem. His skill in drawing three-dimensional diagrams enabled him to prepare materials for class and laboratory to illustrate examples of local geology or to present standard geological situations in a simplified and understandable form Such material was changed from year to year and kept thoroughly up to date.

As senior professor in his department Doctor Goldthwait served several terms as department chairman. He favored the rotating chairmanship system of administration and approved highly of election of younger members of the staff to the chairmanship, sometimes very soon after they had achieved professorial status. His advice and counsel were never forced upon staff members by sole virtue of his senior position but were always available and freely given. He was always ready to state his views, but not ready to insist upon things being done his way. According to one of his colleagues, "... Many a time when things looked rather rough he would come forward with some witty remark which took the edge off the onus of the work to be done."

Field work always occupied an important place in any course Doctor Goldthwait taught. His students became thoroughly familiar with the local geology of the Hanover region and the adjoining part of the Connecticut Valley. Many of his students were taken longer distances, and they were always encouraged to make still longer independent trips. He is reported to have often outwalked his students and to have outclimbed them on some of the more strenuous trips. His students saw him not only as a thoroughly competent academic and laboratory geologist, but as one who was engaged in application of geology to public affairs within the State and over the continent of North America. It would have been difficult not to have learned geology thoroughly from such a man.

When the Dartmouth Outing Club was organized in 1911 Doctor Goldthwait was one of the most enthusiastic faculty supporters and was most active in the work of the club. He coached the first Dartmouth ski teams and for many years acted as judge at winter sports meets and winter carnivals. He skated regularly for many years, mainly on Occum Pond which was just below the family home on Occum Ridge. The whole family took part in this recreation, and the sons learned to skate at an early age. Often on winter Saturdays the whole family went on cross-country ski trips. At noon a fire was built and lunch was prepared in a sheltered spot. Sometimes other faculty families joined in these winter ski picnics. Doctor Goldthwait was quietly proud that his sons went on to be point winners in national competition in winter sports meets—the older son, Richard, in skiing and the younger son, Lawrence, in skating.

During World War I, Doctor Goldthwait served as captain in the army, where he was in charge of the map room in the office of the Chief of Staff in Washington. He returned to civilian life soon after the end of the war and in later years was little inclined to say much about his military service. In World War II his early thorough training and continuing use of fundamental sciences enabled him, after a short review, to teach war-time courses in mathematics and physics. He did not think of these classes as solely a necessary duty but as a privilege which enabled him to take active part in the training of men for military service.

When Doctor Goldthwait came to New Hampshire in 1908 he was not unacquainted with its geology, for the family had long had a pleasant summer place on the shores of Lake Wentworth in the central part of the State. He once told the writer with much amusement that in 1908 C. H. Hitchcock explained carefully that a disadvantage to a young man in coming to Hanover would be the fact that the geology of New Hampshire and Vermont was all "worked out" and that a newcomer must realize his research would necessarily be far-removed. However, not discouraged by Hitchcock's gloomy estimate of the barrenness of New Hampshire geologically, he immediately started work on glacial and postglacial features of the Connecticut valley and soon after on White Mountain erosion levels and glacial geology. He reached the conclusion that cirque cutting on Mt. Washington antedated the last ice sheet. His conclusion was strengthened by new evidence discovered many years later by his son Richard in a detailed study of Mt. Washington. His work on drift dispersal, boulder trains, and direction of ice movement soon made him familiar with all parts of the State. This knowledge increased rapidly after the beginning of his work as geologist for the New Hampshire Highway Department and he was the only thinkable person to write the *Geology of New Hampshire* which appeared in 1925 as "Handbook No. 1" of the New Hampshire Academy of Science. As Hitchcock's huge three-volume work had long been out of print and its concepts outmoded a general summary of the geology of the State on modern lines was most desirable. In a little volume of less than 100 pages Doctor Goldthwait briefly summarized the bedrock geology and devoted most of the volume to surface features. The book, while not neglecting the work of others, was actually a report on his studies of almost 20 years. He supplied the photographs, made the sketches which enlivened its pages, and prepared the two colored maps of the State showing bedrock and glacial features. It is probable that he read all the proofs and supervised the page make-up of the carefully printed volume. The book was eagerly received in all quarters and had a steady sale by the Academy. It is now out-of-print and becoming a collector's item. Doctor Goldthwait had begun the revision of *Geology of New Hampshire* with the assistance of his two sons, Lawrence and Richard. The State plans to publish the revised edition as a memorial. Included in the volume will be a large map, now in press, of the surficial geology of New Hampshire by Lawrence Goldthwait which had been prepared under the supervision of Doctor Goldthwait during the past two years.

Early in 1917, only a few years after Doctor Goldthwait came to Hanover and had become familiar with the geology of the State, especially through his work on glacial and geomorphic features, his advice was sought by the State Highway Department in a study of road materials. Towns (townships) had great autonomy in road building, even of roads which formed, or were to form, a part of the State highway network. Often unsuitable materials were used, although more satisfactory material was available. Through his studies of dispersal of drift materials Doctor Goldthwait had a knowledge of which gravels were high in resistant rocks and which in nonresistant rocks. Gravels of differing resistance are often close together in New Hampshire. The varying rock types exhibited in the stone walls over short distances were often a rough index of gravel composition to be expected. Mr. F. E. Everett, State Highway Commissioner, became convinced that geological advice would be valuable and Doctor Goldthwait was asked to take part in the highway program. Mr. Everett has written:

"I recollect that when the Doctor first came to the Department, the Division Engineers were called in for a conference and he outlined to them what he was attempting to do. The boys were skeptical; they had a feeling that they knew more about the local materials for road building than anyone coming in from outside. However, after one trip with the Doctor, they were convinced that he knew what he was talking about. After that, whenever a new job came up, there was an immediate call for Professor Goldthwait."

It would be expected that engineers would understand the usefulness of geological work and give co-operation. More remarkable was the reaction of town "road agents" who were generally farmers or other men who worked only part time supervising "state aid" and town-road construction within their own towns. Doctor Goldthwait often appeared at district meetings to explain, for about five minutes, the gravel survey program. At the close of the session these men clustered about him, some to greet a long-time friend and other to press invitations upon him to visit their towns to see just the kind of features he had been discussing and to call his attention to some bedrock or glacial feature which they believed he should examine. As a result, he and his assistants were welcomed during the field season, and the geological work was firmly supported at all levels.

After such a favorable start Doctor Goldthwait always devoted some time during each year to Highway Department work. Gravel deposits were mapped on existing maps and through his advice and encouragement the State, through the Highway Department, co-operated with the topographic mapping program of the United States Geological Survey. New Hampshire was one of the first states to be completely mapped with modern topographic maps, for the mapping was continued until all earlier and inaccurate maps were replaced by new maps of highest standard. As each topographic map was completed, mapping of gravel deposits began immediately, usually on the preliminary sheets. As Doctor Goldthwait was given a free hand, the mapping was not confined to gravel, but included all the glacial geology. Mr. Everett fully supported all phases of the glacial studies, and through complete study, in many towns which seemed almost hopeless, road materials were found which would never have been discovered by looking solely for gravel. Many a glacial spillway, high on a mountain side, was followed to its outlet and gravel deposits discovered. When a new map appeared Doctor Goldthwait would mark all notches or creases that might be spillways, and it was astonishing how often the places that appeared favorable to him actually had deposits of scientific and economic importance.

For many field seasons Doctor Goldthwait enlisted the assistance of colleagues and advanced students at Dartmouth and invited the young members of the newly organized department of geology at the University of New Hampshire to take part in the program. The newcomers were instructed not to confine themselves too closely to strictly economic deposits but to become familiar with the general geology of the region being mapped. Doctor Goldthwait organized the field work so that when on Saturdays his field men brought their samples taken during the week to the Highway testing laboratory at Concord he was often also in Concord for a brief conference. His critical comments on the week's work were brief, penetrating, and to the point, but so fair and kindly that one's performance improved constantly. Doctor Goldthwait also arranged that expense and salary accounts be specially received each Saturday morning by the chief auditor of the Department and a check immediately issued. The field men regarded with deep respect and great admiration such a man who could not only point out on a map details of geology not always apparent to those who had covered the ground, but could also cut red tape so that accounts of the geologists were handled and paid in a matter of minutes by the chief auditor in person.

The published reports of the Highway Department from 1918 always included a brief report by the chief geologist. These included summaries of the gravel-survey work, bedrock localities suitable for quarries, the progress of topographic mapping and its importance to all highway and other State works. In later years the geological work also included work on ground-water supplies. Detailed maps of all towns and districts were provided for the State office and division offices. The continued value of the work is indicated in a recent letter from Commissioner Everett:

> "Just this summer, we have been building Route 25, through Sandwich and the available gravel was pretty well exhausted. Our Materials Engineer got out the gravel map and located a red spot that on the ground was entirely surrounded by heavy growth. After clearing the woods away a good supply of gravel was uncovered.
>
> "This is just one case where Professor Goldthwait's work has been of inestimable value to the Department. Everyone in the Department held him in great respect and appreciated his knowledge of New Hampshire geology and the good he accomplished in locating material to carry out our extensive construction programs."

In connection with his Highway Department studies and other work in New Hampshire, Doctor Goldthwait was intimately familiar with the bedrock studies of Professor Marland P. Billings and his Harvard students. He was impressed by their completeness and great detail and realized that the data could only be presented adequately by colored maps on a topographic base. If means could be found to publish the maps in color they would have not only great scientific value but would also be very useful to the Highway Department and other State agencies as well. On Doctor Goldthwait's advice the Highway Department published these maps in groups of three quadrangles, and they continue to appear (in later years jointly sponsored by the Highway Department and the State Planning and Development Commission). From the beginning they were of the highest quality, and Doctor Goldthwait and the Highway Department took justifiable pride in this support of the bedrock mapping program in the State, now so far along toward completion.

A part of Doctor Goldthwait's work for the State was in connection with erosion-control defences along the sea coast. His studies here led him to think again of the problem of movement of the modern strand line. Popular belief, supported by many geologists, was that the New England coast was not stable. He had in 1914 determined that no coastal movement had taken place in historic time in Nova Scotia. In a series of papers in 1936 and 1937 he carefully examined the evidence in northern New England and vigorously presented his belief in vertical stability in modern times.

In his earlier work in New Hampshire, Doctor Goldthwait attempted to find the lines of halt (and possible readvance) of the last ice sheet from the evidence of moraines and outwash. He welcomed, encouraged, and secured support for the varved-clay studies of Antevs as an aid in the solution of the general problem. Classical theory required, during ice retreat in wide valleys and basins with a northerly slope, the presence of large open lakes, as had been the case on a larger scale in the Great Lakes basins which his earlier work had made so familiar to him. The gravel and sand deposits of the coastal region had earlier been interpreted as deposits made along successive portions of a retreating ice margin, many of them built to water level of a higher late glacial sea. As detailed work year after year by Doctor Goldthwait and his assistants in the Highway Department provided more and more complete maps of gravel and other glacial deposits he expected that additional morainic areas would be discovered and that the position of successive ice margins would become clear. Such new areas were not discovered. It became apparent that the outwash often had ice contacts on all sides. No extensive lake deposits were found in the basins. The coastal deposits appeared to be more and more similar to the ice-contact deposits of the inland valleys. From the time of his work in 1903 and 1904 on glacial lake Sudbury in Massachusetts, Doctor Goldthwait was acquainted with the concept of ice retreat from a hilly region by downwastage of only weakly active or stagnant ice, but for 30 years he was inclined to minimize its actuality except in very small and limited areas. His often-expressed attitude during these years was that the evidence was not compelling, and certainly not conclusive. About 1935 Doctor Goldthwait, deliberately avoiding any preconceived ideas of either "normal" retreat or stagnation, collected all the immense amount of data he had, and analyzed them by means of maps, charts, and profiles to determine what they showed about mode of ice disappearance. He invited the most critical comment on data and method from his associates and assistants and from other glacial geologists in New England. In his analysis he refused to accept the weight of "authority" unsupported by evidence. Late in 1938 his analysis was presented in a paper, *The uncovering of New Hampshire by the last ice sheet*. Written in his usual clear and compact style, his conclusions are best set forth in his own words in the introduction to the paper:

"In the hilly interior of the State, our mapping of ice-marginal kames and kame terraces which outnumber all other types of gravel deposits, and of marginal stream channels cut across till slopes, indicates a long continued lowering of the surface of the ice while its thin and irregular edge shifted unevenly from contour to contour. Altho deglaciation in general seems to have proceeded northward or northwestward, locally it was toward the east or west, according to the direction of the upland slopes. In this rough terrain the ice seems nowhere to have had a well-defined 'front' transverse to the ice motion of and more active stages of glaciation. How long and how general the ice lay stagnant in lateglacial stages it is not yet possible to say. Recessional moraines, if present at all, are mere scraps, not in line. Local groups of sharply defined ridges, formerly thought to mark halts in the retreat of the margin of an ice lobe, seem now more likely to be crevasse fillings, formed at unknown distances within the wasting ice-field and wholly without seasonal or climatic significance. . . . More particularly in the lowlying coastal district of southeastern New Hampshire, detailed mapping of the late-glacial deposits has led to a complete revision of views regarding conditions of ice removal, marine submergence, and crustal upwarping."

It may well be that this paper is one of the most important he ever wrote. The possession by a man almost 60 of such flexibility, such ability to consider a problem without bias induced by 35 years of work, and courage to put so much of his life's work in a new framework, is a measure of the man.

In 1934 the State Planning Board (later State Planning and Development Commission) formed a committee on geology to explore the need and desirability of a State geological survey. Doctor Goldthwait was of course the most prominent member of the committee but insisted on the election of another as chairman. It was discovered that the geological work in New Hampshire was proceeding more effectively through the harmonious co-operation and support of the State Highway Department,

Dartmouth College, the University of New Hampshire, the Billings group at Harvard, the Academy of Science, and other agencies, than in some other States which had formal State surveys. Everyone realized that Doctor Goldthwait was responsible for this favorable state of geological affairs, and care was taken to preserve it. The committee became in effect an advisory and co-ordinating board, and the geological publications it began to issue were in addition to, but not in competition with, other agencies working on geological problems. Often the committee, especially through Doctor Goldthwait's intimate knowledge of current geological activity, was able to encourage work in progress and to provide for publication. The same system is in effect today, although now the chief geologist of the Mineral Resources Committee of the State Planning and Development Commission performs the duties of State geologist and is now known by that title. The "Goldthwait system" of cordial co-operation of State and private agencies toward solution of geological problems shows the fine but often unrealized organizing ability of its inventor and should provide a lesson for other States far richer in available funds but with a far smaller proportion of their areas mapped topographically and geologically to the highest standards.

During a part or all of the field seasons of 1909–1916 and again in 1925 and 1926 Doctor Goldthwait was geologist for the Geological Survey of Canada. Studies of glacial geology, physiography, and abandoned shore lines were carried on in Ontario, Quebec, New Brunswick, Nova Scotia, and the Magdalen Islands. His later studies were particularly of the St. Lawrence area. His reports on the shore lines of Algonquin and Nipissing age in Canada began to appear in 1909 in Summary Reports and Memoirs of the Survey, and his broader conclusions were summarized in journal articles. He was responsible for much of the physiographic parts of Guidebooks 1 and 3 of the 12th International Geological Congress held in Canada in 1913. In Memoir 140, *Physiography of Nova Scotia*, which appeared in 1924, although the survey had been completed by 1914, an intensive description and analysis of glacial erosion and glacial deposition were given. An analysis of marine erosion of the present day then led to a description and history of the raised beaches of the province. He particularly considered the problem of shore line movement in historic times, and after a long and careful examination of the evidence reached the conclusion that the coast had been stable through historic time.

Doctor Goldthwait was aware of man's activities in relation to the geomorphic features and the glacial and postglacial deposits which were being studied. This led him to the preparation of occasional geographic papers, and he was active in promoting and aiding the work of others in geographic studies as diverse as varved clays in the Connecticut valley, drowned forests in the Maritime Provinces and New England, and the re-establishment of the meteorological station on Mt. Washington. His geographic contributions have already been more fully evaluated in the Geographical Review, Volume 38, no. 3, 1948, pages 504–505.

Doctor Goldthwait was married to Edith Dunnel Richards of Newtonville, Massachusetts, on June 25, 1906. Their two sons, Richard Parker (b. 1911) and Lawrence (b. 1914) each received a bachelor's degree from Dartmouth and went on to graduate study in geology, each specializing in glacial geology. This choice of a vocation was made without pressure or insistence from their parents, but, once the choice had been made, quiet and effective encouragement was given. The positions attained by the sons, the older as professor of geology at Ohio State University and the younger as assistant professor of geology at the University of Maine, were a matter of quiet gratification to Doctor and Mrs. Goldthwait. Mrs. Goldthwait died on August 10, 1947, after an illness of several years. James Walter Goldthwait died suddenly and unexpectedly in his own home in Hanover on December 31, 1947, probably from exertion in trying to clear a way for his car from his garage to the street through the unusually heavy snow which had fallen in a year-end storm.

Many honors came to Doctor Goldthwait. He was elected Fellow of the Geological Society of America in 1909. He was a Fellow of the Association of American Geographers and of the American Academy of Arts and Sciences. He was one of only three honorary members of the New Hampshire Academy of Science, of which he had been a charter member and which he served as president in 1931–1932. Dartmouth conferred upon him the honorary A.M. degree in 1911, and the University of New Hampshire the Sc.D. degree in 1945. He was a member of Phi Beta Kappa, Sigma Xi, and Gamma Alpha. He was starred in "American Men of Science" beginning with the fourth edition in 1927.

In conversation Doctor Goldthwait was not taciturn, but he expressed himself clearly and completely with an economy of words. Many an important conference was concluded in fifteen minutes or even less, but with no hint of haste, pressure, or brusqueness. His papers before the State Academy of Science were the highlights of the annual meetings. They always dealt with substantial topics and were clearly presented to a general scientific audience forcefully but unhurriedly—but they never were longer than eight minutes, often shorter. His writing, which reads easily, is in a compact, economical style, carefully developed over the years by careful revision and removal of every unnecessary word. He maintained that "hard writing makes easy reading." He could sometimes say more in an abstract than others in a whole paper.

Doctor Goldthwait was in many ways the essence of a New Englander. He had a keen and often pointed humor carefully trained to stop short of being biting or sarcastic. He made little jokes on himself, and his best stories, often very funny indeed, were about himself in an amusing situation rather than of some one else. He was of course very well known over the State to such diverse groups as engineers, teachers, officials, winter-sports enthusiasts, and mountain climbers. No one could ever forget this man with ruddy complexion and prematurely white hair, in gray tweeds who could make such pithy remarks on whatever subject was under discussion, but who could also listen so attentively to the opinions of a student, a selectman of a small town, or a State governor. Fair and just in his estimate of people and situations, his faith in his students and associates was not expressed so much in words as in that quiet support which made the man involved outdo himself.

It is obvious that Walter Goldthwait made significant contributions of lasting national and international importance to glacial geology; that his influence as a teacher of undergraduates in general and of those who went on to become geologists in particular will continue for a very long time; and that his work for the State of New Hampshire has set a pattern of scientific co-operation and contribution. But it is those who worked with him who realize most the depth of his influence in science, in method, and in harmonious living and working.

BIBLIOGRAPHY OF JAMES WALTER GOLDTHWAIT

(1903) (And Ellsworth Huntington) The Hurricane fault in the Toquerville district, Utah. Journal of Geology, volume 11, pages 46–63.

(1904) (And Ellsworth Huntington) The Hurricane fault in the Toquerville district, Utah. Harvard College, Museum of Comparative Zoology, Bulletin 42, pages 199–259.

(1905) The sand plains of glacial lake Sudbury. Harvard College, Museum of Comparative Zoology, Bulletin 42, pages 263–301.

(1906) Correlation of the raised beaches on the west side of Lake Michigan. Journal of Geology, volume 14, pages 411–424.

(1907) The abandoned shore lines of eastern Wisconsin. Wisconsin Geological Survey, Bulletin 17, 134 pages.

(1908) A reconstruction of water planes of the extinct glacial lakes in the Lake Michigan basin. Journal of Geology, volume 16, pages 459–476.

(1908) Intercision, a peculiar kind of modification of drainage. School Science and Mathematics, volume 8, pages 129–139.

(1908) (And W. W. Atwood) Physical geography of the Evanston-Waukegan region. Illinois Geological Survey, Bulletin 7, 102 pages.

(1909) Physical features of the Des Plaines Valley. Illinois Geological Survey, Bulletin 11, 103 pages.

(1909) (And R. C. Jacobson) Preliminary report on measurements of altitude of the Algonquin and Nipissing shore lines in Ontario. Canada, Geological Survey, Summary Report, 1908, pages 112–114.

(1910) An instrumental survey of the shore lines of the extinct lakes Algonquin and Nipissing, in southwestern Ontario. Canada, Geological Survey, Memoir 10, 57 pages.

(1910) Isobases of the Algonquin and Iroquois beaches and their significance. Geological Society of America, Bulletin, Volume 21, pages 227–248.

(1911) Raised beaches of southern Quebec. Canada, Geological Survey, Summary Report, 1910, pages 220–233.

(1911) The twenty-foot terrace and seacliff of the lower St. Lawrence. American Journal of Science, volume 32, pages 291–317.

(1912) Records of postglacial changes of level in Quebec and New Brunswick. Canada, Geological Survey, Summary Report, 1911, pages 296–302.

(1913) Glacial cirques near Mount Washington. American Journal of Science, volume 35, pages 1–19.

(1913) Following the trail of ice sheet and valley glacier on the Presidential Range. Appalachia, volume 13, pages 1–23.

(1913) Excursion in eastern Quebec and the maritime provinces; physiography; Quebec and vicinity, physiographical notes; Riviere du Loup, the postglacial marine submergence; Bic, the postglacial marine submergence; Chaleur Bay, physiographic note. International Geological Congress, XII Session, Canada, Guide Book number 1, pages 16–24, 48–51, 66–67, 77–79, 119–120.

(1913) The upper marine limit at Montreal; the upper marine limit at Covey Hill and vicinity. International Geological Congress, XII Session, Canada, Guide Book number 3, pages 119–126.

(1914) Remnants of an old graded upland on the Presidential Range of the White Mountains. American Journal of Science, volume 37, pages 451–463.

(1914) Supposed evidences of subsidence of the coast of New Brunswick within modern time. Canada, Geological Survey Museum Bulletin 2, pages 45–67.

(1914) Marine shore lines in southeastern Quebec. Canada, Geological Survey, Summary Report, 1912, pages 357–359.

(1914) Marine submergence at Montreal, Covey Hill, and Rigaud Mountain. Canada, Geologica Survey, Summary Report, 1913, page 211.

(1914) Physiography and surficial geology of Nova Scotia. Canada, Geological Survey, Summary Report, 1913, pages 244–250.

(1915) The occurrence of glacial drift on the Magdalen Islands. Canada, Geological Survey, Museum Bulletin 14, page 11.

(1915) Physiography and surface geology of Nova Scotia. Canada, Geological Survey, Summary Report, 1914, pages 102–103.

(1915) The origin of Lost River and its giant potholes. Science, new series, volume 41, pages 834–836.

(1916) Glaciation in the White Mountains of New Hampshire. Geological Society of America, Bulletin, volume 27, pages 263–294.

(1916) Evidence for and against the former existence of local glaciers in Vermont. Vermont State Geologist, Report 10, pages 42–73.

(1922) Preface; Explanation of the map illustrating the recession of the last ice sheet from New England and New York; and folded map, *in* Antevs, E., The recession of the last ice sheet in New England. American Geographical Society, Research Series, number 11, pages vii–x, 104–107.

(1924) Physiography of Nova Scotia. Canada, Geological Survey, Memoir 140, 179 pages.

(1925) The geology of New Hampshire. New Hampshire Academy of Science, Handbook no. 1, 86 pages.

(1927) A town that has gone downhill. Geographical Review, volume 17, pages 527–552.

(1928) The gathering of floods in the Connecticut River system. Geographical Review, volume 18, pages 428–445.

(1928) Preface to Antevs, E., The last glaciation. American Geographical Society, Research series, number 17, pages vii–viii.

(1934) (And C. J. Lyon) An attempt to cross-date trees in drowned forests. Geographical Review, volume 24, pages 605–614.

(1938) Vertical stability of the coast at Marblehead, Massachusetts. American Journal of Science, 5th series, volume 35, pages 81–93.

(1938) (And F. C. Kruger) Weathered rock in and under the drift in New Hampshire. Geological Society of America, Bulletin, volume 49, pages 1183–1197.

(1938) The uncovering of New Hampshire by the last ice sheet. American Journal of Science, 5th series, volume 36, pages 345–372.

(1945) (As member of National Research Council Committee, responsible for New England States) Glacial map of North America. Geological Society of America, Special Papers, number 60.

(1948) Mineral composition of New Hampshire sands. New Hampshire State Planning and Development Commission, Mineral Resource Survey, Part 9, 7 pages.

REPORTS AS GEOLOGIST, NEW HAMPSHIRE STATE HIGHWAY DEPARTMENT

(1918) [Unsigned] Geological note in Purrington, W. F., Report of State Highway chemist, New Hampshire, State Highway Department, Seventh Biennial Report, pages 14–16.

(1920) (And W. F. Purrington) New Hampshire gravels. New Hampshire, State Highway Department, Eighth Biennial Report, pages 207–243.

(1922) Report of the State Highway Department, Ninth Biennial Report, pages 18–23.

(1924) Report of the State Highway geologist. New Hampshire, State Highway Department, Tenth Biennial Report, pages 27–31.

(1926) Native Road Materials in New Hampshire. New Hampshire, State Highway Department, Eleventh Biennial Report, pages 27–40.

(1928) Report of the State Highway geologist. New Hampshire, State Highway Department, Twelfth Biennial Report, pages 27–30.

(1930) Report of the State Highway geologist. New Hampshire, State Highway Department, Thirteenth Biennial Report, pages 25–29.

(1932) Report of Highway geologist. New Hampshire, State Highway Department, Ninth Annual Report, pages 13–17.

(1933) Report of Highway geologist. New Hampshire, State Highway Department, Tenth Annual Report, pages 16–18.

(1934) Report of the geologist. New Hampshire, State Highway Department, Eleventh Annual Report, pages 12–13.

(1935) Report of the geologist. New Hampshire, State Highway Department, Twelfth Annual Report, pages 14–16.

(1936) [Unsigned] Geological work. New Hampshire, State Highway Department, Thirteenth Annual Report, pages 13–15.

(1937) [Unsigned] Geological work. New Hampshire, State Highway Department, Fourteenth Annual Report, page 20.

(1938) [Unsigned] Geological work. New Hampshire, State Highway Department, Fifteenth Annual Report, pages 15–16.

(1939) [Unsigned] Geological work. New Hampshire, State Highway Department, Sixteenth Annual Report, pages 24–26.

(1940) [Unsigned] Geological work. New Hampshire, State Highway Department, Seventeenth Annual Report, pages 20–22.

(1941) [Unsigned] Geological work. New Hampshire, State Highway Department, Eighteenth Annual Report, pages 14–15.

(1942) [Unsigned] Geological work. New Hampshire, State Highway Department, Nineteenth Annual Report, pages 17–18.

MEMORIAL
TO
VINCENT CHESTER SHEPPS
(1928-1967)

MEMORIAL TO VINCENT CHESTER SHEPPS

(1928–1967)

GEORGE W. WHITE
The University of Illinois, Urbana, Illinois

Vincent Chester Shepps was born December 21, 1928, in Lebanon, Pennsylvania. He died August 25, 1967 in Harrisburg. Dr. Shepps is survived by his wife Avrill Brockelman Shepps, whom he met as a fellow graduate student at the University of Illinois, and by their son Anton C. His parents and two brothers also survive.

After receiving the B.S. degree in 1950 from Franklin and Marshall College, he entered the University of Illinois where he successively held appointments as graduate assistant, research assistant, and University Fellow. He received the M.S. degree in 1952 and the Ph.D. in 1955. His master's thesis on sedimentary petrology of some glacial tills in the Allegheny plateau was immediately published. His doctoral dissertation was a detailed study of the glacial geology of an area in northwestern Pennsylvania and formed a large part of *Pennsylvania Geological Survey Bulletin G-32*, of which he was senior author.

His earliest field experience was in the summer of 1949 as assistant to Dr. J. H. Mackin in his investigation for the U. S. Geologic Survey of the Iron Springs, Utah, area. In the summer of 1950 he worked as field assistant to Dr. Jacob Freedman with a U. S. Geological Survey party in Michigan. In the summer of 1951 he was field assistant on the Illinois Survey, working with Dr. P. R. Shaffer in glacial studies in northwestern Illinois. In 1951 and 1952 he was research assistant for the Pennsylvania Geological Survey, working on field studies for his doctoral dissertation. During a part of 1954 he was geologist with the Military Branch of the U. S. Geological Survey, and after two years service in the United States Army, mainly in Germany, he became geologist on the Pennsylvania Geological Survey in January 1957.

His first work on the Pennsylvania Survey involved both field and laboratory studies. His abilities in organizing projects and in producing superior text and illustrations were soon recognized. He organized, led, and wrote the elaborate guidebook for the 24th Annual Field Conference of Pennsylvania Geologists in 1959 and organized the Northwestern Pennsylvania Glacial Geology field trip for the 1960 Pittsburgh meeting of the Geological Society and again prepared an elaborate field guide. During much of 1959 and 1960 he was deeply involved in revising the Geological Map of Pennsylvania and producing it at a new and larger scale. He was not only a superb draftsman but a real artist, and he was solely responsible for training the drafting personnel in the newer techniques of map scribing and color-separation procedures. He directed the production of the geological map from the integration of existing maps with field maps of more recent date and the reconciling of "boundary discontinuities." This map, which appeared in late 1960, was widely regarded at that time as the best state geologic map. It is to be noted that his projects "marched" with dispatch and that the bulletins and maps which he produced from his own work or from the compilation of others appeared in printed form very promptly.

The great success of his work with the geologic map and with the other publications led to his appointment in 1961 as senior staff geologist and editor of the Survey. In this position he was responsible for the publication of all Survey geologic maps and reports. At the same time he was able to continue with his own research. His work on quadrangle mapping, necessary for reconciling certain points for the new map of Pennsylvania, was of the highest order and is now in process of publication.

Pennsylvania was one of the first states to have an official earth-science program and curriculum for the public schools. Dr. Shepps took an active and important part in organizing this program; he edited publications for the projects and wrote two of them himself. His booklet *Pennsylvania and the Ice Age* was a "best seller" of the Survey and more than 40,000 copies have been distributed. It is still being issued at a rate of over 500 copies per month. His advice on publications for new earth-science curriculums was widely sought in other quarters.

Dr. Shepps' work in earth-science education was recognized by his appointment as American Geological Institute visiting geoscientist at several eastern colleges and universities. He was a consultant in earth-science education to many high schools. In this work his unusual clarity, logic, and enthusiasm of expression became widely recognized.

Dr. Shepps' research as a student and as a professional geologist was marked by scientific imagination and application of innovations to solutions of problems. He was unusually resourceful in field work and in laboratory and statistical analysis of petrologic data, and persuasive in writing and graphic presentation of these data. He constructed well-reasoned theories to account for the phenomena he observed. Among his contributions were the discovery of the drumlinoid features of northwestern Pennsylvania, continuity of composition of single till sheets and the differences between them in the northwestern Allegheny plateau, and the existence and wide extent of the Lavery Till of late Wisconsinan age. As a consequence of his knowledge of glaciology he was asked to write a description of glacial deposits for a forthcoming encyclopedia (*see* Bibliography). In addition to his skill and productivity as a geologist, he was also an outstanding editor, administrator, and artist. He was elected to Fellowship in The Geological Society of America in 1965.

Dr. Shepps' notable success in his editorial work with the Survey and his prominent position in earth-science education led to an invitation in 1966 from the Pennsylvania Department of Public Instruction to take charge of the publication activities of this organization which has an annual budget in the millions of dollars. As director of this new Bureau of Publications he was responsible for overseeing the work of editors of over 500 publications issued each year. His talents immediately began to be apparent in systematizing and reorganizing the publication program of the department. A skillful writer and illustrator himself, he was able to encourage and promote good work and to stimulate the members of his staff to greater achievement. Had he been able to continue in this position his outstanding organizational ability would have been soon recognized with greater opportunities in public administration.

Although his illness developed in early 1967 and was diagnosed in April as lung cancer, he persevered with courageous will power to continue his work until only a few days before his death in August. He was thus able to complete projects in process and to leave his geological and educational projects in such shape that they can be continued and completed by others.

Dr. Shepps was an accomplished silversmith and an artist in enamels. His work was widely exhibited and won many awards. He was recognized as one of the leading craftsmen in silver and enamel in Pennsylvania and in 1963 was president of the Pennsylvania Craftsman's Guild. Mrs. Shepps, a recognized artist in enamels, shared his artistic interests and they worked together on many projects.

It is not often that such a combination of outstanding enthusiasm, ability, and productivity as a research geologist, administrator, editor, and artist is combined in one man. He was, and remains, an inspiration to all who knew him.

In preparing this memorial I have been given helpful information by Mrs. Avrill B. Shepps and staff members of the Pennsylvania Geological Survey.

BIBLIOGRAPHY OF VINCENT CHESTER SHEPPS

1952 (with White, G. W.) Characteristics of Wisconsin tills in northwestern Ohio (Abstract): Geol. Soc. America Bull., v. 63, p. 1388.

1953 Correlation of the tills of northeastern Ohio by size analysis: Jour. Sed. Petrology, v. 23, p. 33–48.

1955 The glacial geology of a part of northwestern Pennsylvania: Ph.D. thesis, Illinois Univ. Dissertation Abstracts, v. 15, p. 1832–1833.

1957 (with White, G. W.) Glacial deposits of northwestern Pennsylvania (Abstract): Geol. Soc. America Bull., v. 68, p. 1902–1903.

1958 "Size factors," a means of analysis of data from textural studies of tills: Jour. Sed. Petrology, v. 28, p. 482–485.

1959 The glacial geology of Crawford and Erie Counties: Field Trip A, Guidebook to 24th Annual Field Conference of Pennsylvania Geologists, p. 2–11.

—— Glacial geology of northwestern Pennsylvania: Geol. Soc. America Guidebook for Field Trips, Pittsburgh Meeting, p. 167–188.

—— (and White, G. W., Droste, J. B., and Sitler, R. F.) The glacial geology of northwestern Pennsylvania: Pennsylvania Geol. Survey Bull. G-32, 4th series, 59 p.

1960 (with Gray, Carlyle, and others) Geological map of Pennsylvania: Pennsylvania Geol. Survey Map 1. (Supervising editor and designer of this map.)

1962 Pennsylvania and the Ice Age: Pennsylvania Geol. Survey Educ. Ser. no. 6, 33 p.

—— Glaciers, unit 10 *in* Geology and earth sciences sourcebook: New York, American Geol. Institute, Holt, Rinehart and Winston, p. 179–198.

1963 *Editor*, Symposium on Middle and Upper Devonian stratigraphy of Pennsylvania and adjacent states: Pennsylvania Geol. Survey Bull. G-39, 301 p.

1966 The point system of jurying: Ceramics Monthly, p. 24–25.

—— The point system of jurying: Outlook, v. 7, no. 2, p. 2–3.

—— Jurying by point system: Handweaver and Craftsman, v. 17, no. 3, p. 7.

1967 (and Tomikel, J. C.) The geography and geology of Erie County, Pennsylvania: Pennsylvania Geol. Survey Inf. Cir. 56, 64 p.

—— Sections on "Glacial deposits, kames, eskers, drumlins, till, tillite, and till fabric," *in* Fairbridge, Rhodes, *Editor*, Encyclopedia of Earth Sciences: New York, Rinehold, in press.

MEMORIAL
TO
J. ERNEST CARMAN
(1882-1966)

MEMORIAL TO J. ERNEST CARMAN
(1882–1966)

GEORGE W. WHITE

Joel Ernest Carman was born August 13, 1882, on a farm in Handley, Iowa, 2½ miles southwest of St. Charles in Madison County and about 25 miles southwest of Des Moines. He died January 11, 1966. He was the fourth of seven children of Jacob Waters Carman, a Civil War veteran. "Joel Ernest" was a common name in the Carman family and was always used in the form "J. Ernest." He graduated from St. Charles High School in 1897. He entered Simpson College, Indianola, Iowa, that fall, but it was necessary for his mother to establish residence in Indianola as the college would not admit students under 16 who did not live with their parents. His detailed daily record and personal account books, which he began in 1897 and continued to 1962, show total expenses of $1,002.59 for the four college years.

Following Ernest Carman's graduation from college in 1901, he taught for the year 1901–02 at the Hattel District #4 School, South Township, Madison County, Iowa, a school of all 8 grades having 33 pupils. One of his pupils of that year states her high opinion of him as a teacher and speaks of the admiration and affection which the pupils had for him. In the spring and summer of 1902 he attended Iowa State Normal School at Cedar Falls, and after further study during the summer of 1903, received the degree M.Di. (Master of Didactics). He taught science, and coached football and track in 1902–03 at the Hawardin, Iowa High School and in 1904 at the Council Bluffs High School. He had great success as an athletic coach, his football teams in particular achieving records that brought him state-wide prominence.

In spite of Ernest Carman's success as an athletic coach his real interest was in science and particularly in earth science. In order to increase his knowledge he took the University of Chicago geology field trip through the West in the summer of 1904 and made the decision to take further work in geology. In 1904–05 he taught in a high school in the mornings and took courses at the University of Chicago in the afternoons to prepare for graduate work. He held graduate fellowships at Chicago from 1905 to 1908, when he completed the residence requirements for the Ph.D. In the summers of 1905 and 1906 he worked for the Illinois and Iowa Geological Surveys in investigations along the Mississippi Valley and in 1907 for the U. S. Geological Survey in Wyoming. As travel in Wyoming was by horse-drawn vehicles, partly over country without roads, his earlier farm experience and great strength made him valuable not only as a geologic aid, but also in the work of transport of the camp and field equipment.

In his graduate work at the University of Chicago, Ernest Carman had as teachers among others T. C. Chamberlin and R. D. Salisbury. It is little wonder, therefore, that his attention was particularly drawn to Pleistocene geology and his first large research project was a field study in 1905 and 1906, jointly supported by the Iowa and Illinois Surveys, of the Pleistocene geology of the Mississippi Valley between Savanna, Illinois, and Davenport, about 15 miles into Iowa and the same distance into Illinois. Arthur C. Trowbridge was his assistant in the second field season. As Professor Salisbury was consulting geologist of the Illinois Geological Survey and in charge of the preparation of educational bulletins, he supervised the field work and guided the publication of the report in 1909 as *Bulletin 13* of the Illinois Survey. The area of the report extends from the Driftless Area to below the junction of the Rock River with the Mississippi, an area where several drift sheets of several ages come together and overlap with a complex series of drainage changes caused by the different ice advances. This complicated area is still a subject for continued investigation. All the later workers speak highly of Dr. Carman's observations and explanations of the data in *Bulletin 13*, which is a pioneering work on the interpretation of the Pleistocene geology of the upper Mississippi Valley, particularly of the Rock Island area. From personal discussion with him in later years it is probable that some of his theories were too far "advanced" for the times and the printed text is somewhat more conservative than the original draft. At least he always spoke approvingly of the newer conceptions of later workers of the sequence and extent of the different ice advances into the region. As the bulletin was designed as one in the Educational Series, it contains introductory material on the bedrock geology and on the general geomorphology and principles of landform development. The careful organization of this material, as well as of the glacial part, and the lucid explanation and the clear diagrams are typical of all his publications and of his teaching methods throughout his whole career.

Field work for his doctoral dissertation, a study of the Pleistocene geology of northwestern Iowa, was begun in 1909 and continued in 1910, 1911, and 1913 and the report was completed and the Ph.D. awarded in 1915. This report, published by the Iowa Geological Survey in 1917, was a detailed study of the western part of the Des Moines glacial lobe and of the older drift from the Altamont moraine westward to the state line. In the 1917 report all the drift west of the Altamont moraine was interpreted as Kansan in age. Field work after 1917, when new exposures allowed more detailed study, led him to the conclusion that part of the drift was later than Kansan and was interpreted as Iowan. The revised report was published in 1931 by the Iowa Survey and has remained a definitive work for this part of the state until recent times, when many miles of superhighway cuts and deep auger borings have furnished many more data for stratigraphic analysis.

At the conclusion of his course work in 1908, Ernest Carman was invited to the University of Cincinnati, where he was the second man to be appointed to the Department of Geology after the establishment of the department by N. M. Fenneman in 1907. His first appointment was as museum assistant in Geology, but he was made instructor in 1909 and had advanced to assistant professor by 1915, when he was appointed professor of historical geology and curator of the Museum of Geology at

The Ohio State University, to fill the vacancy created by the untimely death of Professor Charles S. Prosser.

While in Cincinnati, although Carman was writing on the report of Pleistocene geology of northwestern Iowa, he became familiar with the famous exposures of Ordovician rocks in the Cincinnati area. As these rocks had been studied in detail by many persons, he recognized that a more promising field for stratigraphic investigations was in the carbonate rocks above the Cincinnatian, which outcrop to the north, and his interest was drawn to the Monroe group of rocks of Silurian and Devonian age. Upon moving to Columbus he began work on the stratigraphy of the Monroe group of rocks, beginning with those exposed in the southern part of the outcrop in Highland County. His field work was supported by the Geological Survey of Ohio and he spent each summer in the field from 1917 to 1940, with the exception of the World War I years. Investigation in Highland County soon showed that the Monroe rocks were better exposed to the north and the section was more complete, so he gradually extended his work northward to the boundary of the state in Lucas County. He spent several summers in the Toledo region making detailed investigations on the stratigraphy and structure and extended his work across the state line into Michigan. His work on the Monroe rocks led to several publications, and a long manuscript on the geology of Lucas County was essentially completed by 1941. New outcrops in quarries and other large excavations, and cores from extensive drillings were studied after 1941, and since he felt that further additions were necessary he was unwilling to have this manuscript published without incorporating the new data, a task on which he continued to work.

In his work in Lucas County he discovered the Holland Quarry Shale, a thin Early Devonian unit in a depression in the Raisin River Dolomite (Silurian), underlying the Sylvania Sandstone (Middle Devonian). From the Shale he "obtained a large collection of ostracoderms, early fishes, eurypterids, and some plant remains." In 1956 the collection was presented to the Field Museum and the forms studied in detail: R. H. Denison described part of the fauna in "Fishes of the Devonian Holland Quarry Shale," (*Fieldiana-Geology*, v. 11, 1961, p. 555–618) and E. N. Kjellesvig another part in "Eurypterids of the Devonian Holland Quarry Shale of Ohio," (*ibid.*, p. 79–98).

In 1918 and 1919 Dr. Carman was on the overseas YMCA staff as recreation director of a rest camp in France. One of his YMCA associates was R. A. Daly. After the armistice Dr. Carman visited some of the famous European geological localities in the Alps. Characteristically, one of his aims was to climb Mont Blanc; although the weather was unfavorable, this would be his only opportunity, so he made the ascent.

After the death of J. A. Bownocker, Professor Carman was appointed head of the Department in 1929, a position he held until 1944. Almost all of this time was a trying period for academic administrators; the Great Depression began in 1929 and universities were just recovering from its effects when the onset of the World War II period brought problems of student enrollment, staff members in military or other war service, travel and equipment shortages and many others. During all these years Professor Carman gave long and conscientious attention to university and department problems, continued his full teaching program and spent each summer in the field.

He retired as professor of geology in 1952, at which time his portrait was presented by alumni and associates to hang beside those of earlier distinguished staff members on the walls of Orton Library of Geology.

Professor Carman was president of the Ohio Academy of Science 1943–45 and was the only person to serve the Academy as president for two consecutive terms. His presidential address, "Geologic Interpretation of the Scenic Features of Ohio," was long months in preparation and is actually a fine summary of Ohio geology and its relation to surface features. Two of his students assisted in the preparation of special maps and sections. The illustrations were presented as slides shown by two projectors (at a time when this method of presentation on two screens was very novel), and the address was given without mention of changing of slides because the presentation had been carefully rehearsed and the student assistants recognized all of the cues for changes. As *Reprint Series 3* of the Geological Survey of Ohio, this address continues to be one of its popular publications.

Professor Carman's classes were always most carefully prepared year after year. The material was meticulously outlined and presented in sequential fashion. In his graduate courses he encouraged his students to prepare their material for seminar presentation in the same clear way. He encouraged discussion of new and controversial ideas. He particularly appreciated new and recent foreign references, which he carefully checked and then incorporated into his own notes.

His famous course at Ohio State was the graduate course in stratigraphy of Ohio which was given on weekends in the field, studying rock units from Ordovician to Permian and taking some account of Pleistocene features. Each student's weekly report was critically read and edited for text and illustrations and always returned before the next weekend so that the student could sharpen his geological vision and take advantage of the comments when writing the next report. The course ended with a searching final examination. On one of these weekend trips in October, 1923, Professor Carman was assisting his students to remove a particularly fine fossil from one of the Cincinnatian limestone layers in an outcrop in Kentucky, just across the Ohio River from Cincinnati. A fragment of steel from his hammer hit his eye. That a fragment had actually entered the eye was not realized for many days and it was then too late to save the sight of the eye. He adjusted to the loss of binocular vision and never mentioned his handicap except when in later years features shown on airphoto stereo pairs were discussed with him.

Those who knew Professor Carman in his later professional career little realized his earlier athletic prowess as player and coach. Although he was unusually young for a college student, he was well over six feet tall and rugged. In his senior year at Simpson, while only 18 years old, he was an outstanding member of the track team as a hurdler and established a college record for the high jump which was not broken for many years. In his senior year he was end on the football team and in the annual game with Simpson's traditional rival, Penn College, Ernest Carman won the game with a 70-yard run. The game was recorded in detail in the school paper and of course Carman became a campus hero. In his two high school positions after graduating from college he was track and football coach. His football teams were so successful that he immediately became a state figure in athletics and could have made a career in this field. From his teaching methods, one can be sure his teams were thoroughly drilled in fundamentals and that their high success came from a few plays precisely executed, but with one or two unexpected and unlooked for "different" ones that could go for long gains. While attending the University of Chicago as a graduate student he played tackle on the varsity team, one of the country's outstanding teams of that time. (Graduate students were eligible for such participation in those days.) It was little known that he maintained his interest in varsity sports and was an attendant at university football games. Only a very few people knew of his technical interest because, unlike most fans, he did not discuss his opinions unless seriously pressed for them. His analysis of line-play was especially penetrating and he would have made a valuable scout for a college or professional coaching staff.

Professor Carman was generally formal in manner and not much given to small talk, but was always willing to discuss problems and situations with students and staff members. But with a small group in the field, and particularly when he and one other were looking at geology, he was discovered to have a fund of humorous stories about his earlier geological experiences—on some occasions he could even be persuaded to recount experiences in athletics and mountain climbing. His great generosity to all good causes, including assistance to many students, was carefully concealed and only the students he helped and associates who were solicitors for Community Fund, Red Cross, or church knew this facet of his character. In his will he provided a $25,000 fund to Simpson College to endow a scholarship for a graduate from a Madison County high school and an equal amount to The Ohio State University for support of geology department activities.

Ernest Carman married Gertrude Hancox on Thanksgiving Day, 1908. She was the youngest daughter of Andrew Hancox, a Methodist minister who was pastor at the Carmans' church in St. Charles, 1897–1900. Gertrude's older brothers and sisters were fellow students of Ernest at Simpson College, which is about 20 miles west of St. Charles, and she herself graduated from Simpson in 1905. Mrs. Carman died January 25, 1942. Professor Carman is survived by his sister, Mrs. Grace (W. H.) Breeding of St. Charles, Iowa; his son Ralph, an insurance executive of Wilbraham, Massachusetts, and two grandsons and a granddaughter.

In preparing this memorial I have been given helpful information by Mr. Ralph H. Carman, by staff members of the Geology Department of The Ohio State University and of the Geological Survey of Ohio, and by Professor A. C. Trowbridge. I am particularly indebted to Professor Ruth M. Jackson, Archivist of Simpson College, who searched early records and conducted extensive personal interviews in Iowa to secure information on Professor Carman's early years. Extensive data from these and other sources, which could only be summarized in this memorial, have been deposited in the University Archives of The Ohio State University.

BIBLIOGRAPHY OF J. ERNEST CARMAN

1909 The Mississippi valley between Savanna and Davenport: Ill. Geol. Survey, Bull. 13, 95 p.

1913a Notes on the Nebraskan drift of the Little Sioux Valley in Cherokee County, Iowa: Iowa Acad. Science, Proc., v. 27, p. 231–236.

1913b The Wisconsin drift plain in the region about Sioux Falls, South Dakota: Iowa Acad. Science, Proc., v. 27, p. 237–250.

1917 The Pleistocene geology of northwestern Iowa: Iowa Geol. Survey, v. 26, p. 232–445.

1922 Some subsurface rock channels and cavities filled with glacial material: Ohio Jour. Science, v. 22, p. 125–128.

1927a The Monroe division of rocks in Ohio: Jour. Geology, v. 35, p. 481–506.

1927b Fossil footprints from the Pennsylvanian system in Ohio: Geol. Soc. America Bull., v. 38, p. 385–395.

1927c The rock section at the O'Shaughnessy Dam (Scoito River, Delaware County, Ohio): Ohio Jour. Science, v. 27, p. 289–294.

1928 The O'Shaughnessy Dam and Reservoir: Discussion: Am. Society of Civil Engineers Proc. v. 54, p. 1325–1327.

1929a The geology of the Columbus region: Chapter II, p. 5–9, in "Columbus, an analysis of a city's development," published by the Columbus Chamber of Commerce.

1929b (and Schillhahn, E. O.). The Hillsboro sandstone of Ohio: Jour. Geology, v. 38, p. 246–261.

1930 Drainage changes in the Toledo region: Ohio Jour. Science, v. 30, p. 187–193.

1931 Further studies on the Pleistocene geology of northwestern Iowa: Iowa Geol. Survey, v. 35, p. 15–193.

1934 (with Stout, Wilber). Relationship of accumulation of oil to structure and porosity in the Lima-Indiana field: Problems of Petroleum Geology: Am. Assoc. Petroleum Geologists, p. 521–529.

1936 Sylvania sandstone of northwestern Ohio: Geol. Soc. America Bull., v. 47, p. 253–266.

1946 The geologic interpretation of scenic features in Ohio: Ohio Jour. Science, v. 46, p. 241–283. (Also as Ohio Geol. Survey Reprint Ser. No. 3).

1947 Geologic section of the Chillicothe test-core: Ohio Jour. Science, v. 47, p. 49–54. (Also as Ohio Geol. Survey Rept. Inv. 2).

1960 The stratigraphy of the Devonian Holland Quarry Shale of Ohio: Fieldiana-Geology, v. 14, No. 1, p. 1–5.

memorial
to

Harold Rollin Wanless
1898-1970

by

George W. White

THE
GEOLOGICAL SOCIETY
OF AMERICA

NOVEMBER, 1971

MEMORIAL TO HAROLD ROLLIN WANLESS (1898-1970)

GEORGE W. WHITE
University of Illinois, Urbana, Illinois

Harold Rollin Wanless was born December 5, 1898, in Chicago, Illinois. He died in Champaign, Illinois, June 3, 1970. He was the son of William and Rhoda Tanner Wanless. The father, a promising young lawyer, was in San Francisco at the time of the 1906 earthquake, and he lost his life in that disaster. His mother taught nature study in the Chicago schools and Harold, from his earliest days, learned to recognize birds, flowers, and rocks, and was practically brought up with a binocular in his hand. He learned photographic methods early and experimented with and improved upon various kinds of cameras; he was one of the first to use color photography.

After graduation from Waller High School, in 1916, he was awarded the Princeton Club of Chicago Fellowship. He was an undergraduate at Princeton from 1916 to 1920, when he received the B.Sc. in geology with high honors. During World War I he was a member of the Princeton detachment of the Student Army Training Corps from September to December, 1918, and rose to the rank of sergeant. Since he came to Princeton with professional qualifications as a photographer he soon became a staff member of the *Princeton Photographic Magazine* and the *Princeton Pictorial.* Harold was often assigned to record some of the activities of distinguished visitors and thus met many very famous people.

Following graduation, he remained at Princeton where he was appointed University Fellow in 1920-1921, Reader in Geology in 1921-1922, and Warfield Scholar in 1922-1923. His master's thesis, *Geology of the Rosendale Cement District, Ulster County, New York*, was presented in 1921. It is an example of his meticulous attention to detail and of his use of all these details in analyzing the geology of a region, in extracting general principles, and in determining geological history. The report consists of two large volumes; the first of 279 typed pages includes a detailed description of scores of sections and deals in great detail with the stratigraphy, paleontology, mineralogy, economic geology, tectonic history, historical geology, and paleogeography of the region. The illustrations are assembled in a second large volume of 123 superb photographic plates, many of them folding, mainly of strata and of tectonic and sedimentary structures, but some of topography and some of portraits of the Princeton staff and students of historical importance. The labeled stratigraphic outcrop plates show many classic exposures which probably no longer exist.

Beginning with the field season of 1920, Wanless worked in the Big Badlands of South Dakota with Professor W. J. Sinclair. This field work continued in the summers of 1921 and 1922, under the direction of Professor Sinclair, and provided the material for his doctor's thesis presented in 1923. The thesis, which dealt with the lithology and stratigraphy of the White River Beds of South Dakota, was published in two long papers in the *Proceedings of the American Philosophical Society*. He was able to determine that the sediments, with their rich fossil content, were not lacustrine deposits, but were "river-channel, freshwater, pond, sheet flood, and eolian elements," and from these drew conclusions about the source of the material and its method of transportation, and about the conditions under which the White River faunas lived and evolved. This relation of sedimentary character to the lateral and vertical changes of the deposits was to be found again and again as he studied Pennsylvania deposits in Illinois and throughout the United States, and indeed the world.

In addition to the scientific reports presented in his thesis, and the two long journal articles, he prepared for his own personal records an elaborate account of his experiences, illustrated with drawings and photographs. His field assistant was J. B. Mawdsley and, in addition to Professor Sinclair, he was associated in his field and laboratory work from time to time with Professors A. H. Phillips, R. W. Chaney, A. H. Pillsbury, and W. B. Scott. His personal account, preserved in a large, handsomely bound volume, provides interesting biographical material on these men.

Upon the completion of the Ph.D. in 1923, the new Dr. Wanless was invited to join the staff of the University of Illinois Department of Geology as Instructor, at a salary of $1800. He was promoted to the rank of Associate in 1925, a rank that existed at that time between Instructor and Assistant Professor. He became Assistant Professor in 1929, Associate Professor in 1937, and Professor in 1945. He was Acting Head of the Department in the academic year 1946-1947. He became Professor Emeritus in 1967, after 44 years of active service.

When Dr. Wanless came to Illinois, he participated in the large beginning courses in physical geology and historical geology, and for many years taught the course in geology for students in the College of Agriculture, with great success. He also taught vertebrate paleontology in the early years. Soon after he came to Illinois, he began to teach a course in field geology and in geological maps, and was very successful in imparting his own meticulous field methods to many generations of students. When air photos became generally available in the 1930s, Professor Wanless was in the forefront in developing the use of these for geological purposes. As an expert photographer himself, he understood not only the benefits, but the limitations of the method.

As his earlier interests were in sedimentation and sedimentology, it was natural that he began to teach courses in sedimentation and sedimentary petrology and continued the sedimentation part throughout most of his academic career. He divided the work between the theoretical classroom instruction in sedimentation and the laboratory work in sedimentary petrology. Many field trips to local and distant localities were included, and his students learned from actual experience to collect material for analysis in the laboratory.

Graduate students soon sought to work under his direction and, especially after World War II, his office, library, and laboratories always had several students working throughout the day and far into the night. More than 128 students were awarded master's and doctor's degrees under his direction. Many of these men are now in positions of importance in industry, government, and universities, both on this continent and in Europe and Asia.

Soon after he came to Illinois he began work with the Illinois State Geological Survey. His work was first that of quadrangle mapping, but he soon found in the Pennsylvanian rocks variations of the same kinds with which he was so familiar in the Oligocene beds of South Dakota, and he was to become the leading authority on conditions of Pennsylvanian sedimentation in the Illinois Basin. He and J. Marvin Weller, then paleontologist on the Illinois Survey, collaborated on the classic 1932 paper on cyclic sedimentation which established the term "cyclothem" in geological literature. The importance of his work in cyclic sedimentation is well expressed in a letter from Dr. K. C. Dunham, F.R.S., Director of the Institute of Geological Sciences:

> I first met Harold at the International Congress in London in 1948, where we were both taking part in a symposium on Rhythm in Sedimentation. It was thrilling to do so, for he was joint author, with J. M. Weller, of the term *cyclothem*, so aptly descriptive of the rhythmic unit characteristic of Pennsylvanian sediments from West Virginia through Illinois to Kansas. The term could equally be applied, it seemed to me, to the Visean and Namurian sediments first described by Westgarth Forster as the Lead Measures in northern England; and it has become an accepted part of descriptive terminology in this country also. His brilliant studies of the varying emphasis and nature of the cyclothems when viewed on a continent-wide basis have greatly illuminated an important branch of sedimentology, giving perspective to processes which were of great extent in late Paleozoic times throughout much of what is now the northern hemisphere. They also served to provoke and crystallise thought about the factors controlling this widespread and remarkable form of accumulation.
>
> Then later, as a visiting professor at the University of Illinois, I had the great pleasure of being for a short spell his colleague. I pay tribute to his mastery as a teacher, both in his quiet talks with his students in his office, and with a small party of graduates in the field, when he would pursue with untiring zeal the Carboniferous beds from Champaign to the Mississippi itself.

Professor Wanless found in work in western Illinois that the Paleozoic sedimentary rocks were overlain by varying thicknesses of Pleistocene deposits. He soon began to sort these out, as well as to work on the bedrock geology. He was able to establish the existence of Nebraskan deposits in Illinois and to differentiate western Kansan from eastern Illinoian drift. He also contributed to establishing a stratigraphic sequence of Kansan, Illinoian, and Wisconsinan deposits. This work was a distinct contribution to Pleistocene stratigraphy, history, and sedimentation. As a stratigrapher he was keenly aware of the differences in lithologic units and if he had concentrated in this field he would soon have become known as an outstanding authority on Pleistocene stratigraphy, sedimentation, and sedimentary petrology.

He was able to demonstrate to those graduate students fortunate enough to be admitted to his field course, Geology of the Upper Mississippi Valley, the whole sequence from the Precambrian rocks of southern Wisconsin to the Mesozoic and Tertiary rocks in extreme southern Illinois. The Pleistocene material covering the bedrock was also demonstrated, not as an annoying concealment of the bedrock, but as a very important entity of its own and its morphology and stratigraphy were carefully pointed out.

Professor Wanless's knowledge of areal geology came from his many travels through the United States. His spring field course in southern Illinois and northern Kentucky demonstrated the relations of rocks and the structures there to the topography. His field course for beginning students to Tennessee and North Carolina was based on his intimate knowledge of all the rocks which were seen along the route to the Great Smoky Mountain National Park and the return by a different route.

In the summers of 1945 and 1947 he held an appointment as Professor at the University of Michigan and took part in the Summer Field Camp of that university in Wyoming. Some of the Illinois students who accompanied him to this camp collected material for graduate theses under his direction. He was very anxious to see a University of Illinois field camp established and was the first director of the Illinois Field Camp at Fort Lewis, Colorado, near Durango, in 1949, and continued in this capacity through 1950 and 1951.

Professor Wanless used sabbatical leaves in 1938-1939, 1948-1949, 1958-1959, and 1965 to enlarge his studies of Pennsylvanian strata and of cyclic sediments throughout North America, in Australia, South Africa, and Europe (especially Holland and Great Britain). His work was supported by a succession of grants from The Geological Society of America, the National Science Foundation and the University of Illinois Research Board. (His original requests were usually so modest and the proposed budgets so frugal that it was generally necessary to revise them in the department office before final transmittal, in order to bring them in line with the probable real cost of the project!) His reports to the University on his activities during a leave were fully documented and detailed meticulously the successful accomplishment of his plans.

As Professor Wanless's work in the Pennsylvanian rocks developed over a wider and wider area, and as more and more subsurface information, as well as outcrop information, became available, he and his students were able to show the wide extent of many of the members of the various formations and to trace them through alluvial, paludal, and marine facies. They began construction of a series, eventually to include scores, of paleogeographic maps of almost separate instants in Pennsylvanian time. These maps depict surfaces, channels, and materials, and show why swamp deposits, which became coal, were localized in the places where they were found. It was possible to show why the coal occurred where it did and why certain filled channels existed which contained fluids, some of which were hydrocarbons. Some of the maps were precise enough to show pointbars and even natural levees along channels. The relation of this work to the beginning paleotectonic map program of the U.S. Geological Survey, and ultimate connection of Wanless with that work, is described in the following communication from E. D. McKee:

In the early 1950s the U. S. Geological Survey was developing plans for a new project to be known as the Paleotectonic Map program. This program involved the systematic gathering and recording of all available data on the history of each geologic system and the preparation of a series of isopach, lithofacies, environmental, and other maps to portray these data. It constituted a synthesis for the entire United States and for each of the principal subdivisions of geologic time. Also in the early 1950's, but quite independently, Harold Wanless, of the University of Illinois, had become deeply engrossed in the study of environmental and tectonic features of the Pennsylvanian System of the United States and was actively engaged in the preparation of a series of maps illustrating results of his investigations.

At the annual meeting of the Geological Society of America in 1953, I had the privilege of seeing some of Harold Wanless' newly prepared maps of the Pennsylvanian and of hearing him discuss his plans and progress in developing others. The maps were novel and stimulating and indicated many new and little-recognized features. As I was at that time about to accept a post in Denver with the U. S. Geological Survey to develop the newly established Paleotectonic Map program, I lost no time in arranging a meeting with Wanless in order to discuss problems of mutual interest and benefit in the field of paleotectonics.

During our conference at the Geological Society meeting, it soon became apparent that because of similar and overlapping interests in programs, it would be desirable to combine forces on the Pennsylvanian part of the Geological Survey project. Harold Wanless therefore became a part-time geologist of the Survey and began a work program that was to last until his death in 1970 and that was to constitute a major job in the compilation of data and in analyzing the history of the Pennsylvanian System. By 1970 he had virtually completed his part of the study—covering all rocks of this age east of the Mississippi, plus Iowa and Missouri—and had prepared 6 isopach maps, 5 lithofacies maps, a paleogeologic map, a suprageologic map, numerous cross-sections, and interpretive maps. The amount of detailed work and painstaking analysis that went into the study of this vast region during those 17 years was tremendous but is a real monument and tribute to the man who followed it through to completion.

In later years Professor Wanless's interest in photo-geology turned to the study of changes in coastlines, as shown by airphotos taken over a period of years. These changes of coastal, deltaic, and lagoonal features helped explain some of the features revealed by his outcrop and subsurface studies of Pennsylvanian rocks. The study of modern shorelines led to cooperation with Dr. Francis Shepard, his former associate at the University of Illinois. During 1968, Professor Wanless spent six months at the Scripps Institution of Oceanography working with Dr. Shepard on a book describing the coasts of the United States, including Alaska and Hawaii. He was in California again from February to October, 1969, during which time the book, *Our Changing Coastlines*, was completed and sent to the publishers, McGraw-Hill. The book which appeared in May, 1971, deals with coastal descriptions, origin of features, and particularly the changes in the evolving coastlines.

Dr. Wanless was elected to Fellowship in The Geological Society of America in 1931. He was also a member of American Association of Petroleum Geologists, Paleontological Society, American Geophysical Union, American Society of Photogrammetry, Association of Geology Teachers, American Association for the Advancement of Science, Illinois State Academy of Science (President 1937-1938), Illinois Geological Society, American Ornithological Society, Wilson Ornithological Club, Sigma Xi (President, Illinois Chapter, 1964-1965), Champaign County Audubon Club (President, 1942), Urbana Exchange Club (President, 1953), and Princeton Club of Chicago. He was a long-time member of the Society of Economic Paleontologists and Mineralogists and was elected to honorary membership in 1970. The award was presented posthumously to his son at the Calgary meeting on June 23, 1970. However, as the officers knew of his serious illness, the Certificate of Award was sent to him early in May and was proudly displayed by him to the many visitors to his hospital room.

Professor Wanless was often invited to give lectures at universities and before professional groups, not only in the United States and Canada, but also overseas in Australia, South Africa, Holland, Great Britain, and elsewhere. He was American Association of Petroleum Geologists Distinguished Lecturer in 1958, and American Geological Institute Visiting Scientist in 1959, 1961, and 1966. His lectures were models of organization and were accompanied by expertly prepared illustrations. He was one of the very early users of color photographs to record and illustrate features he was discussing.

Professor Wanless's personal library, which filled two large rooms, included bound volumes of the standard journals, usually from the first volume. Government survey publications and monographs on coal-containing strata came to him from all over the world. Because of his world-wide correspondence, his reprint collection, meticulously catalogued and cross-indexed, was truly enormous. He encouraged his students, both undergraduate as well as graduate, to use his library and reference cards freely, which contributed to a continuously close association with his students. Rarely has a man been so loved by his students and associates. He was utterly selfless in his consideration for others and in sometimes going not only the "second mile," but the third and fourth for a student with some personal or scientific problem. His faith and confidence almost always paid off and more than one "problem-child" is now a successful geologist, who may not fully realize how he was snatched from the brink of disaster by Professor Wanless. Some of his students knew of his international reputation and to them, especially foreign students, it was almost shocking to find that the famous man was so considerate. Others less experienced geologically, who knew him first as a kind and helpful man, were hardput to realize, and realize they did eventually, that he was also a great scholar.

It is pleasing to record that some feeling of this admiration and affection was exhibited to Harold Wanless in his lifetime. In 1949, in honor of his 25th anniversary as a member of the Department of Geology at Illinois, he was presented with a purse and a volume of letters and greetings from 153 of his former students. At the 40th anniversary of his service at Illinois in 1963, alumni, students and staff joined in a quiet, intimate dinner to celebrate this anniversary. At this time so many letters were received from former students that it required two bound volumes to contain them when they were presented to him. At the dinner he was presented with a beautifully engrossed scroll which included this statement:

> Geologists will be forever indebted to you for your eminent and voluminous contribution to your science. These intellectual benefactions are epitomized for us by your original and provocative investigation of cyclothemic sedimentation and by your enormous labor of love in unraveling the incredibly complex mysteries of the paleogeography and paleocology of the Pennsylvanian System. Your close associates remember with gratitude your boundless friendship, benevolent counsel, unpretentious humanity, and selfless service to the University and community. They are grateful and proud for having had an opportunity to have had some share in the life of you and your devoted wife, Grace.

At his retirement in 1967, the alumni presented him with a bronze plaque of appreciation and told him that funds had been collected for a "Wanless Room" and for a portrait. In the remodeling of the Natural History Building for the Geology Department a large room was to be especially fitted and furnished in a very "non-classroom" style, with kitchen attached, as a lounge and conference room for students and staff. Although he was not able to attend the dedication on November 8, 1969, and the unveiling of the portrait (reproduced here), it is a matter of comfort that he was able to see the completed room a few months before he died.

Scores of alumni have written of their personal experience with Professor Wanless in the field, in the laboratory, and elsewhere. Many speak of the part-time student jobs he secured for them and for his concern, which extended to providing personal financial assistance. His associates realized he was a very generous man, but the full extent of his generosity only has become apparent after his death, as many former students have commented upon it with personal examples. Brief excerpts from only a few of the letters are given to show the differing aspects of his association with students and former colleagues and to contribute a picture of this many-sided teacher, scholar, and citizen:

> He talked to students as if the student was contributing something to the conversation that was worth hearing . . . . Down inside he was a very tough and strong man I think, but his manners were so considerate and gracious that one seldom realized the extent of that part of his nature.

> He always had time to discuss personal or geological matters with students and to give his suggestions and criticisms, no matter how busy he was . . . . He was one of the most unselfish individuals I have ever known.

The association with Dr. Wanless was responsible for many of the students choosing to major in geology. There are many geologists in this country today who were committed to the profession by their early association with Dr. Wanless. It will be a long time before another of his equal will pass this way again.

Just after the War, as a student who arrived at midyear I felt I faced a huge and impersonal new world . . . . Dr. Wanless's first few statements [that year he was acting head of the Geology Department] put me completely at ease, and from his memory, not from the files, he discussed an orderly summary of my background, a list of the geology courses I had completed, and gave a straightforward recommendation as to the work I should plan to take . . . . I sensed right then a man who would have time for me no matter what other pressures were on his mind.

As a new student in a different country, he immediately proceeded to make me feel at home by personally introducing me to other students and other members of the department. He made sure also, that my wife and I met people outside the university, many of whom we still count as friends.

I was met at the Champaign airport by Professor Wanless who drove me to my lodgings and helped me with my baggage. This was quite unexpected because in my own country the student is supposed to serve his teacher in all possible ways . . . . I can hardly believe it if any professor in a Chinese university would do the same to a student as Dr. Wanless did for me. This is what I came to admire in American education and in the great personality of Dr. Wanless . . . . He was a friendly, generous, and kind scholar.

By 1937, Section E of the American Association for the Advancement of Science had been languishing with very low attendance. At the request of the officers Wanless organized a Symposium for the 1937 Indianapolis meeting on "Oil in the Pennsylvanian of the Eastern Interior Basin." The papers offered a sweep of Pennsylvanian stratigraphy, structure, and paleontology that attracted the largest audience section we had known in years . . . . That Symposium started the Section on Geology on the upgrade. It was a notable contribution, not only to the subject matter, but also to an organization that needed help.

His most impressive traits were his boundless energy and enthusiasm and his desire to have his students see as clearly as he himself saw . . . . He was a master of field work and his powers of observation and deduction were legendary . . . . When I last saw him in 1968, as a speaker before a large group of Illinois graduates in Louisiana, although he was under medical care for the malignancy which later claimed his life, he was the very picture of silver-haired vigor and pink-cheeked enthusiasm.

Harold Wanless's dedication to geology was considered by many to make him a driving hazard—especially in the day of superhighways. Anyone who had ridden with him knows that he enjoyed pointing out geologic features, including details of an outcrop, even though he was driving past them at 60 mph. Graduate students and associates used a variety of strategems to supply Dr. Wanless with such a mass of maps to follow that it became necessary for someone else to drive . . . . In order to conserve student resources he picked low-priced hotels, the quality of which became legendary throughout the years and provided memories and anecdotes still shared by many students.

Dr. Wanless was at his best as a teacher in the field . . . . His field course, "The Geology of Upper Mississippi Valley" was his best. He literally knew every outcrop in the Upper Mississippi Valley from personal experience. He challenged the students in the field . . . to make their own interpretation from their own observations . . . . Of my courses in geology this was undoubtedly one of the most rewarding.

The impact of the Wanless method of teaching flowed from making lucid the well-established subject matter, but the varied views of the major workers in each field were cited and expounded so that a balanced feel for the unresolved was imparted. His illustrative materials must represent the best in existence.

Many students and associates have commented on his prodigious memory:

He had total recall and did not have to refer to notes to recall the smallest detail of field sections, which he had seen at any place in the world.

His memory was probably photographic and he could reproduce from memory whole pages of text and of notes, and even reproduce large portions of topographic maps from memory.

He remembered the names of all of his students from the earliest years—their names, families, and interests. He was most helpful in finding jobs for the students.

For many years, at the annual meeting of The Geological Society and of the American Association of Petroleum Geologists, Illinois alumni met for a luncheon, sometimes with attendance of as many as 100 people. Harold would ask each graduate to stand and would promptly mention his business affiliation, where he was located, and make a few pertinent remarks about his family. We marveled at his tremendous memory and naturally it enhanced our affection for him.

I remember him particularly for his tremendous ability to recall information in journals, seeming to be able to cite author, title, page number, and paragraph at a moment's notice . . . . He somehow found time to go through in detail each manuscript submitted by a student, making constructive suggestions and comments which materially improved his presentation . . . . His method of teaching was to pose a question that made you think, rather than asking

> questions which required parroting back answers . . . . His interests were much broader than geology, for one 6 a.m. I met him unexpectedly on a woodland trail bird-watching. He knew his birds and flora nearly as well as he knew rocks . . . .
>
> From my experience in the field course "Geology of the Upper Mississippi Valley," I began to appreciate the man behind the fine reputation . . . . He had almost total recall of detailed geologic characteristics of the various outcrops visited throughout the entire upper Mississippi Valley. In spite of the whirlwind of activities, there was always time to answer over again the many questions posed by his sometimes not-so-apt students.

While a student at Princeton, Harold Wanless was a member of the Gateway Club, where the Rogers brothers of Philadelphia were fellow members. Through them he met their sister, Miss Grace Rogers. This acquaintanceship eventually resulted in their marriage on August 7, 1926. Mrs. Wanless accompanied her husband on many of his travels and participated in some of his field work. Later, their son, Harold Rogers Wanless, who was born in 1942, accompanied his parents on their travels and to the Illinois Summer Field Station; he learned to recognize rocks, minerals, and fossils along with the letters of the alphabet. It was only natural for their son to engage in geology from his earliest years, and later to apply his great ability as a swimmer and diver to research in marine geology.

A slowly developing malignancy, first discovered in 1964, was kept under control by medication and a series of operations so that Professor Wanless continued his teaching until retirement in 1967; he then continued his research and writing at an even increased pace until early 1970. Even after he entered the hospital, he continued to revise manuscripts and correct proofs until a few days before his death on June 3, 1970.

Thousands of students knew Professor Wanless as an outstanding teacher and hundreds of "his own" students in geology testify to his influence on their personal lives and professional careers. His scholarship was recognized in North America and overseas through honors and appointments. Above all he was a man who was gentle, modest, infinitely patient, scrupulously honest, a man of dedication and transparent integrity. His scientific and personal influence will persist through many generations, as his students in turn pass on these attitudes to their students and associates.

In preparing this memorial I have been aided by information from Mrs. Wanless and from many alumni who are former students of Professor Wanless. H. B. Willman and C. A. Chapman were helpful in providing data and in suggesting certain improvements in organization and expression. Harold Rogers Wanless has helped in the preparation of the Bibliography. A great deal of material which could only be summarized here, including scores of letters from alumni, will be deposited in the University of Illinois Archives.

BIBLIOGRAPHY OF HAROLD ROLLIN WANLESS

1922 Notes on sand calcite from South Dakota: Am. Mineralogist, v. 7, p. 83-86.

—— Lithology of the White River sediments: Am. Philos. Soc. Proc., v. 61, p. 184-203.

1923 The stratigraphy of the White River beds of South Dakota: Am. Philos. Soc. Proc., v. 62, p. 190-269.

1927 Oil possibilities of the Alexis quadrangle, Mercer and Warren Counties, Illinois: Illinois Petroleum, v. 8, p. 1-7.

1928 Pleistocene and recent history of Alexis quadrangle, Illinois: Illinois Acad. Sci., Trans., v. 20, p. 254-260.

1929 Geology and mineral resources of the Alexis quadrangle, Illinois: Illinois Geol. Survey Bull. 57, 230 p.

—— Nebraskan till in Fulton County, Illinois: Illinois Acad. Sci. Trans., v. 21, p 273-282.

1931 Pennsylvanian cycles in western Illinois: Illinois Geol. Survey Bull. 60, p. 179-193.

—— Pennsylvanian section in western Illinois: Geol. Soc. America Bull., v. 42, p. 801-812.

—— The question of a Pennsylvanian overlap in the Rock Island region: Illinois Acad. Sci. Trans., v. 24, p. 331-340.

1932 (with J. M. Weller) Correlation and extent of Pennsylvanian cyclothems: Geol. Soc. America Bull., v. 43, p. 1003-1016.

1935 (with F. P. Shepard) Permo-carboniferous coal series related to southern hemisphere glaciation: Science, v. 81, p. 521-522.

1937 Sea level and climatic changes related to late Paleozoic cycles: Geol. Soc. America Bull., v. 47, p. 1177-1206.

—— Coals of the eastern interior basin: Proc. Ind. Coal Min. Instit., p. 28-33.

1938 Geological records of a rhythmic nature: Illinois Acad. Sci. Trans., v 31, p. 7-14.

1939 (with J. M. Weller) Correlation of minable coals of Illinois, Indiana, and western Kentucky: Am. Assoc. Petroleum Geologists Bull., v. 23, p. 1374-1392.

—— Correlation of minable coals of Illinois, Indiana, and western Kentucky: Am. Inst. Mining & Metallurgical Engineers, Ann. Mtg., Class F. Coal Div., p. 1-14.

—— Pennsylvanian correlations in the eastern interior and Appalachian coal fields: Geol. Soc. America Spec. Paper 17, 130 p.

1940 The use of color slides as an aid in geologic teaching: Illinois Acad. Sci. Trans., v. 33, p. 171-172.

1942 (with J. M. Weller, L. M. Cline, and D. G. Stookey) Interbasin Pennsylvanian correlations, Illinois and Iowa: Am. Assoc. Petroleum Geologists Bull., v. 26, p. 1585-1593.

1944 (with R. C. Moore and Pennsylvanian Sub-Committee National Research Council) Correlations of Pennsylvanian formations of North America: Geol. Soc. America Bull., v. 55, p. 657-706.

—— (with D. W. Franklin) Pennsylvanian stratigraphy of part of southern Indiana: Illinois Acad. Sci. Trans., v. 37, p. 85-92.

1946 Pennsylvanian geology of a part of the southern Appalachian coal field: Geol.

Soc. America Mem. 13, 162 p.

1947 Introduction to symposium on Pennsylvanian problems: Jour. Geol., v. 55, p. 184-193.

—— Regional variations in Pennsylvanian lithology: Jour. Geol., v. 55, p. 237-259.

1948 Memorial to Terence Thomas Quirke: Illinois Acad. Sci. Trans., v. 41, p. 119.

—— Memorial to Thomas Edmund Savage: Illinois Acad. Sci. Trans., v. 41, p. 121.

1950 Selection of aerial photographs for teaching geology, and aerial photograph collection: Photogrammetric Engineering, v. 16, p. 796-806.

—— Late Paleozoic cycles of sedimentation in the United States: International Geological Congress, Report of 18th Session, London, Part IV, p. 17-28.

1951 (and Jacqueline Patterson) Cyclic sedimentation in the marine Pennsylvanian of the southwestern United States: Troisieme Congre pour L'avancement des Etudes de Stratigraphique et de Geologie du Carbonifere Compte Rendu, Heerlen, Netherlands, v. 2, p. 653-664.

—— Outstanding aerial photographs in North America: Am. Geol. Inst. Rept., No 5, 87 p.

1953 Evidence of eustatic shifts in sea level in the Pennsylvanian of the southwestern United States: Tulsa Geol. Soc. Digest, v. 20, p. 65-69.

—— Development of methods and materials for teaching photogeologic interpretation: Selected papers on photogeology and photointerpretation: Comm. on Geophysics and Geography, Research and Development Board, Washington, p. 163-205, 36 figs.

1954 Studies of field relations of coal beds: Second Conference on the Origin and Constitution of Coal, Crystal Cliffs, Nova Scotia, June 1952, p. 148-175.

1955 (with R. L. Belknap, and Helen Foster) Paleozoic and Mesozoic rocks of Gros Ventre, Teton, Hoback, and Snake River Ranges, Wyoming: Geol. Soc. America Mem. 63, 90 p., 13 pl.

—— Pennsylvanian rocks of eastern interior basin: Am. Assoc. Petroleum Geologists Bull., v. 39, p. 1753-1820.

1956 Classification of the Pennsylvanian rocks of Illinois as of 1956, with correlation chart by Raymond Siever: Illinois Geol. Survey Circ. 217, 14 p.

—— Problems of the Pennsylvanian of the United States: Tulsa Geol. Soc. Digest, v. 24, p. 56-61.

1957 Relations between Pennsylvanian rocks of the eastern interior and northern midcontinent coal basins: Kansas Geol. Soc., Guidebook 21st Field Conference, p. 85-91.

—— (with J. C. Mueller) Differential compaction of Pennsylvanian sediments in relation to sand-shale ratios, Jefferson County, Illinois: Jour. Sed. Petrology, v. 27, p. 80-88.

—— (with W. G. Ziebell, E. A. Ziemba, and A. Carozzi) Limestone texture as a key to interpreting depth of deposition: International Geological Congress, 20th Session, Mexico, Sec. 5, p. 65-82.

—— Geology and mineral resources of the Beardstown, Glasford, Havana and Vermont quadrangles: Illinois Geol. Survey Bull. 82, 233 p.

1958 Pennsylvanian faunas of the Beardstown, Glasford, Havana and Vermont quadrangles: Illinois Geol. Survey Rept. Inv. 205, 59 p.

—— Water and agriculture in the United States of America, some relations: The

Living Earth, Sydney, Australia.

1960 Evidences of multiple late Paleozoic glaciation in Australia: International Geological Congress, 21st Session, Copenhagen, pt. 12, Regional Paleogeography, p. 104-110.

—— (with R. M. Kosanke, J. A. Simon, and H. B. Willman) Classification of the Pennsylvanian strata of Illinois: Illinois Geol. Survey Rept. Inv. 214, 84 p.

1961 Depositional basins of some widespread Pennsylvanian coal beds in the United States: Nova Scotia Res. Council, 3rd Conf. on Origin and Constitution of Coal, June, 1956, p. 94-128.

1962 Pennsylvanian rocks of eastern interior basin: Am. Assoc. Petroleum Geologists Spec. Vol., Pennsylvanian system in the U. S., p. 4-59.

1963 (with J. B. Tubb, D. E. Gednetz, and J. L. Weiner) Mapping sedimentary environments of Pennsylvanian cycles: Geol. Soc. America Bull., v. 74, p. 437-486.

1964 Pennsylvanian period and system: Lexicon of major geologic time units of the world: International Commission on Stratigraphy, v. 8, Congres Geologique International, p. 1-64.

—— Local and regional factors in Pennsylvanian cyclic sedimentation: Kansas Geol. Survey Bull. 169, p. 593-606.

1965 Environmental interpretation of coal distribution in the eastern and central United States: Proc. Illinois Mining Institute, p. 19-36.

—— (with M. T. El-Ashry) Shoreline features and their changes: Photogrammetric Engineering, v. 33, no. 2, p. 184-189.

—— Aerial stereo photographs: T. N. Hubbard Scientific Co., Northbrook, Illinois, 92 p.

1968 Photo interpretation of shoreline changes between Capes Hatteras and Fear (North Carolina): Marine Geology, v. 6, p. 347-379.

1969 Eustatic shifts in sea level during the deposition of late Paleozoic sediments in the central United States: West Texas Geol. Soc. Symposium on Cyclic Sedimentation, p. 41-54.

—— Marine and non-marine facies of the Upper Carboniferous of North America: Compte Rendu 6e Congres Intern. Strat. Geol. Carbonif. Sheffield, 1967, v. 1, p. 293-336.

—— (with J. R. Baroffio and P. C. Trescott) Conditions of deposition of Pennsylvania coal beds: Environments of coal deposition: Geol. Soc. America Spec. Paper 114, p. 105-142.

1970 Late Paleozoic deltas in the central and eastern United States: Deltaic sedimentation, modern and ancient: Soc. Econ. Paleontologists and Mineralogists Spec. Pub. 15, p. 215-245.

1971 (with F. P. Shepard) Our changing coastlines: McGraw-Hill Book Company, New York.

In addition to the books, monographs, and journal articles listed, 36 abstracts; 19 reviews; 27 laboratory manuals and field trip guides, some containing over a hundred pages; and 6 Illinois Academy of Science reports are included in the complete bibliography, which will be filed in the University of Illinois Archives.

History of Geology

An Arno Press Collection

Association of American Geologists and Naturalists. **Reports of the First, Second, and Third Meetings of the Association of American Geologists and Naturalists, at Philadelphia, in 1840 and 1841, and at Boston in 1842.** 1843

Bakewell, Robert. **An Introduction to Geology.** 1833

Buckland, William. **Reliquiae Diluvianae:** Or, Observations on the Organic Remains Contained in Caves, Fissures, and Diluvial Gravel. 1823

Clarke, John M[ason]. **James Hall of Albany:** Geologist and Palaeontologist, 1811-1898. 1923

Cleaveland, Parker. **An Elementary Treatise on Mineralogy and Geology.** 1816

Clinton, DeWitt. **An Introductory Discourse:** Delivered Before the Literary and Philosophical Society of New-York on the Fourth of May, 1814. 1815

Conybeare, W. D. and William Phillips. **Outlines of the Geology of England and Wales.** 1822

Cuvier, [Georges]. **Essay on the Theory of the Earth.** Translated by Robert Kerr. 1817

Davison, Charles. **The Founders of Seismology.** 1927

Gilbert, G[rove] K[arl]. **Report on the Geology of the Henry Mountains.** 1877

Greenough, G[eorge] B[ellas]. **A Critical Examination of the First Principles of Geology.** 1819

Hooke, Robert. **Lectures and Discourses of Earthquakes and Subterraneous Eruptions.** 1705

Kirwan, Richard. **Geological Essays.** 1799

Lambrecht, K. and W. and A. Quenstedt. **Palaeontologi:** Catalogus Bio-Bibliographicus. 1938

Lyell, Charles. **Charles Lyell on North American Geology.** Edited by Hubert C. Skinner. 1977

Lyell, Charles. **Travels in North America in the Years 1841-2.** Two vols. in one. 1845

Marcou, Jules. **Jules Marcou on the Taconic System in North America.** Edited by Hubert C. Skinner. 1977

Mariotte, [Edmé]. **The Motion of Water and Other Fluids.** Translated by J. T. Desaguliers. 1718

Merrill, George P., editor. **Contributions to a History of American State Geological and Natural History Surveys.** 1920

Miller, Hugh. **The Old Red Sandstone.** 1857

Moore, N[athaniel] F. **Ancient Mineralogy.** 1834

[Murray, John]. **A Comparative View of the Huttonian and Neptunian Systems of Geology.** 1802

Parkinson, James. **Organic Remains of a Former World.** Three vols. 1833

Phillips, John. **Memoirs of William Smith, LL.D.** 1844

Phillips, William. **An Outline of Mineralogy and Geology.** 1816

Ray, John. **Three Physico-Theological Discourses.** 1713

Scrope, G[eorge] Poulett. **The Geology and Extinct Volcanos of Central France.** 1858

Sherley, Thomas. **A Philosophical Essay.** 1672

Thomassy, [Marie Joseph] R[aymond]. **Géologie pratique de la Louisiane.** 1860

Warren, Erasmus. **Geologia:** Or a Discourse Concerning the Earth Before the Deluge. 1690

Webster, John. **Metallographia:** Or, an History of Metals. 1671

Whiston, William. **A New Theory of the Earth.** 1696

White, George W. **Essays on History of Geology.** 1977

Whitehurst, John. **An Inquiry into the Original State and Formation of the Earth.** 1786

Woodward, Horace B. **History of Geology.** 1911

Woodward, Horace B. **The History of the Geological Society of London.** 1907

Woodward, John. **An Essay Toward a Natural History of the Earth.** 1695